The Chang'E-1 Topographic Atlas of the Moon

Chunlai Li · Jianjun Liu · Lingli Mu · Xin Ren · Wei Zuo

The Chang'E-1 Topographic Atlas of the Moon

2nd Edition 2016

Springer

Chunlai Li
Jianjun Liu
Lingli Mu
Xin Ren
Wei Zuo
National Astronomical Observatories,
Chinese Academy of Sciences (NAOC)
Beijing, China

ISBN 978-3-662-48437-1 ISBN 978-3-662-48439-5 (eBook)
DOI 10.1007/978-3-662-48439-5
Springer Heidelberg New York Dordrecht London

Library of Congress Control Number: 2015957994

1st edition was published bilingual (Chinese and English): 嫦娥一号全月球地形图集 by Compiling Committee of the Chang'E-1 Topographic Atlas of the Moon,

Printed on acid-free paper

Springer Berlin Heidelberg is part of Springer Science+Business Media
(www.springer.com)

Preface

Since the end of the 1950s, a series of unmanned and manned lunar exploration activities have taken place. With the enhancement of China's overall national strength, the development of its technology and continuous technological breakthroughs in aerospace engineering and deep space exploration, China is qualified to carry out lunar exploration. On October 24, 2007, the first lunar probe-Chang'E-1 was successfully launched into space. It entered the earth-moon transfer trajectory on October 31 and successfully started to orbit the Moon on November 5. At 16:49 pm, November 20, the CCD stereo camera on the Satellite opened its "eyes", and began to scan the surface of the Moon. After several days of continuous imaging by integrating the multi-track data, the first lunar image from China's first lunar probe was mosaicked and issued on November 26, which marked the success of China's first lunar exploration. On September 28, 2009, *The Global Chang'E-1 Digital Topographic Map* was evaluated. The results showed that the topographic map had reached the international advanced level in data coverage, plane positioning accuracy, elevation accuracy and spatial resolution.

The Moon carries the dream of mankind for thousands of years. So far, dozens of lunar probes have been launched. In order to help unveil the mysteries of the Moon, scientists used telescopes and probes to obtain data, and produced a variety of lunar topographic maps.

Early lunar topography maps were mainly produced based on photos taken by satellites, such as the 1960 publication of *Atlas Obratnoi Storony Luny (Atlas of the Far Side of the Moon)* containing former Soviet Luna 3 spacecraft. In 1969, Lewis made *The Times Atlas of the Moon*, which divided the moon surface into 110 areas and included almost all the moon images (libration area not included) and contents of the Apollo 11 moon landing. In 1971, David Bowker and Kenrick Hughes, assembling about 675 planar images acquired by Lunar Orbiter, compiled the *Lunar Orbiter Photographic Atlas of the Moon*. This atlas was a lunar global map of the widest coverage including the near side and far side with the highest accuracy at that time, and became an essential reference book for workers in lunar science. In 1999, Eliason et al. produced the *UVVIS 750 nm Basemap* by using the Clementine 750 nm images. In 2004, Ben Bussey and Paul Spudis integrated Lunar Prospector data with Clementine data, and compiled *The Clementine Atlas of the Moon*. In 2007, Phillip used Luna 9 data to compile the *The International Atlas of Lunar Exploration*. In 2010, *The Chang'E-1 Image Atlas of the Moon* was produced by Li Chunlai et al. (at 120 m resolution), which was an image atlas with the most complete coverage among products of equivalent resolution. In March, 2011, *The Kaguya Lunar Atlas: The Moon in High Resolution*, compiled by Motomaro Shirao and Charles A. Wood, used the HDTV (High Definition Television Camera, carried on the Kaguya) images to illustrate 100 of the Moon's most interesting landforms. On March, 15, 2011, JPL released the Wide Angle Camera (WAC, on board the Lunar Reconnaissance Orbiter) global mosaic comprised of over 1,500 lunar images acquired between November 2009 and February 2011. Up to now, JAXA (Japan Aerospace Exploration Agency) has made the TCOrtho (Terrain Camera Orthophoto) available, which was acquired by the Kaguya. On February 6, 2012, China National Defense Technology Industry Bureau published *Chang'E 2 Global Map of the Moon* at *7 m resolution* – high resolution lunar global map of the broadest coverage so far; at the same time, it also published *Chang'E -2 High Resolution Lunar Image Atlas*. Aforementioned maps and atlas include images obtained by the satellite to show the morphology of the Moon vividly. With the development of the photogrammetry and laser ranging technology, it has become an important trend to use the topographic data to compile lunar topographic maps.

Lunar topographic map production began in the 1960s. In the following 50 years a large number of topographic maps had appeared. The first one is the 1 : 1 000 000 *Lunar Astronautical Chart (LAC) Series*, including 38 sheets, which were produced by ACIC (Aeronautical Chart Information Center, United States Air Force) from 1962 to 1967. The LAC portrayed lunar topography by shaded relief with a western illumination, approximate contours (300 meter interval), spot elevations, height differences between level features and crater depths. In 1964, 1 : 5 000 000 *Topographic Lunar Series* were published by ARMY MAP SERVICE, CORPS OF ENGINEERS. Currently, the intermediate and large scale topographic maps mainly employ ground-based telescope, lunar orbiter, and Apollo data. During 1965–1967, the *Apollo Intermediate Chart Series*, including 20 sheets of 1 : 500 000 low-latitude topographic maps with Mercator projection based on ground-based telescope and Lunar Obiter data, was released by the ACIC. In 1971, the *Large Scale Topophotomaps Series* composed of 41 large-scale topographic atlas sheets, was published by Defense Mapping Agency, Topographic Center. Based on data obtained by Apollo 15, the above series contains eight sheets of 1 : 10 000, three sheets of 1 : 25 000 and thirty sheets of 1 : 50 000. During 1973–1980, The *Lunar Topographic Orthophotmaps and Lunar Orthophotomaps Series*, including 215 pieces of 1 : 250 000 maps, were edited by Defense Mapping Agency, Aerospace Center and Lawrence A. Schimerman. This Series is the first comprehensive and continuous mapping to be accomplished on the basis of Apollo Mission 15-17 mapping photographs. During 1976–1980, Defense Mapping Agency Aerospace Center compiled *Lunar Shaded Relief (LSR) Series* in low latitude area, including 11 sheets of 1 : 1 000 000 maps. Later, several editions of 1 : 5 000 000 topographic maps have been published by USGS. *Shaded Relief Map of the Mare Orientale Area of the Moon* was released in 1978. *Shaded Relief Map of the Lunar Far Side, Shaded Relief Map of the Lunar Polar Regions, Shaded Relief Map of the Lunar Near Side* were available in 1980, 1981 and 1992 respectively.

From the beginning of the 21st century, lunar exploration entered another thriving age. Space powers and international organizations embarked on lunar exploration activities aiming at returning to the moon. Launched lunar missions included SMART-1 of ESA, Selene (Kaguya) of Japan, Chang'E-1 and Chang'E-2 of China, Chandrayaan-1 of India, LRO, ARTEMIS and GRAIL of the USA. Planned missions include the Chang'E series of China, the Selene2 of Japan, the Luna Glob of Russia, the MoonLITE of the UK and the Google Lunar X-Prize funded by Google, etc. Data by this generation of lunar exploration missions have improved compared with the historical data in spatial coverage and resolution, and are expected to cast far-reaching influences on lunar cartography. In 2002, Roseik et al produced a new set of 1 : 10 million scale lunar maps (Clementine Color-coded Shaded Relief Map) combining color-coded topography with shaded relief data and nomenclature, in which topographic data are from the Clementine laser altimeter (covering the Moon surface of 79°S-81°N) and photogrammetric data collected from Clementine UVVIS images. In March 2008, JAXA produced Lunar Topographic Map using KAGUYA (SELENE) LALT observation data with 1 km interval contours. Since November 2009, the Japan Aerospace Exploration agency (JAXA) has being releasing DOM (Digital Orthophoto Map) and DEM data with 7.4 meter resolution, obtained by the Terrain Camera. In the last month of 2010, NASA's Goddard Space Flight Center released LOLA Topographic Map, using Digital Evaluation Model (DEM) data obtained by LOLA at resolution of 30 meters. In November, 2011, Arizona State University in Tempe published the GLD100 (Global Lunar DTM 100 m) data with a pixel resolution of 100 meters and *Color Shaded Relief* map. The GLD100 was created by the evaluation data derived from images taken by Wide Angle Camera and altimetry data by LOLA. At present, the National Astronomical Observatories, Chinese Academy of Sciences (The Ground Application System of Lunar Exploration Program) is processing the Chang'E-2 CCD stereo camera data.

Following the publication of *The Chang'E-1 Image Atlas of the Moon*, *The Chang'E-1 Topographic Atlas of the Moon* is based on the lunar global DEM derived from the CCD stereo image data with digital photogrammetry. The spatial resolution of DEM in this *Atlas* is 500 m, with horizontal accuracy of 192 m and vertical accuracy of 120 m. This atlas uses color shaded relief map with contour lines to show the lunar topographic characters. The CE-1 topographic data can provide foundational information for the study of lunar topography, morphology, and geologic structures, as well as lunar evolution research.

The compilation of the Atlas was mainly organized and carried out by the National Astronomical Observatories of the Chinese Academy of Sciences (The Ground Application System of China's Lunar Exploration Program) and Sinomaps Press, making their best effort to meet the scientific, practical, and artistic requirements put forward by the Compiling Committee. However, room for improvement is believed to still exist.

Any criticism and correction on the Atlas is sincerely welcome.

This Atlas is an important research result of our Lunar Exploration Program. It is a collective product of the joint efforts of the five Systems of the Program, including the Spacecraft, the Launch Vehicle, the Launch Site, the Tracking, Telemetry and Control (TT&C) and the Ground Application Systems. It embodies the hard work of various personnel in technology, education, production, and management in the fields of aerospace engineering and topography & remote sensing. This Atlas is presented as a gift to the great of China's Lunar Exploration Program and space industry.

Compiling Committee of *The Chang'E-1 Topographic Atlas of the Moon*

February, 2012

Publication Committee of *Science Achievement Series of China's Lunar Exploration Program*

Consultant Committee	
Honorary Directors	Yongxiang Lu, Qide Han, Gang Wan
Director	Qiufa Chen
Deputy Directors	Mianheng Jiang, Jianlin Cao, Dongkui Liu, Hongguang Niu, Hejun Yin, Xingrui Ma, Qunli Xiong
Members	Jian Song, Guanhua Xu, Jianqi Zhang, Enjie Luan, Jiadong Sun, Ziyuan Ouyang, Bingzhong Chen, Jingshan Jiang, Lehao Long, Weiren Wu
Editorial Board	
Honorary Directors	Enjie Luan, Jiadong Sun
Director	Ziyuan Ouyang
Deputy Directors	Jun Yan, Benzheng Li, Chunlai Li
Members	Zhijian Wu, Hao Hu, Rongqiao Zhang, Xiaohan Liao, Yingjie Yu, Xiaoqun Liu, Guoxiang Ai, Xianlin Liu, Renxiang Wang, Jun Gao, Deren Li, Junyong Chen, Jiayao Wang, Qingxi Tong, Dengyun Yu, Huixian Sun, Weiping Qian, Zhaoyu Pei, Ji Wu, Xiaoyu Hong, Jianyu Wang, Huanyu Wang, Baochang Zhao, Jin Chang, Yongliao Zou, Jianzhong Liu

Compiling Committee of *The Chang'E-1 Topographic Atlas of the Moon*

Chief Cartographer	Chunlai Li
Deputy Chief Cartographers	Jianjun Liu, Lingli Mu, Xin Ren, Wei Zuo
Editors	Lingli Mu, Wei Zuo, Xin Ren, Zhoubin Zhang, Hongbo Zhang, Jianzhong Liu, Yan Su, Weibin Wen, Xiaoqian Wang, Xiaoduan Zou, Xiaoxia Zhang
Data Processing	Xin Ren, Lingli Mu, Jianjun Liu, Fenfei Wang, Xu Tan, Jianqing Feng, Guangliang Zhang, Liang Geng, Liyan Zhang, Lei Zheng, Deqing Kong, Xinying Zhu, Fang Wang, Qiang Fu, Yuan Xiao, Fanlu Wu
Mapping	Lingli Mu, Xiaoduan Zou, Yuxuan Liu, Fenfei Wang, Jingtao Xiao, Lingmin Meng, Xiaoqian Wang, Xingye Gao, Jinjin Zhao, Liping Xing

Part I Introduction of the CCD Stereo Camera

CCD Stereo Camera

CE-1's optical sensor is a three-line-array CCD push-broom camera. The camera is loaded on a big area array CCD detector, reading the 11st, 512nd, and 1013rd rows perpendicular to the direction of flight. The data are regarded as the front, nadir and back image arrays, of which the pixel is 512 columns, and the view angle between adjacent arrays is 16.7°. Based on the predetermined in-orbit flight parameters, the camera's scanning speed is set to be 11.89 f/s, to ensure that the satellite's three-line-array scanning can obtain consecutive image data on front, middle, and back angles (See in Fig. 1.1). The main specifications of three-line-array CCD stereo camera are shown in Tab. 1.1.

Characteristics of Orbits and Attitudes

The design parameters of CE-1 circumlunar orbits and attitudes are as follows (Tab. 1.2):
(1) Orbital data in operation
CE-1's actual dynamic range of orbit altitude is from 195.53 km to 202.07 km, with an average of 197.19 km; the dynamic range of orbital eccentricity is from 0.00017 to 0.012293, with an average of 0.006349; the dynamic range of orbital inclination is from 87.6° to 90.0°, with an average of 88.20°.

CCD Stereo Camera

(2) Attitude data in operation
The attitude is relatively stable: more than 95% of attitude values are less than 0.03° and standard deviation of the value of attitude angle to the Moon is less than 0.02°; 96% of the attitude change rate is less than 0.0016°/s; stability of yaw angles and roll angles in orbit coordinates is better than 0.003°/s (3σ); stability of pitch angles is better than 0.008°/s (3σ).

Data Acquisition and Coverage

Because of the adjustment of the orbit attitude, the maintaining of payloads, the global CCD image was finally fulfilled, after 4 covering cycles, in two operation periods, making the time phase of the image data inconsistent. During the first covering cycle, 228 tracks of data in areas of 70°N–70°S were acquired; in the second period, 247 tracks from 70°N–70°S and 17 tracks from the polar regions were acquired; and in the third period (till January 27, 2008), 14 tracks from 70°N–70°S and 58 tracks from the polar regions were acquired. In the second operation period starting from May 1, 2008. 6 tracks from 70°N–70°S were acquired, fulfilling the data coverage in 70°N–70°S. Data coverage of the polar regions were fulfilled on July 1.

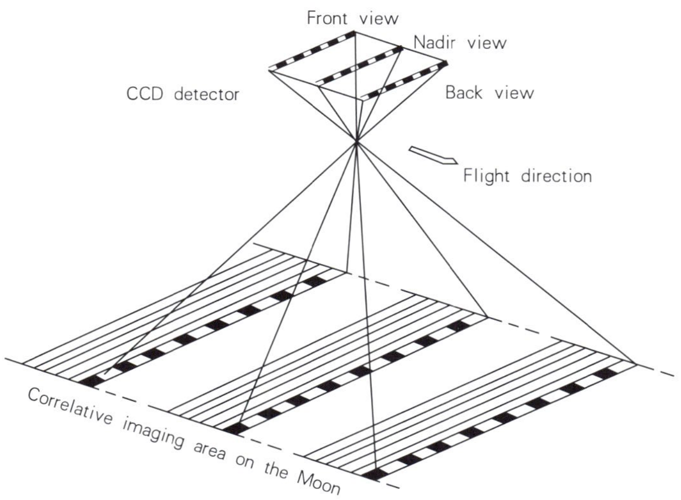

Fig. 1.1 CCD stereo camera imaging process

Table 1.1 The main specifications of CCD stereo camera

Name	Index
Frame frequency	11.89 f/s
Spectrum range	0.5 μm ~ 0.75 μm
Radiometric resolution	8 bit
Focal length	23.33 mm
Pixel per linear array	512
Size of pixel	14 μm × 14 μm
Relative aperture	F/5
Adjacent angle between linear arrays	16.7°
Time of exposure	3.2 ms, 7 ms, 20 ms and 84 ms
MTF	≥ 0.2
S/N ($\rho = 0.2$ $\theta = 60$)	≥ 100
Imaging width above 200 (km)	L = 60 km
Baseline to height ratio	≥ 0.6
Pixel spatial resolution (subastral point)	120 m

Table 1.2 The design parameters of CE-1 circumlunar orbits and attitudes

CE-1 parameter	Value
Orbital altitude (km)	195.464 ± 25
Inclination	90° ± 5°
Eccentricity	0
Orbital period (min)	127.164
Attitude control precision (three dimensional)	± 1°(3σ)
Attitude change rate (three dimensional)	0.01°/s(3σ)

C. Li et al., *The Chang'E-1 Topographic Atlas of the Moon*,
DOI 10.1007/978-3-662-48439-5_1, © Springer-Verlag Berlin Heidelberg 2016

Part II Global Topographic Map of the Moon

Data Receiving

The Ground Application System is responsible for the acquisition of all the image data from CCD camera and other scientific data. Reception of data from Chang'E-1 adopts two radio antennas acquisition systems, i.e. the 50 m-antenna earth station in Miyun Station and 40 m-antenna earth station in Kunming Station. The two antennas adopt parallel working modes and receive the data synchronously to ensure the completeness of data. The data acquisition code rate is 3 Mbps, and satellite-ground link allowance is larger than 6 dB (BER $< 10^{-6}$).

Data Preprocessing

The production of this Atlas adopts data of three-linear-array. The preprocessing procedures include communication channel processing, data source package extracting and sequencing, duplicates removing, image frame format reforming, radiometric correction and photometric calibration.

Radiometric Correction

Radiometric correction aims to correct the systematic or random radiometric distortion in the raw data due to external factors in the process of data acquisition and transmission. Radiometric correction will eliminate or correct the radiometric distortion caused by radiometric errors. The correction includes dark current subtraction, relative calibration, absolute calibration and filling of missing linear array.

Photometric Calibration

Photometric calibration mainly refers to the correction of radiancy of the targets acquired by sensors due to parameters of solar irradiance angle, emergence angle, phase angle and solar distance. With photometric calibration, all the image data will be normalized into the same irradiance parameter, to eliminate the influence of different irradiance conditions on the brightness of the images.

Processing of the Lunar Global Topographic Data

The lunar global topographic data of Chang'E-1 were obtained by the digital photogrammetry of the three-line-array CCD stereo image data. The processing of the global data includes: data organization, image matching, three-line block adjustment, topographic data generation at block scale and topographic data mosaiking at globe scale.

Data Organization

In order to ensure the realization of data processing and the precision of processing results, in the processing of lunar global topographic data, the lunar surface is divided into 202 blocks. Each block spans latitude range of 14 degrees, including about 20 tracks of images. In each block, the original images are cut to ensure that the adjacent block images have lateral overlap (latitudinal direction) of 4 degrees and forward overlap (longitudinal direction) of 2 degrees.

Image Matching

The SIFT (Scale Invariant Feature Transform) and LSM (Least Squares Matching) methods are used to find the matching points in forward, nadir and backward view images. The image matching provides sparse and dense matching points for the block adjustment and DEM generation. Finally, in the three-dimensional digital photogrammetry workstation environment, the data processing technicians check and edit the above points one by one.

Three-Line Block Adjustment

Block adjustment through three-line block independent models is used for block adjustment and global block adjustment to provide processing parameters for the absolute orientation and seamless mosaic of lunar global topographic data.

Topographic Data Generation at Block Scale

At first, the dense matching points and forward intersection method are employed to calculate the elevation, longitude and latitude coordinates of the matching points. Then, the elevation points are interpolated into DEM data of 500 m resolution by Triangulated Irregular Network (TIN). At last, according to the geometric position, the adjacent data are mosaicked into a block DEM data.

Topographic Data Mosaicking at Globe Scale

The global mosaic topographic data are stored with different projections. Mercator Projection with parallel of 35° is used in the range from 70°N–70°S and Azimuthal Projection with parallel of 70° in the range from 60° to the poles. After data quality check, the DEM data of every block are transformed under the same projection system. Subsequently, all of the projected data are mosaicked up. In the polar shadowed regions, the CCD camera can't obtain terrain information. Thus, laser altimetry data of CE-1 are used to fill in the "holes at the pole".

C. Li et al., *The Chang'E-1 Topographic Atlas of the Moon*,
DOI 10.1007/978-3-662-48439-5_2, © Springer-Verlag Berlin Heidelberg 2016

The Color Coded Shaded Topographic Maps of the Moon from China's First Lunar Probe Chang'E-1

North Polar Region

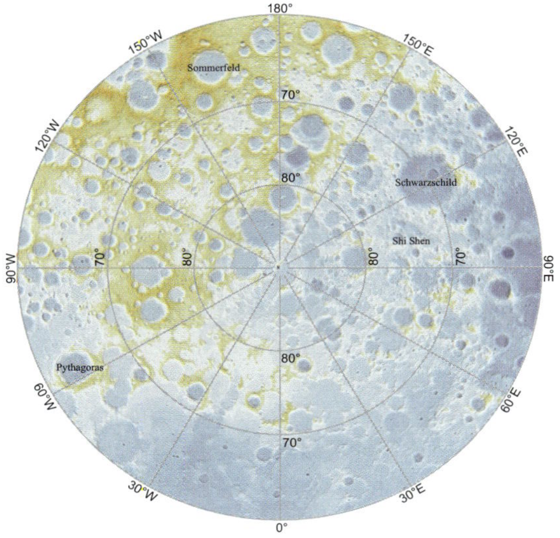

South Polar Region

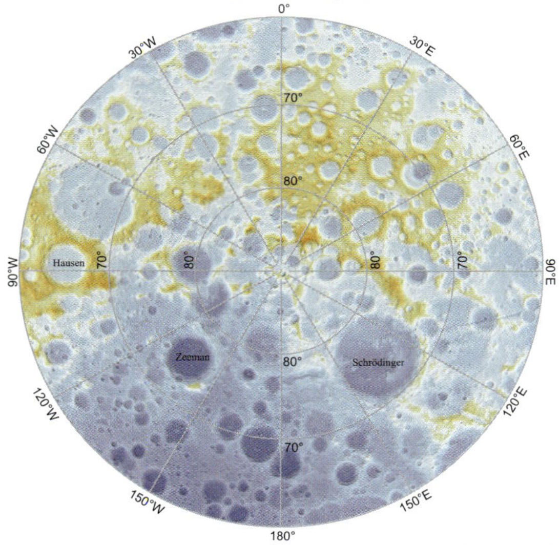

Scale of the polar regions

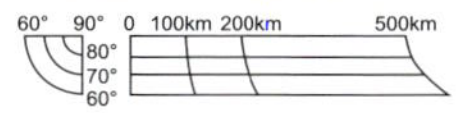

Sketch of the polar regions

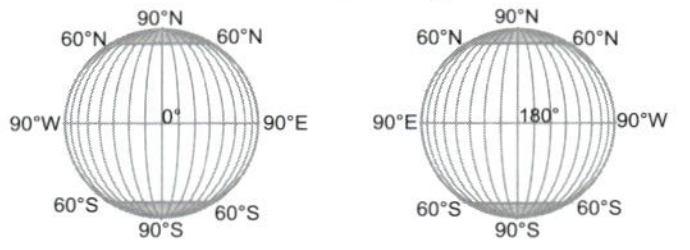

Graduation tints

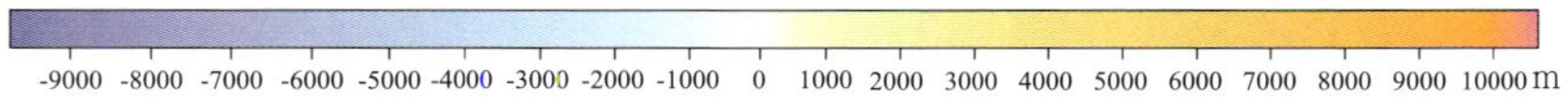

Index

Name	Lat.	Long.	Page
Aitken	16°48'S	173°24'E	148
Apollo	36°06'S	151°48'W	150
Aristoteles	50°12'N	17°24'E	44
Chang Heng	19°00'N	112°12'E	85
Chang-Ngo	12°42'S	2°06'W	118
Ching-Te	20°00'N	30°00'E	80
Compton	55°18'N	103°48'E	47
Copernicus	9°42'N	20°06'W	97
Crookes	10°18'S	164°30'W	109
Fabry	42°54'N	100°42'E	47
Gärtner	59°06'N	34°36'E	31
Gill	63°54'S	75°54'E	190
Giordano Bruno	35°54'N	102°48'E	65
Grimaldi	5°30'S	68°18'W	115
Guanghan Gong	44°7'N	19°31'W	42
Hausen	65°00'S	88°06'W	185
Helmholtz	68°06'S	64°06'E	190
Jackson	22°24'N	163°06'W	69
Kao	6°42'S	87°36'E	123
Kepler	8°06'N	38°00'W	96
Kuo Shou Ching	8°24'N	133°42'W	91
Lents	2°48'N	102°06'W	93
Mare Anguis	22°36'N	67°42'E	82
Mare Australe	38°54'S	93°00'E	162
Mare Cognitum	10°00'S	23°06'W	117
Mare Crisium	17°00'N	59°06'E	82
Mare Fecunditatis	7°48'S	51°18'E	121
Mare Frigoris	56°00'N	1°24'E	29
Mare Humboldtianum	56°48'N	81°30'E	32
Mare Humorum	24°24'S	38°36'W	136
Mare Imbrium	32°48'N	15°36'W	59
Mare Ingenii	33°42'S	163°30'E	166
Mare Marginis	13°18'N	86°06'E	103
Mare Moscoviense	27°18'N	147°54'E	87
Mare Nectaris	15°12'S	35°30'E	140
Mare Nubium	21°18'S	16°36'W	138
Mare Orientale	19°24'S	92°48'W	133
Mare Serenitatis	28°00'N	17°30'E	60
Mare Smythii	1°18'N	87°30'E	103
Mare Spumans	1°06'N	65°06'E	102
Mare Tranquillitatis	8°30'N	31°24'E	100
Mare Undarum	6°48'N	68°24'E	102
Montes Alpes	46°24'N	0°48'W	43
Montes Apenninus	18°54'N	3°42'W	78
Montes Caucasus	38°24'N	10°00'E	60
Montes Jura	47°06'N	34°00'W	42
Montes Pyrenaeus	15°36'S	41°12'E	141
Oceanus Procellarum	18°24'N	57°24'W	75
Ohm	18°24'N	113°30'W	72
Plato	51°36'N	9°24'W	43
Pontécoulant	58°42'S	66°00'E	190
Poynting	18°06'N	133°24'W	71
Pythagoras	63°30'N	63°00'W	27
Rima Sung-Mei	24°36'N	11°18'E	79
Rima Wan-Yu	20°00'N	31°30'W	77
Schrödinger	75°00'S	132°24'E	200
Schwarzschild	70°06'N	121°12'E	22
Shi Shen	76°00'N	104°06'E	22
Sinus Iridum	44°06'N	31°30'W	42
Sinus Medii	2°24'N	1°42'E	99
Sommerfeld	65°12'N	162°24'W	24
Tsu Chung-Chi	17°18'N	145°06'E	87
Tycho	43°24'S	11°06'W	174
Vavilov	0°48'S	137°54'W	111
Wan-Hoo	9°48'S	138°48'W	111
Zeeman	75°12'S	133°36'W	195

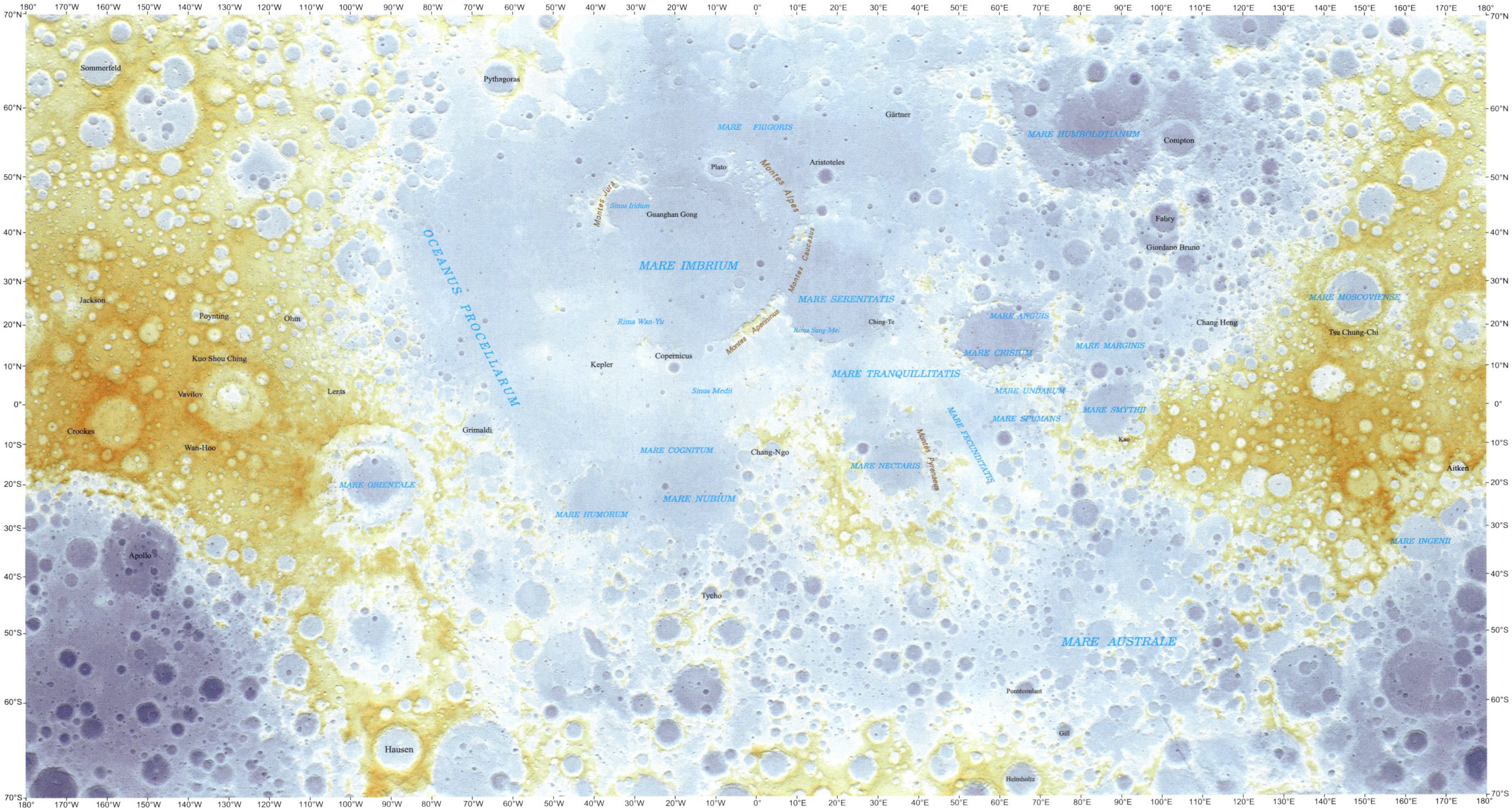

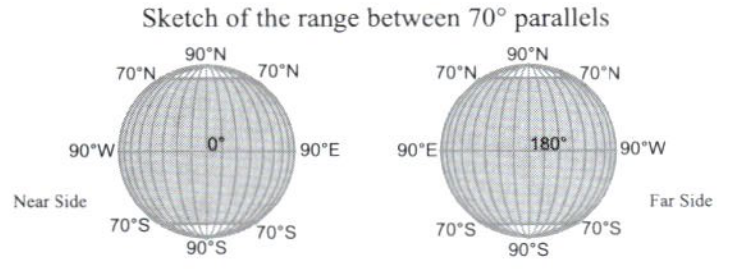

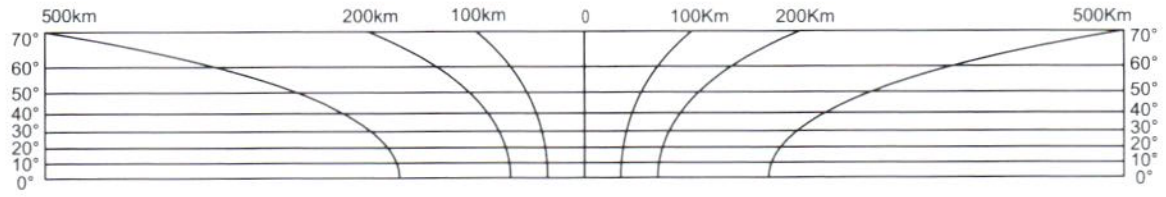

The Color Coded Shaded Topographic Map of the Moon from China's First Lunar Probe Chang'E-1 is made of the Digital Elevation Model (DEM) data, The space resolution of the DEM data is 500 m, with the horizontal accuracy 192 m and the vertical accuracy 120 m. The elevation information is represented by different colors (see the graduation tints).The areas of low altitude below the geoid are shown in blue tones, and those above the geoid in deep yellow tones. The lunar geoid is defined as a spherical surface of radius 1737.4 km. The Mercator projection, adopting in the 35°parallel of both south and north latitude, is used for the right part of the map (the rectangular image), including an area between 70° parallels. The left part of the map (the two pies) displays the polar regions of the Moon, covering the ranges of 60°N–90°N and 60°S–90°S respectively. The azimuth isometric projection, adopting in the 70°parallel, is used for the polar regions.

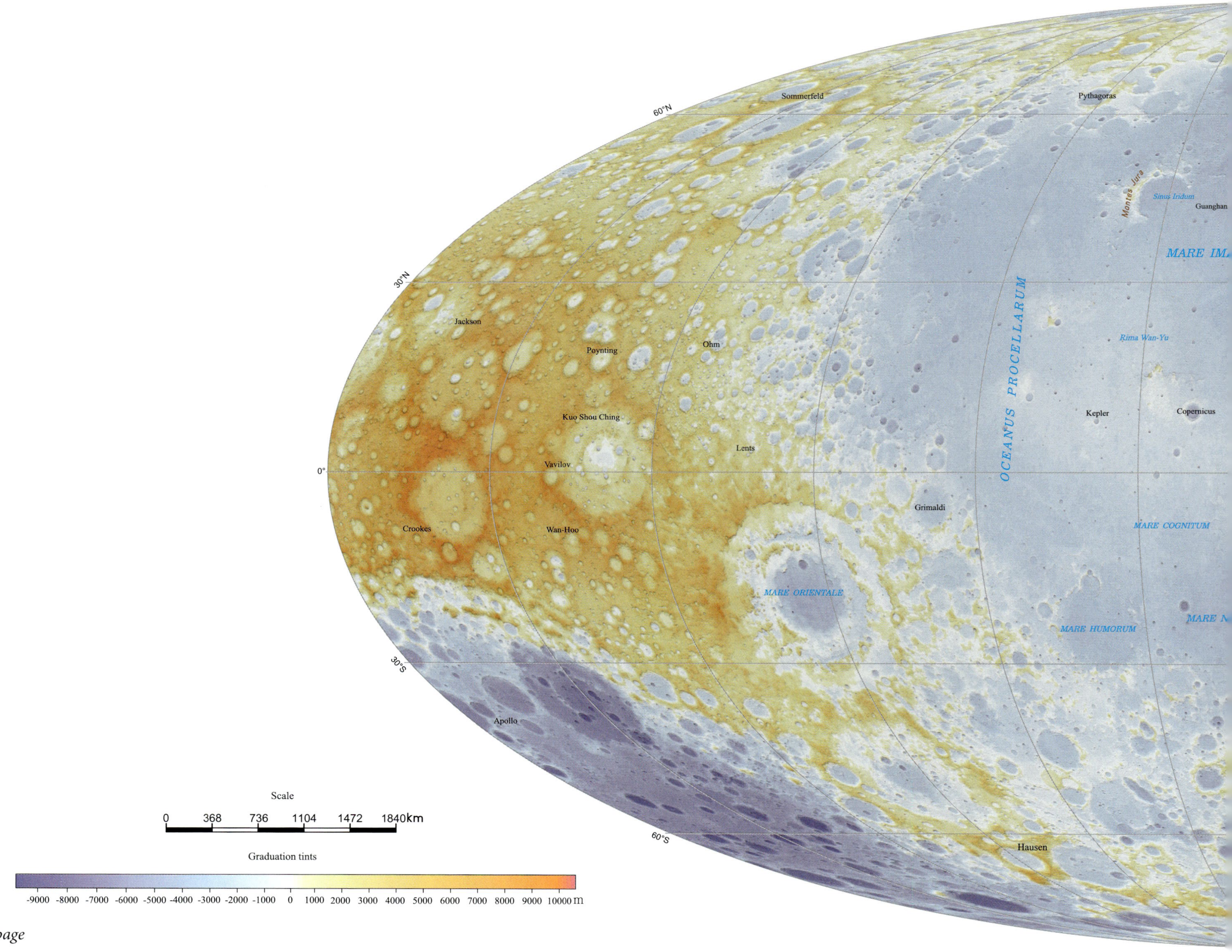

continues on next page

The Color Coded Shaded Topographic Map of the Moon from China's First Lunar Probe Chang'E-1 is made of the Digital Elevation Model (DEM) data. The space resolution of the DEM data is 500 m,with the horizontal accuracy 192 m and the vertical accuracy 120 m. The elevation information is represented by different colors (see the graduation tints). The areas of low altitude below the geoid are shown in blue tones, and those above the geoid in deep yellow tones. The lunar geoid is defined as a spherical surface of radius 1737.4 km. The Mollweide projection is used in this map, covering the whole Moon, from 180°W to 180°E in longitude, and 90°N to 90°S in latitude. In this map, the near side of the Moon (90°W–90°E) is on the center area and the far side is separately surrounding the near side.

Near Side

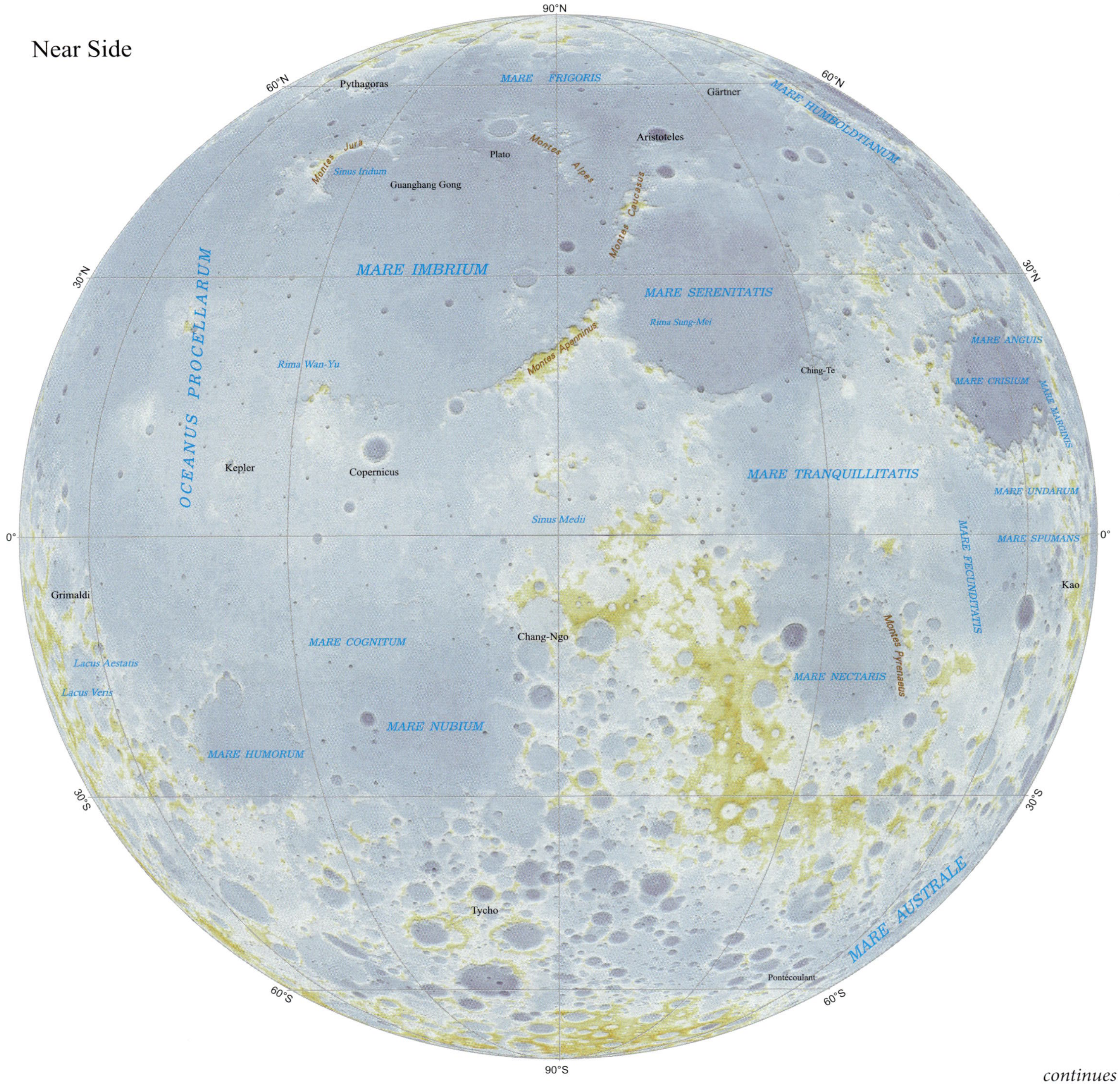

continues on next page

The Color Coded Shaded Topographic Map of the Moon from China's First Lunar Probe Chang'E-1 is made of the Digital Elevation Model (DEM) data, The space resolution of the DEM data is 500 m,with the horizontal accuracy 192 m and the vertical accuracy 120 m. The elevation information is represented by different col ors (see the graduation tints). The areas of low altitude below the geoid are shown in blue tones, and those above the geoid in deep yellow tones. The lunar geoid is defined as a spherical surface of radius 1737.4 km. The Orthographic projection is used in this map, covering the whole Moon, from 180°W to 180°E in longitude, and 90°N to 90°S in latitude. The left image is the near side of the Moon, and the right one is the far side.

Far Side

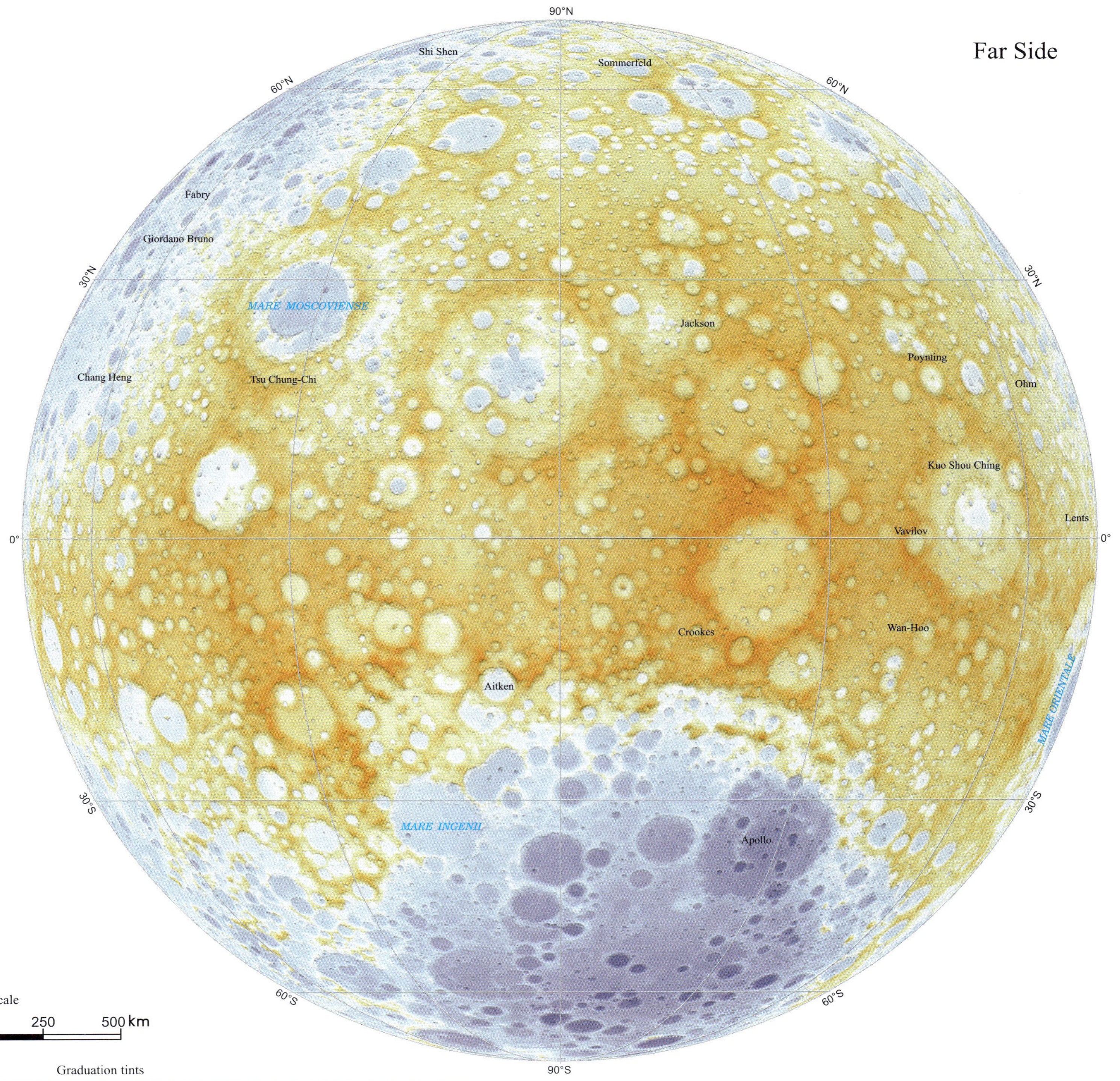

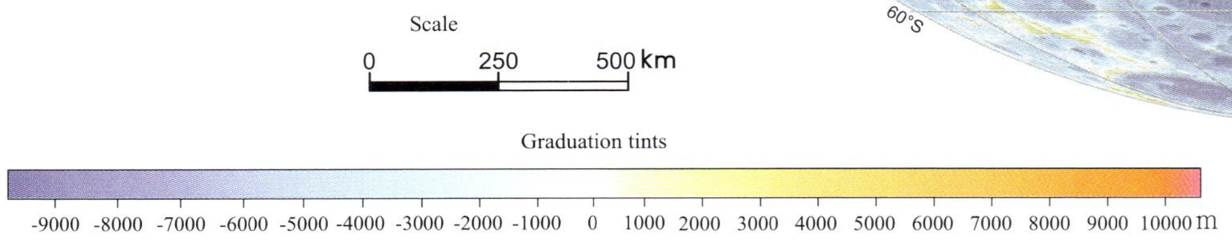

Part III Subdivision Topographic Maps of the Moon

The Chang'E-1 Topographic Atlas of the Moon and *The Chang'E-1 Image Atlas of the Moon* (published in Jun., 2010) use the same subdivision model and projections, which have 188 subdivisions. The subdivision model and projection parameters are described in the ◘ Tab. 3.1. In order to facilitate the inquiry, the page number is marked on the index of the subdivision topographic maps, shown in page 12–13.

In the subdivision topographic maps, color coded shaded relief maps are used for the terrains of the lunar surface, and the contours to show the elevation information. In order to show the topographic features of the lunar surface accurately, the lunar topography is divided into five main types: deep mare, shallow mare, plains, hills and mountains. Therefore, elevation (encoding) color tables are designed according to the topographic types and placed at the bottom of the map. In the shaded relief maps, light comes from north with solar elevation angle of 60–70 degrees. The contours with 500 m intervals are extracted from DEM data. After cartographic generalization, the contours in brown are assigned the same scale as the shaded relief map to ensure matching with the latter. In addition, elevations of the highest and lowest points in each subdivision topographic map are marked out.

The main surface features with prominent topographic characteristics are annotated with 17 types of lunar location names, including Mare, Crater, Rima, shown in the legend. In order to help understand and find the Moon names, location, geometry, and page number of the lunar place names are described in the Appendix.

◘ **Table 3.1** Subdivision Scope and Coding of CE-1 Topographic Maps of the Moon

Code	Latitude range	Longitude range	Projection	Central meridian	Standard parallel
A***	84°N–90°N	360°	Polar azimuthal projection	0°	90°N
B***	70°N–84°N	45°	Lambert conformal conic projection	Average value of the longitude range	77°N
C***	56°N–70°N	30°			63°N
D***	42°N–56°N	24°			49°N
E***	28°N–42°N	20°			35°N
F***	14°N–28°N	18°			21°N
G***	0°–14°N	18°	Mercator projection		7°N
H***	0°–14°S	18°			7°S
I***	14°S–28°S	18°	Lambert conformal conic projection		21°S
J***	28°S–42°S	20°			35°S
K***	42°S–56°S	24°			49°S
L***	56°S–70°S	30°			63°S
M***	70°S–84°S	45°			77°S
N***	84°S–90°S	360°	Polar azimuthal projection	0°	90°S

C. Li et al., *The Chang'E-1 Topographic Atlas of the Moon*,
DOI 10.1007/978-3-662-48439-5_3,

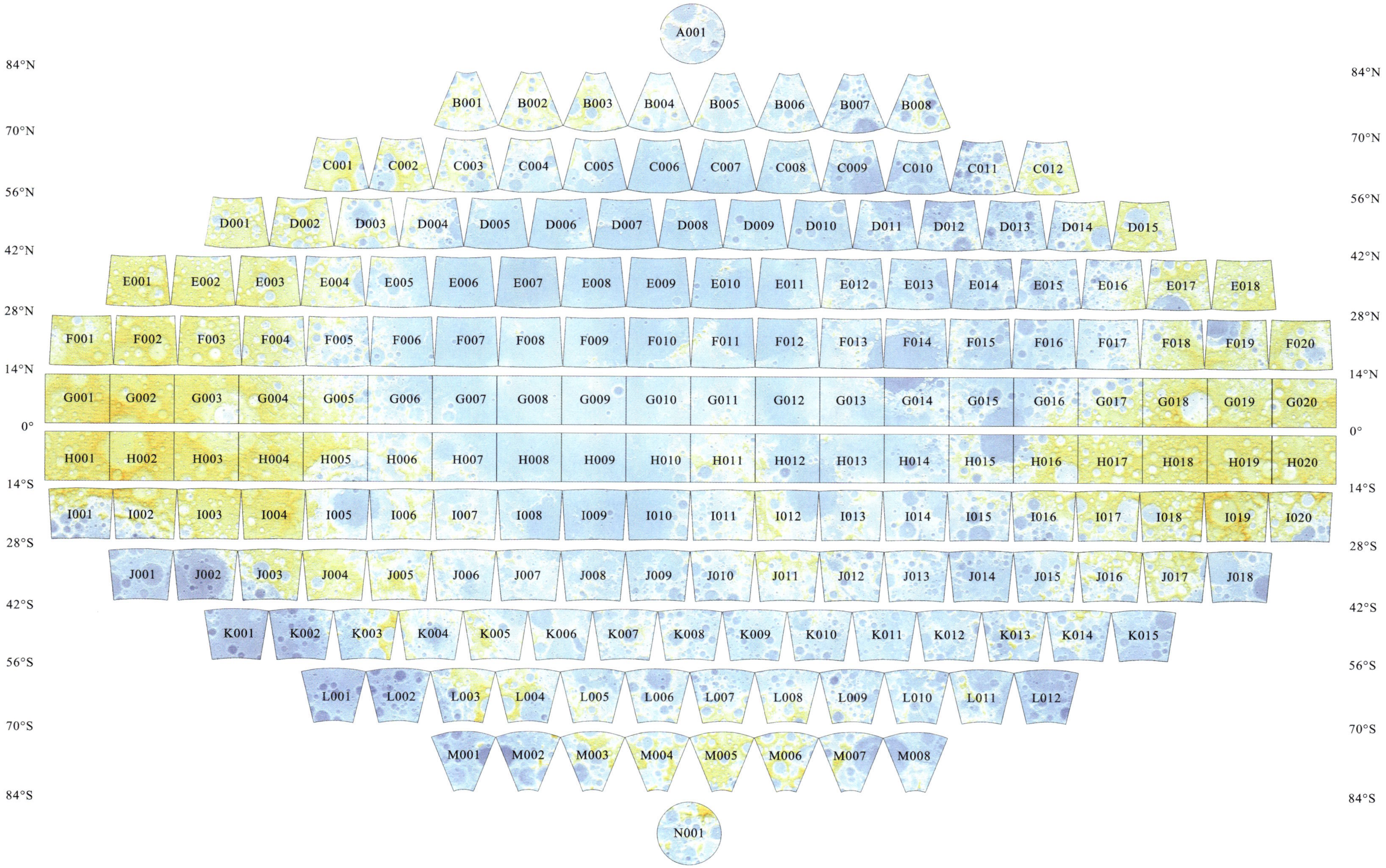

Code of the Subdivision Topographic Maps of the Moon

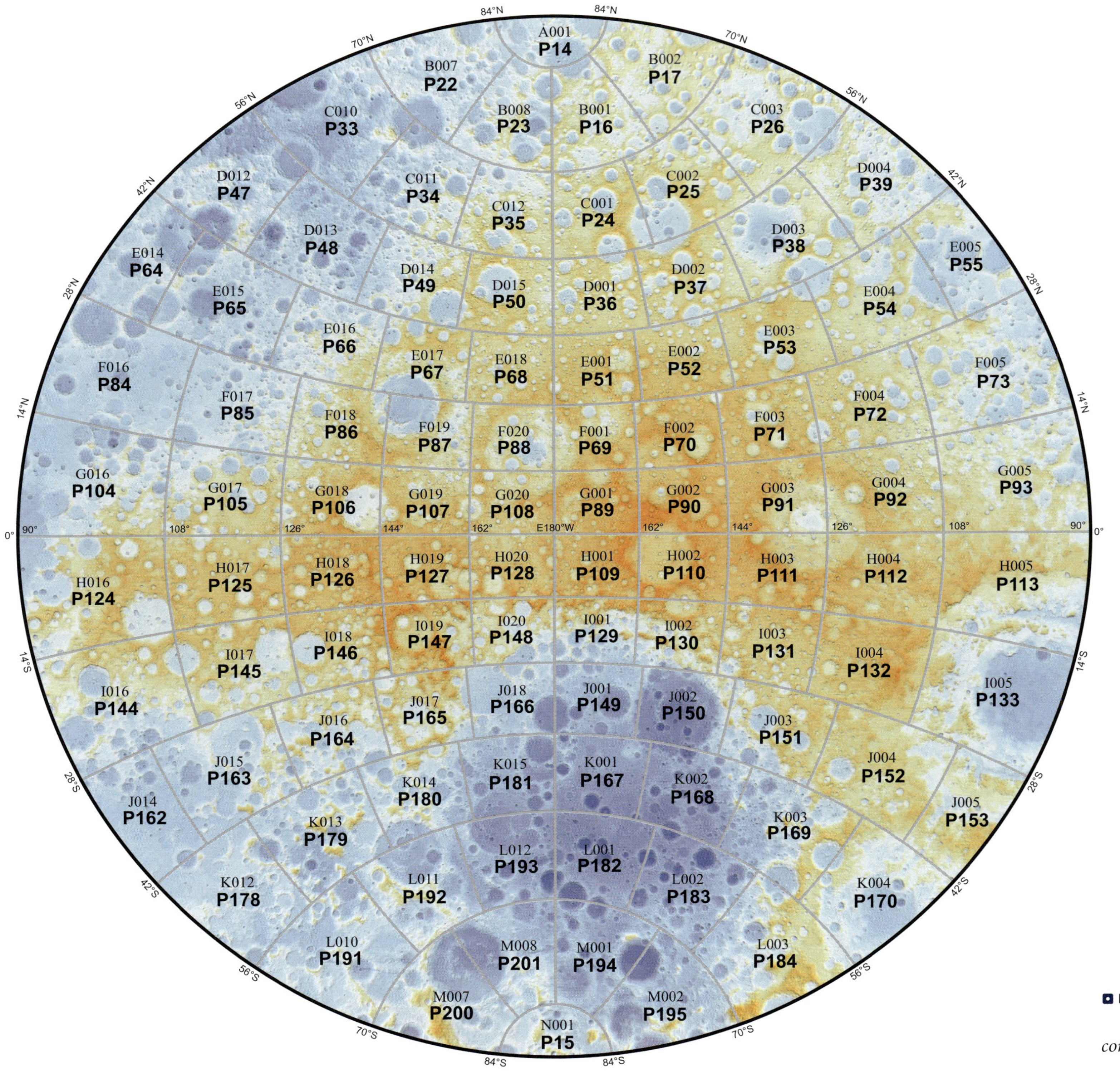

Index of the Subdivision Topographic Maps of the Moon

continues on next page

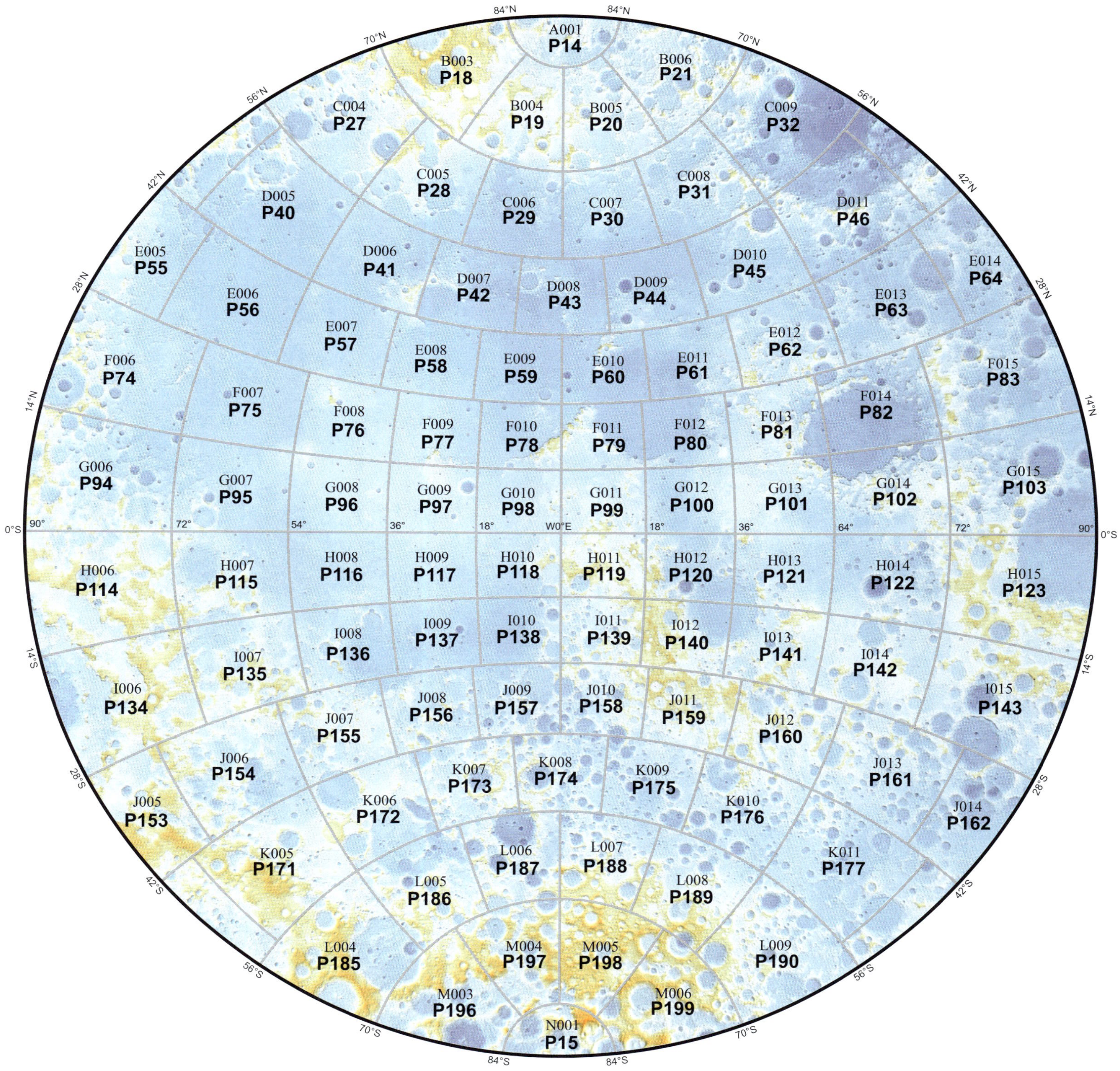
A001 P14
B003 P18
B004 P19
B005 P20
B006 P21
C004 P27
C005 P28
C006 P29
C007 P30
C008 P31
C009 P32
D005 P40
D006 P41
D007 P42
D008 P43
D009 P44
D010 P45
D011 P46
E005 P55
E006 P56
E007 P57
E008 P58
E009 P59
E010 P60
E011 P61
E012 P62
E013 P63
E014 P64
F006 P74
F007 P75
F008 P76
F009 P77
F010 P78
F011 P79
F012 P80
F013 P81
F014 P82
F015 P83
G006 P94
G007 P95
G008 P96
G009 P97
G010 P98
G011 P99
G012 P100
G013 P101
G014 P102
G015 P103
H006 P114
H007 P115
H008 P116
H009 P117
H010 P118
H011 P119
H012 P120
H013 P121
H014 P122
H015 P123
I006 P134
I007 P135
I008 P136
I009 P137
I010 P138
I011 P139
I012 P140
I013 P141
I014 P142
I015 P143
J005 P153
J006 P154
J007 P155
J008 P156
J009 P157
J010 P158
J011 P159
J012 P160
J013 P161
J014 P162
K005 P171
K006 P172
K007 P173
K008 P174
K009 P175
K010 P176
K011 P177
L004 P185
L005 P186
L006 P187
L007 P188
L008 P189
L009 P190
M003 P196
M004 P197
M005 P198
M006 P199
N001 P15
84°N
70°N
56°N
42°N
28°N
14°N
0°S
14°S
28°S
42°S
56°S
70°S
84°S
90°
72°
54°
36°
18°
W0°E
18°
36°
64°
72°
90°

Image Code A001

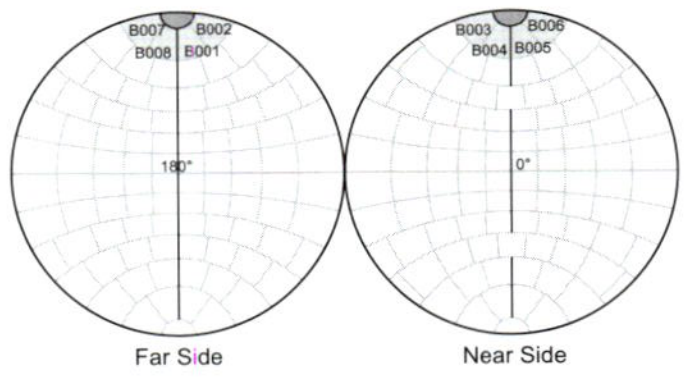

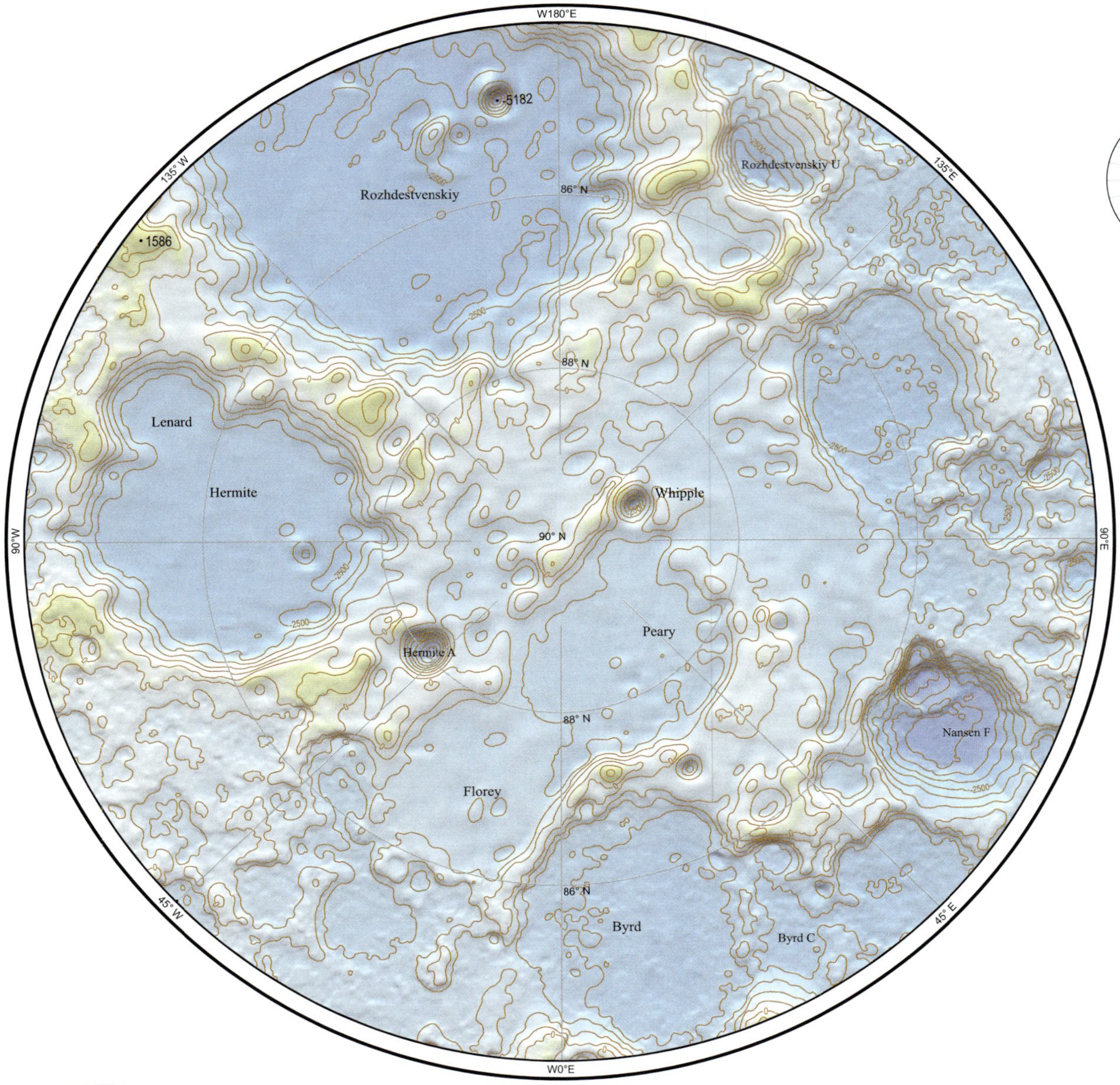

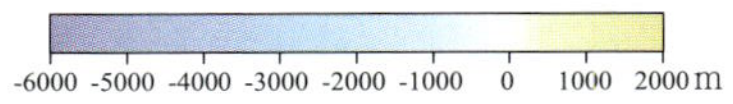

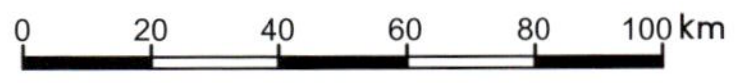

Image Code **N001**

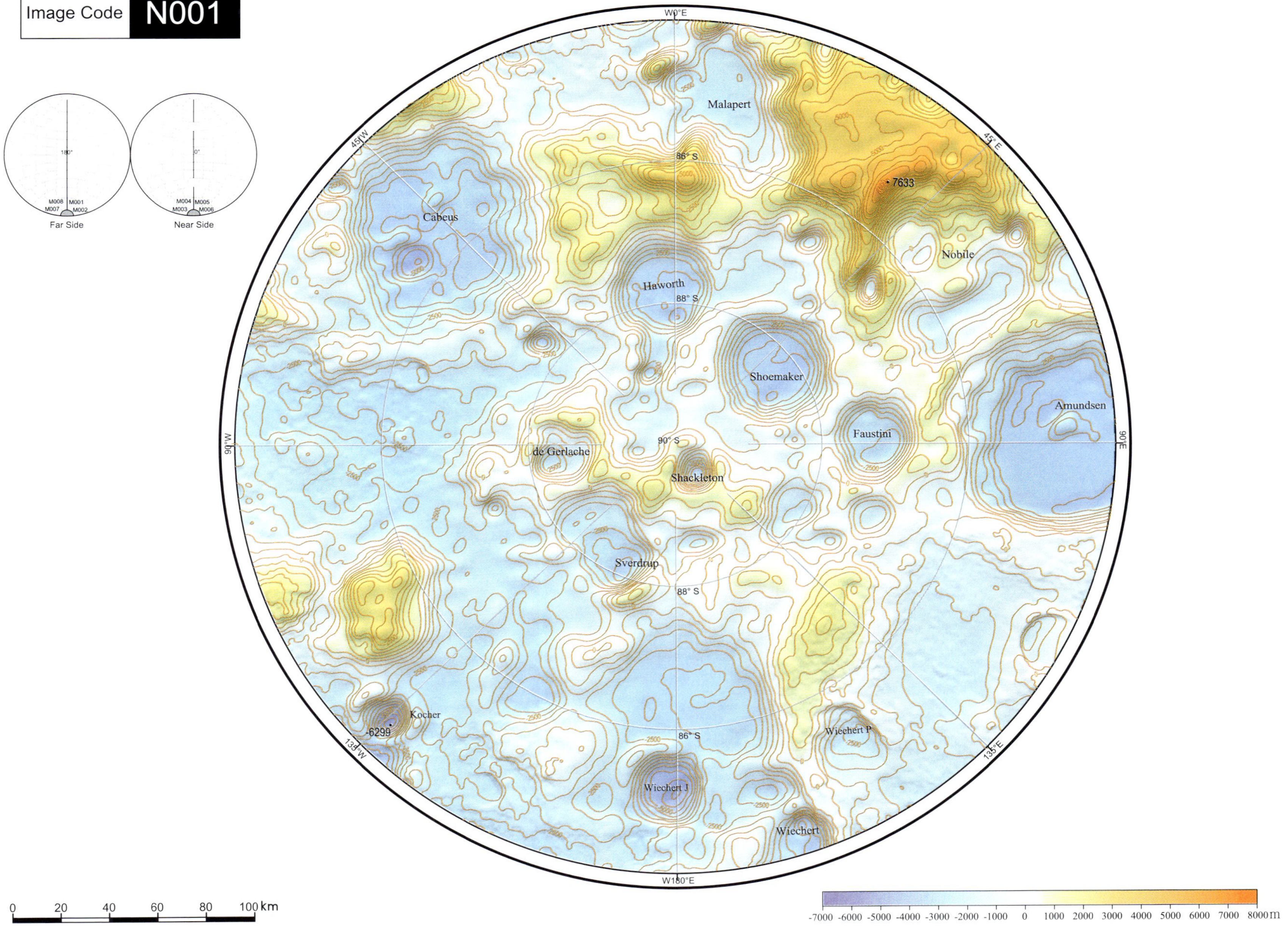

Image Code B001

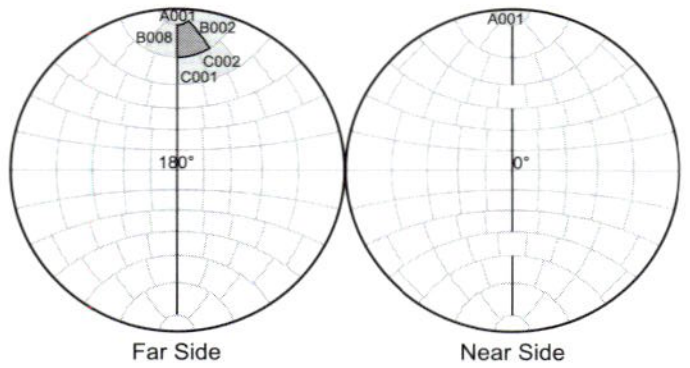

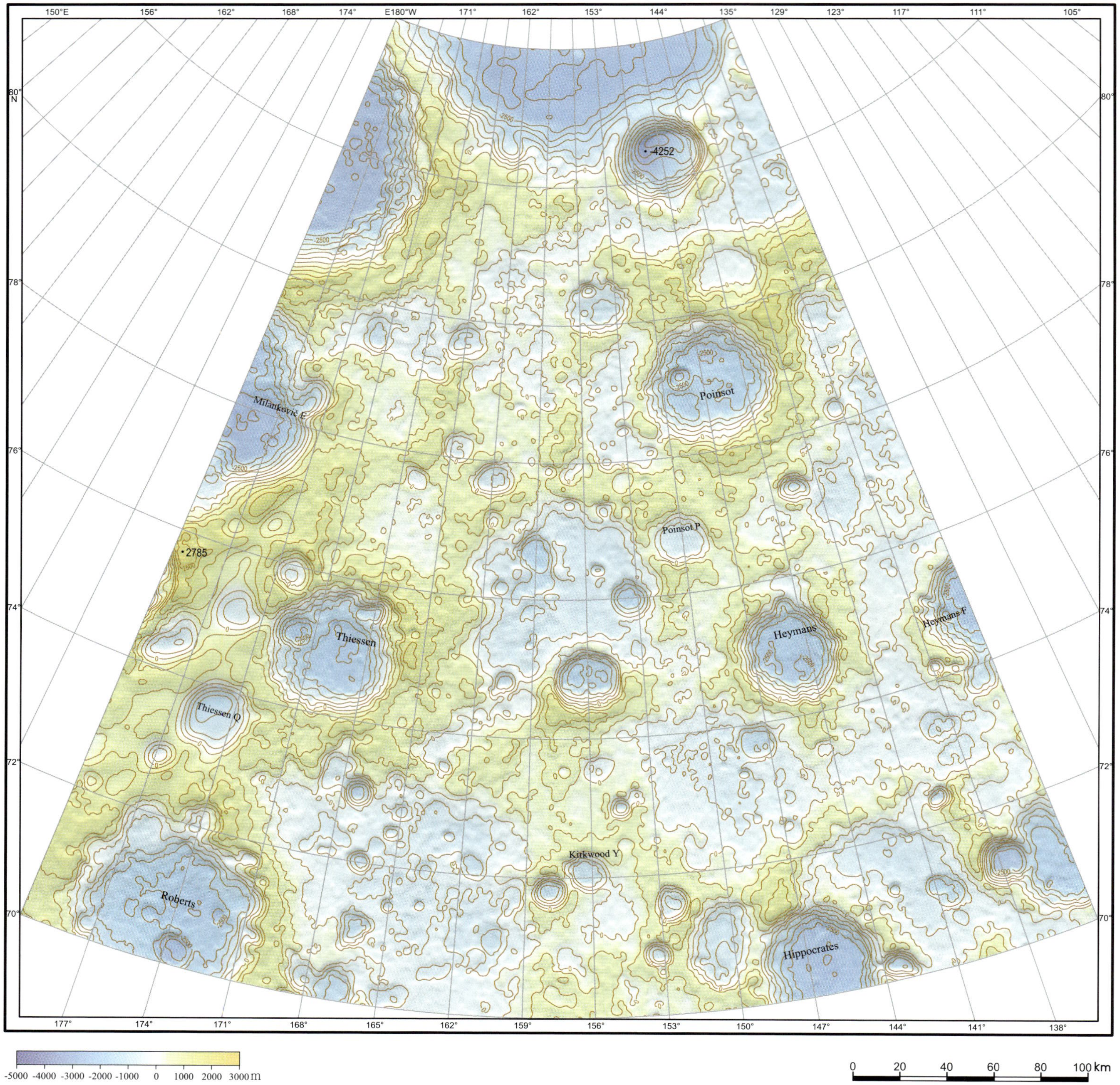

Image Code **B002**

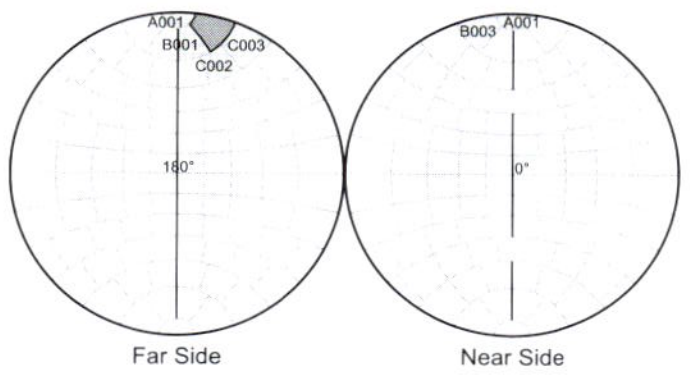

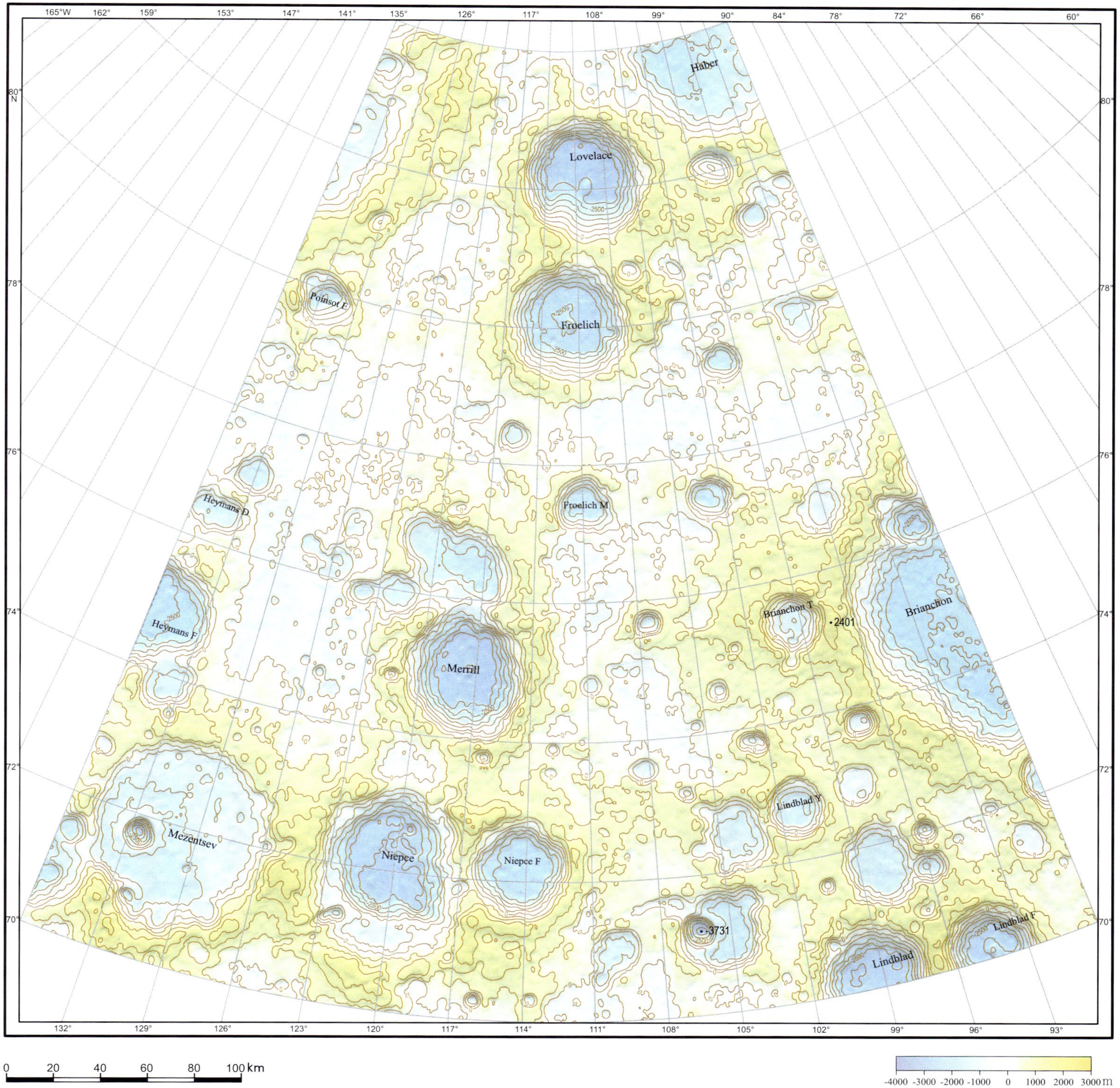

Image Code B003

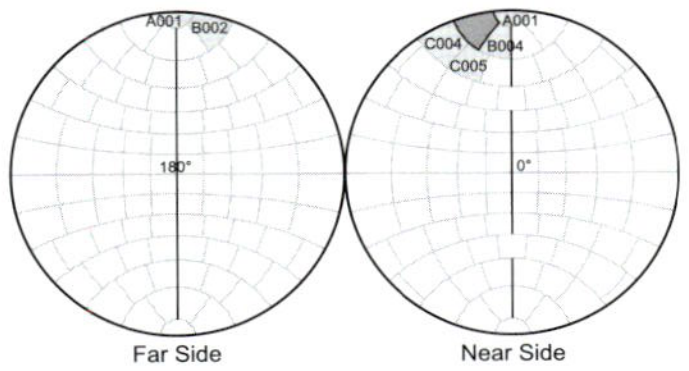

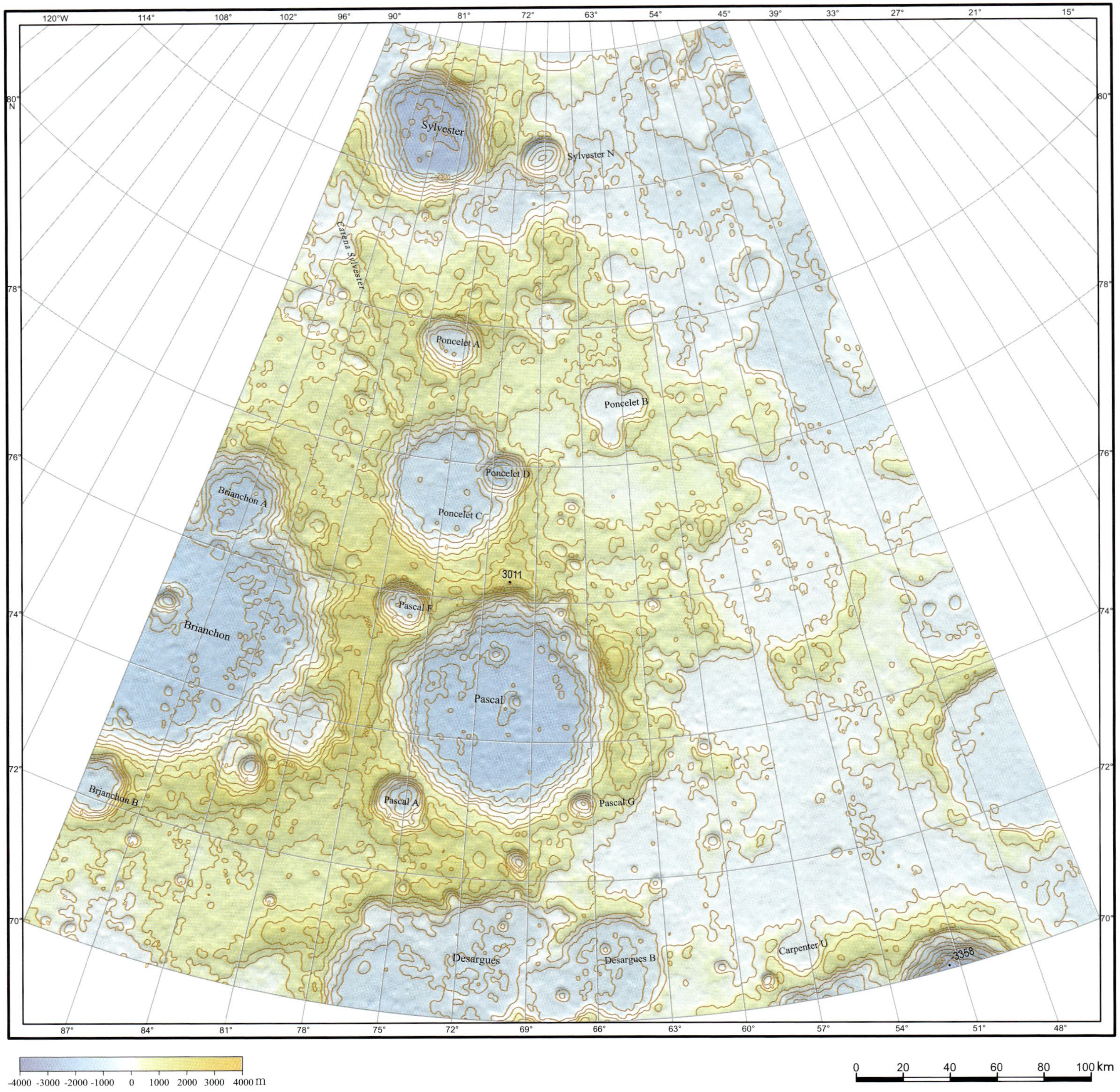

Image Code B004

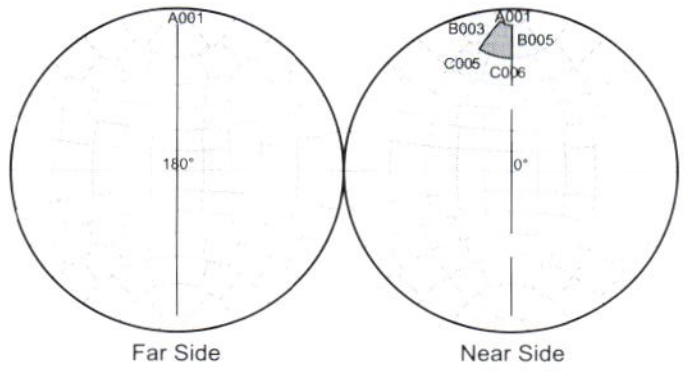

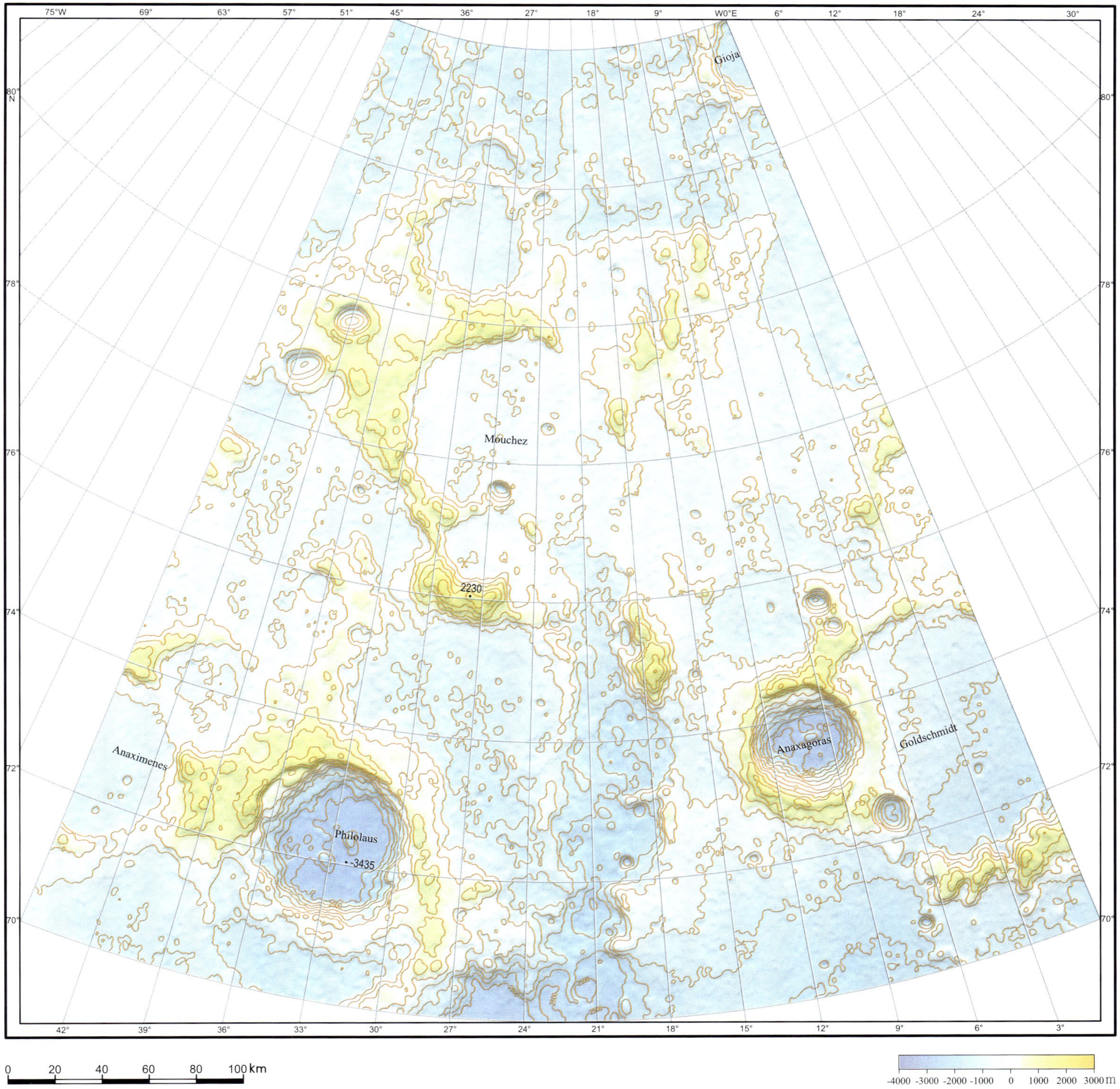

Image Code B005

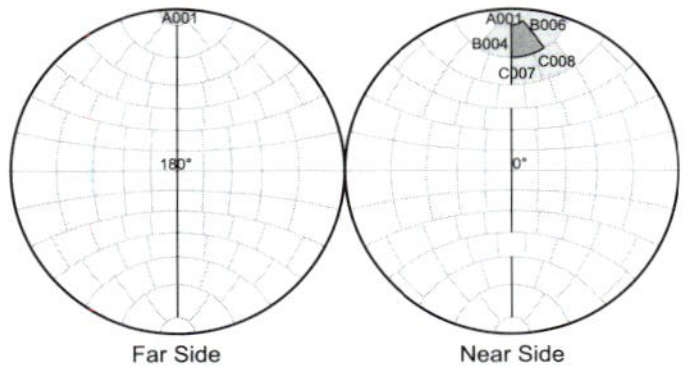

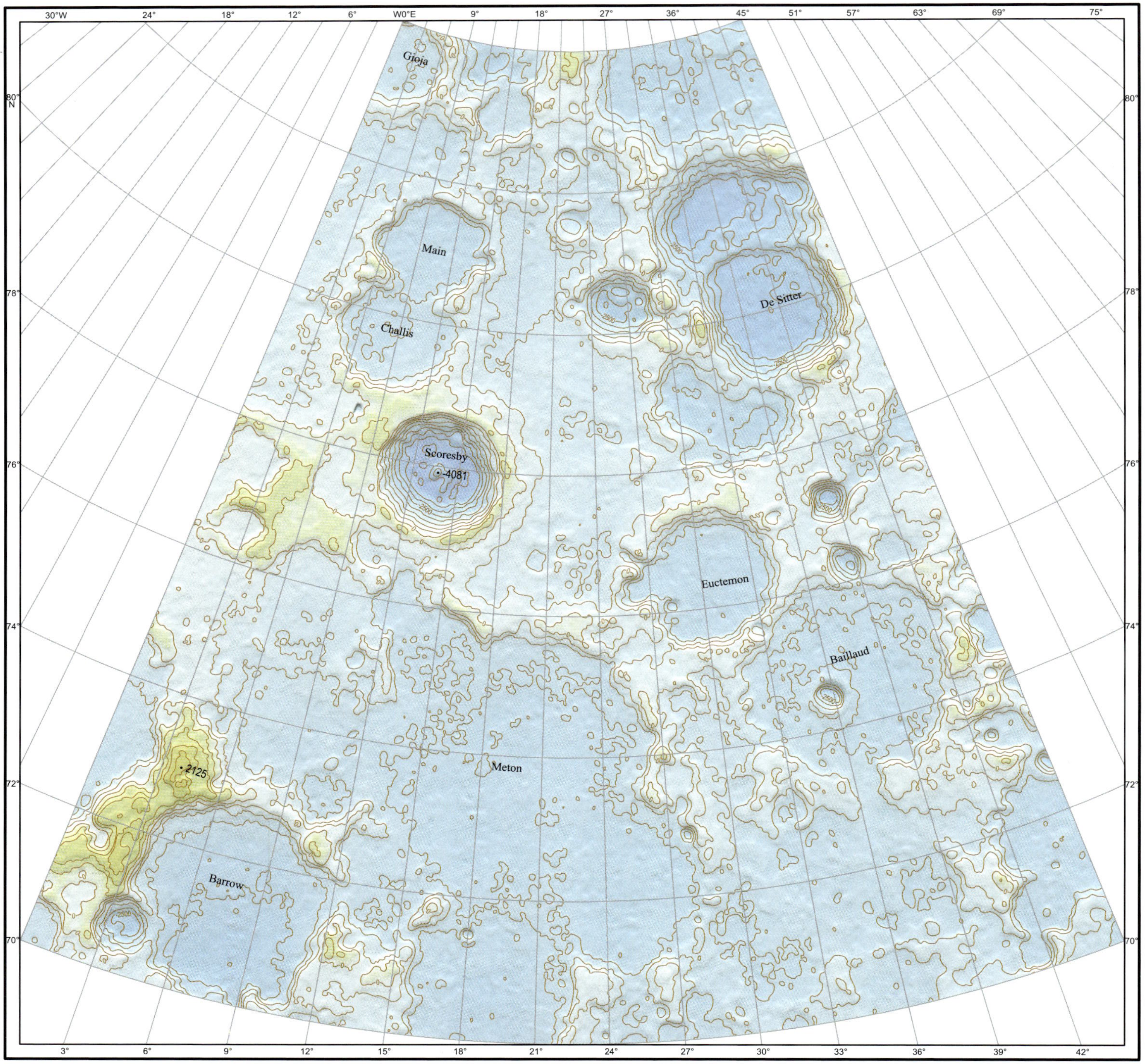

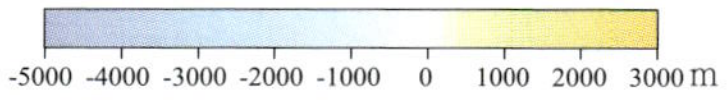

Vallis, Valles

Sinus, Sinūs

Intermediate contour

Index contour

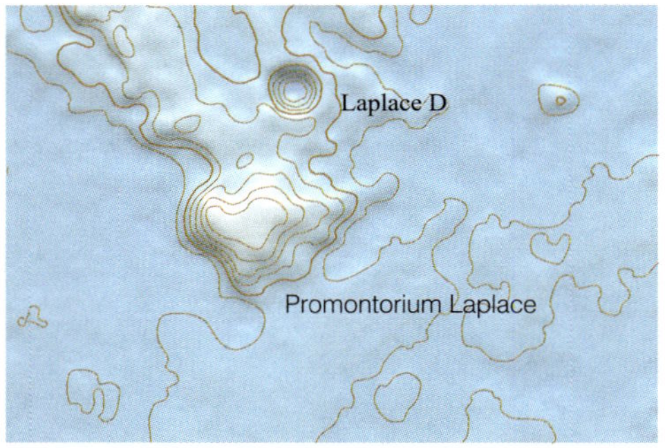

Promontorium, Promontoria

Palus, Paludes

• 5035

Spot elevation

Rima, Rimae

Legend

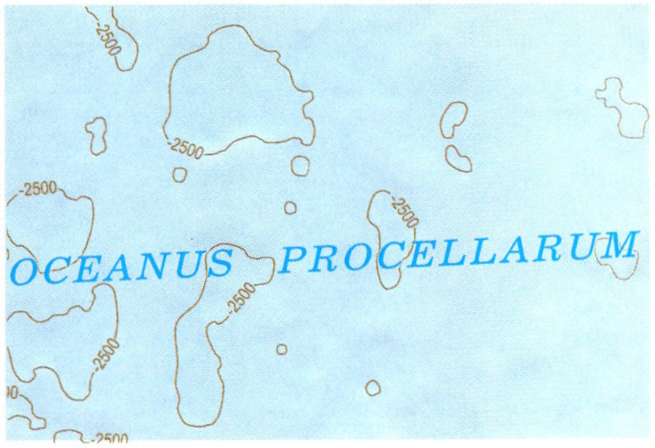

Oceanus, Oceani

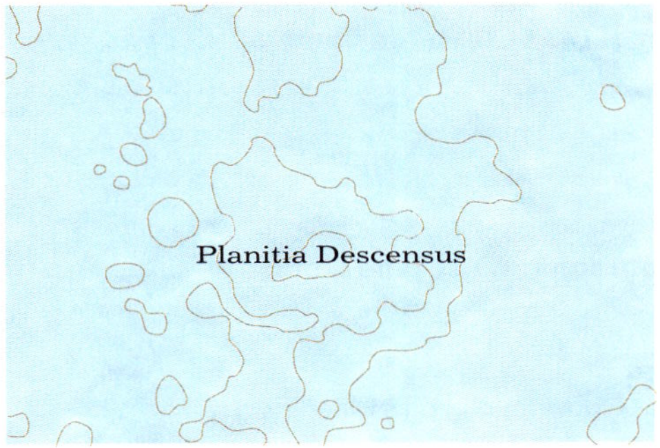

Planitia, Planitiae

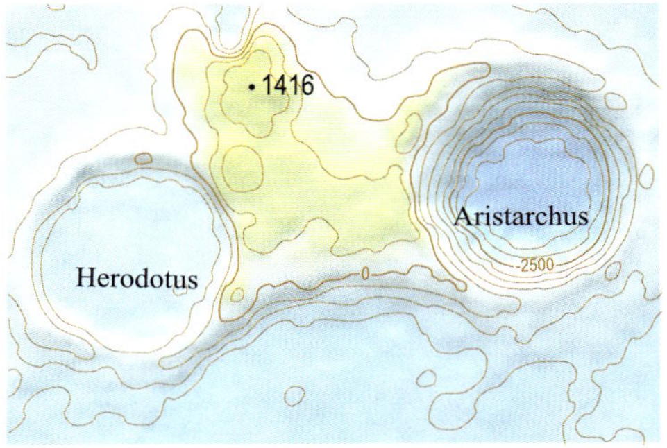

Crater, Craters

Mare, Maria

Mons, Montes

Satellite Feature

Lacus, Lacūs

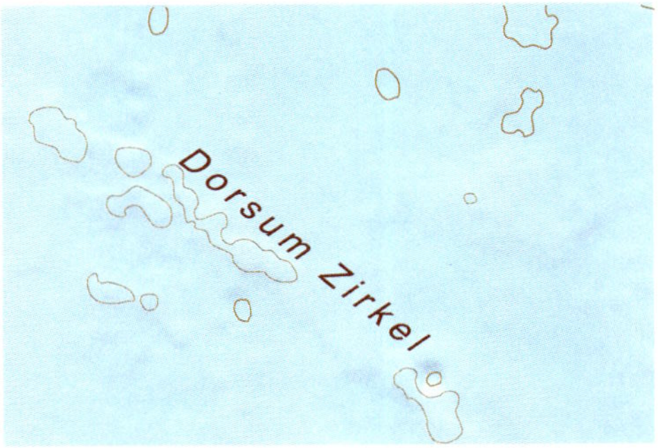

Dorsum, Dorsa

Catena, Catenae

Table of Contents

Part I Introduction of the CCD Stereo Camera 1
CCD Stereo Camera 1
Characteristics of Orbits and Attitudes 1
Data Acquisition and Coverage 1

Part II Global Topographic Map of the Moon 3
Data Receiving 3
Data Preprocessing 3
Processing of the Lunar Global Topographic Data 3
The Color Coded Shaded Topographic Map of the Moon from China's First Lunar Probe Chang'E-1 4

Part III Subdivision Topographic Maps of the Moon 10

Appendix-Gazetteer 203

Image Code B006

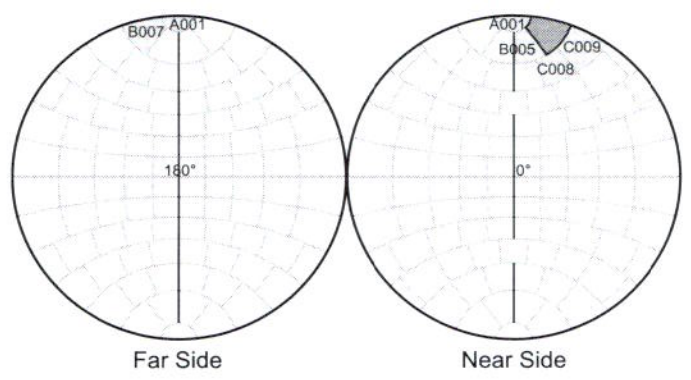

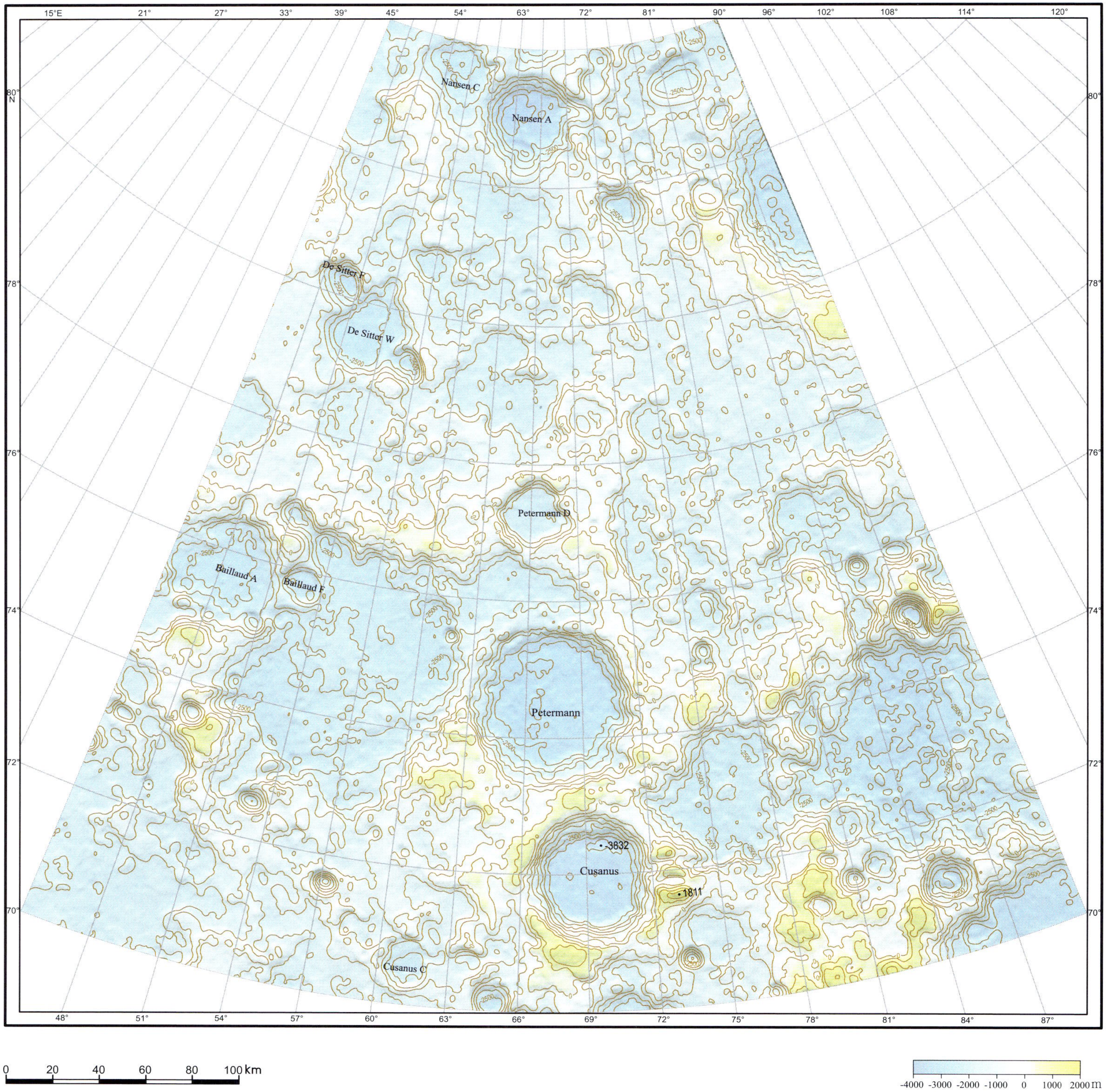

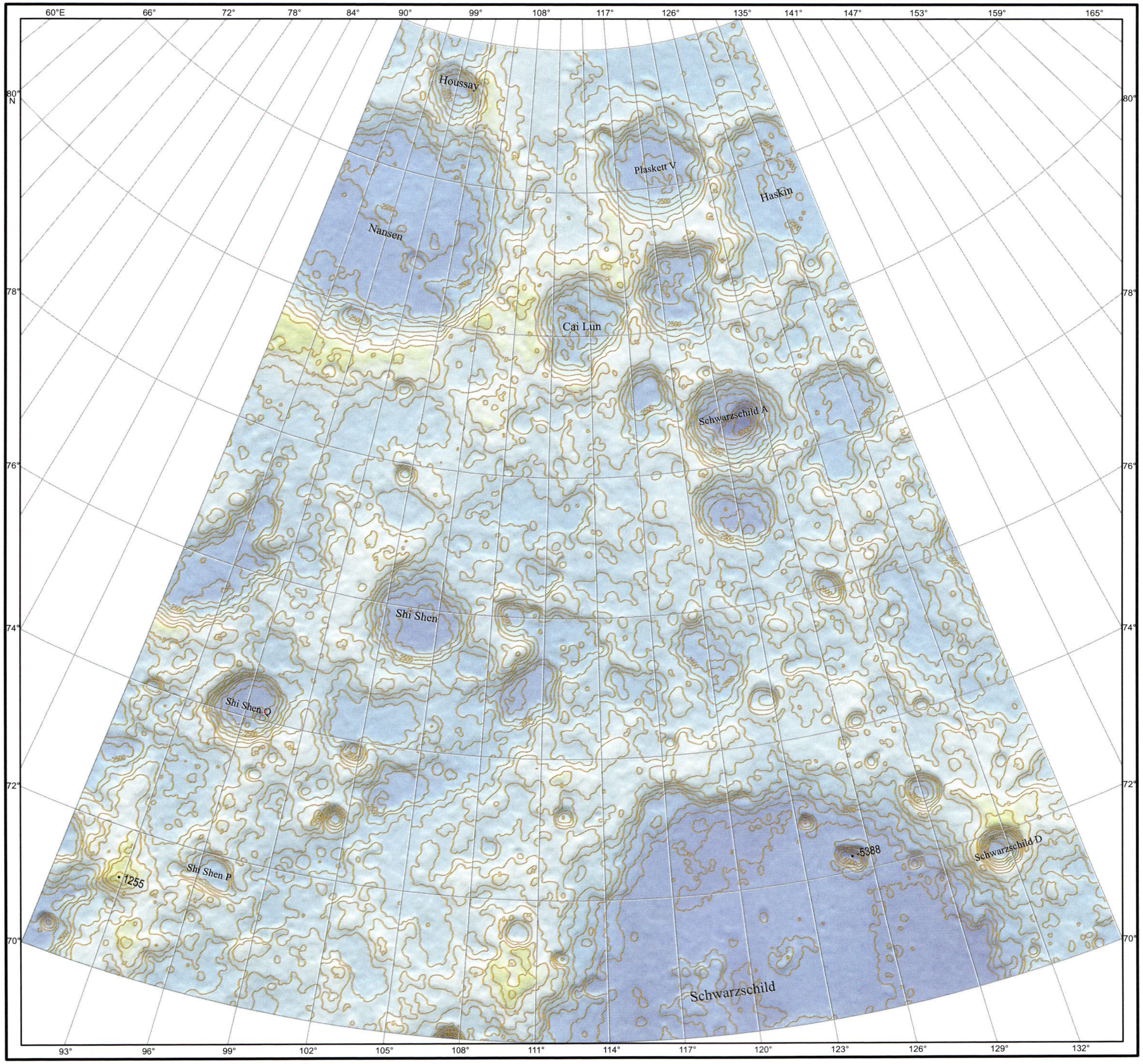

Image Code **B007**

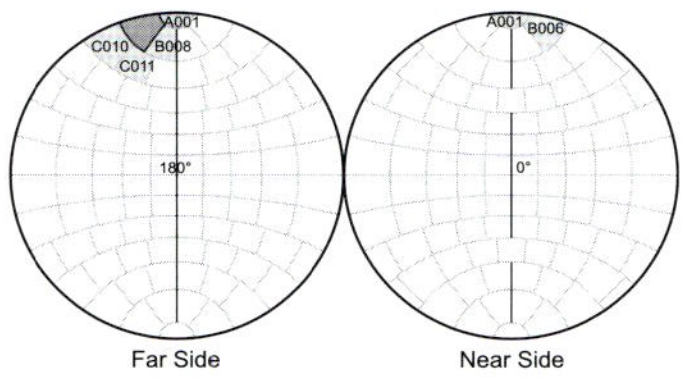

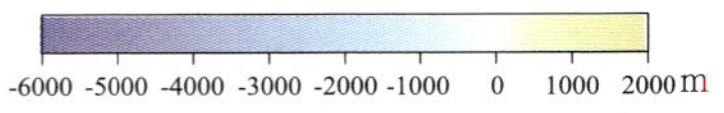

Image Code **B008**

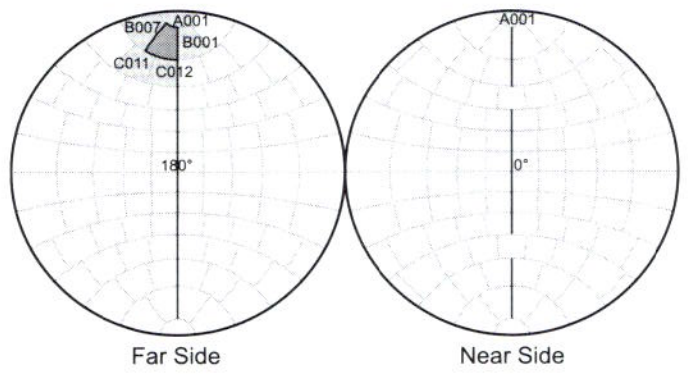

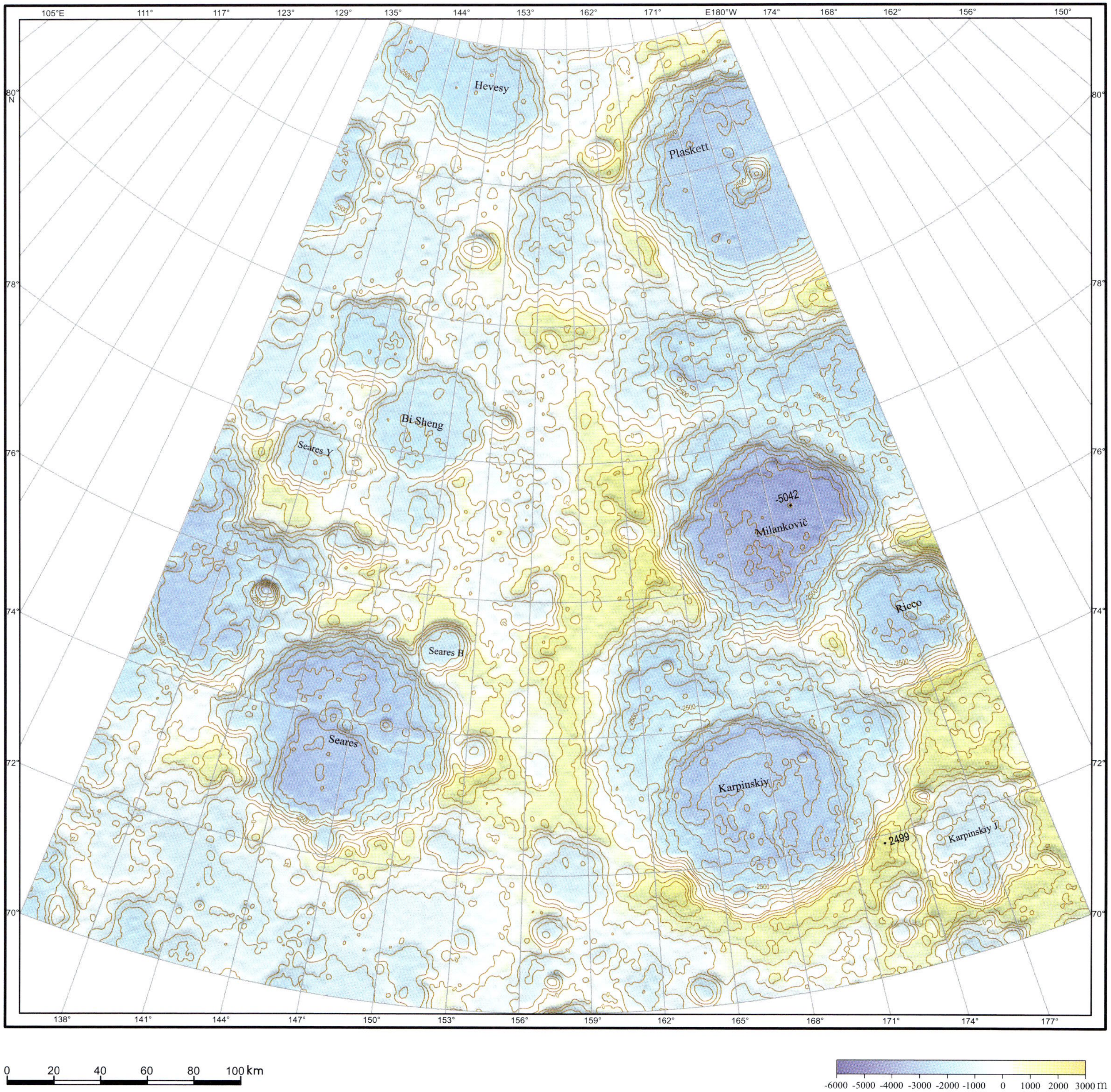

Image Code C001

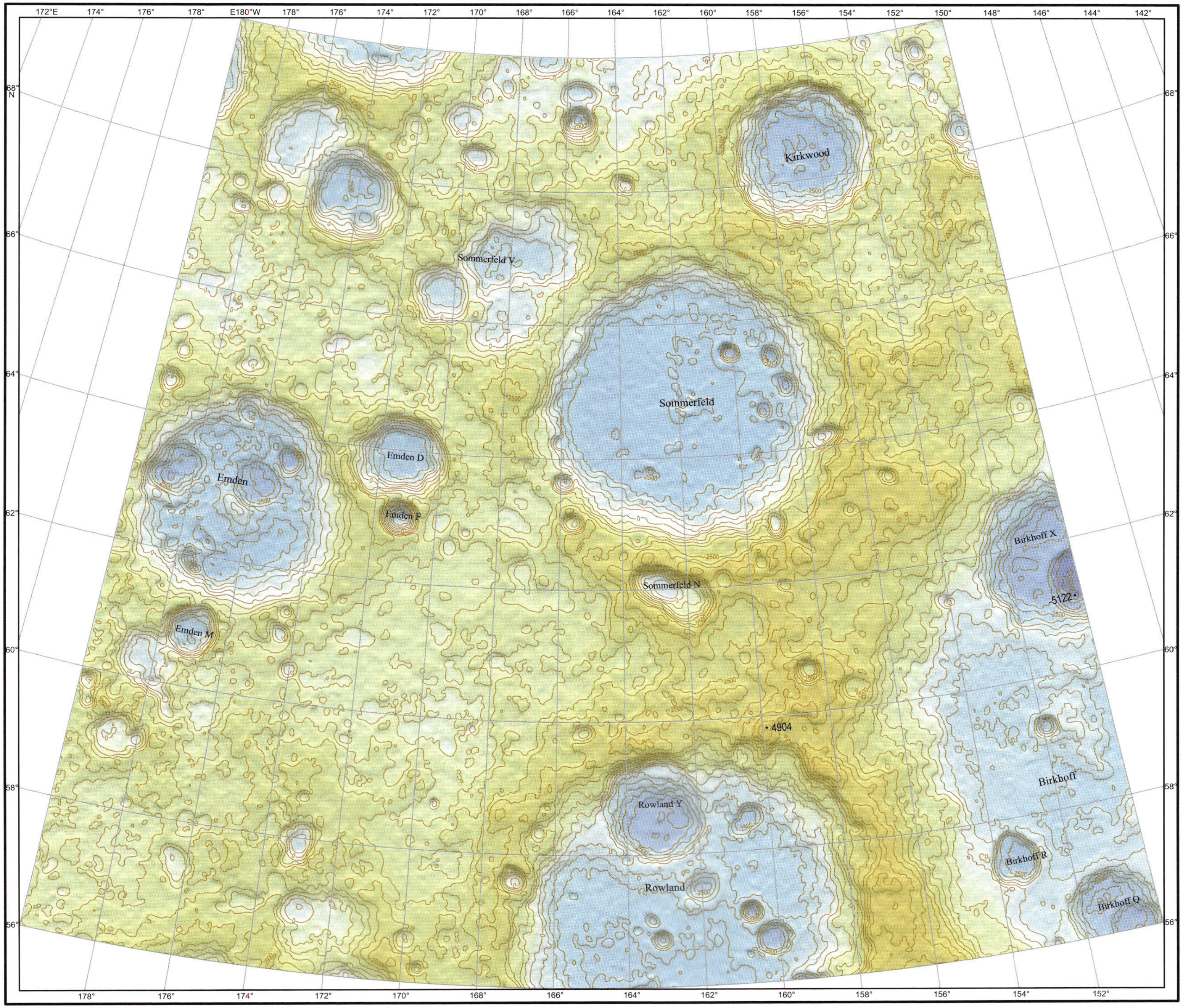

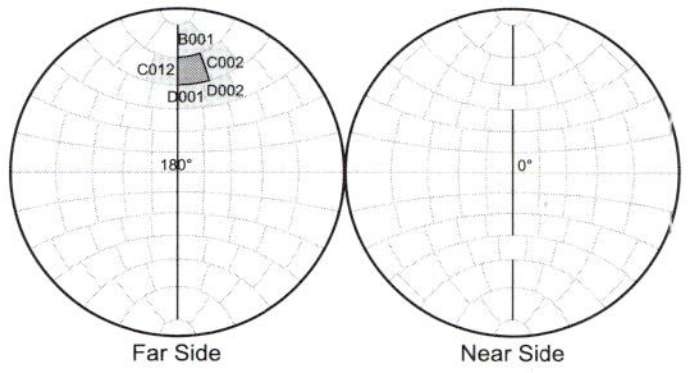

-6000 -5000 -4000 -3000 -2000 -1000 0 1000 2000 3000 4000 5000 m

0 20 40 60 80 100 km

Image Code C002

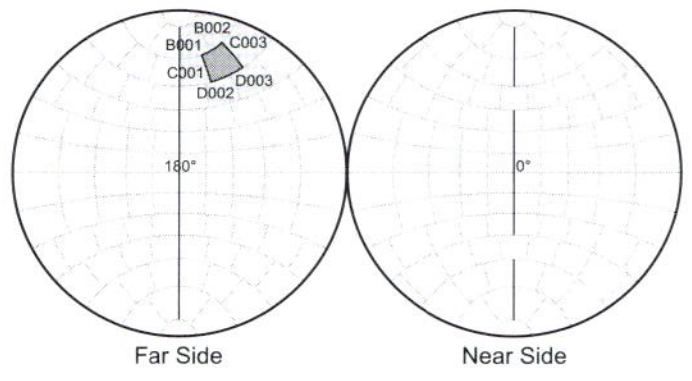

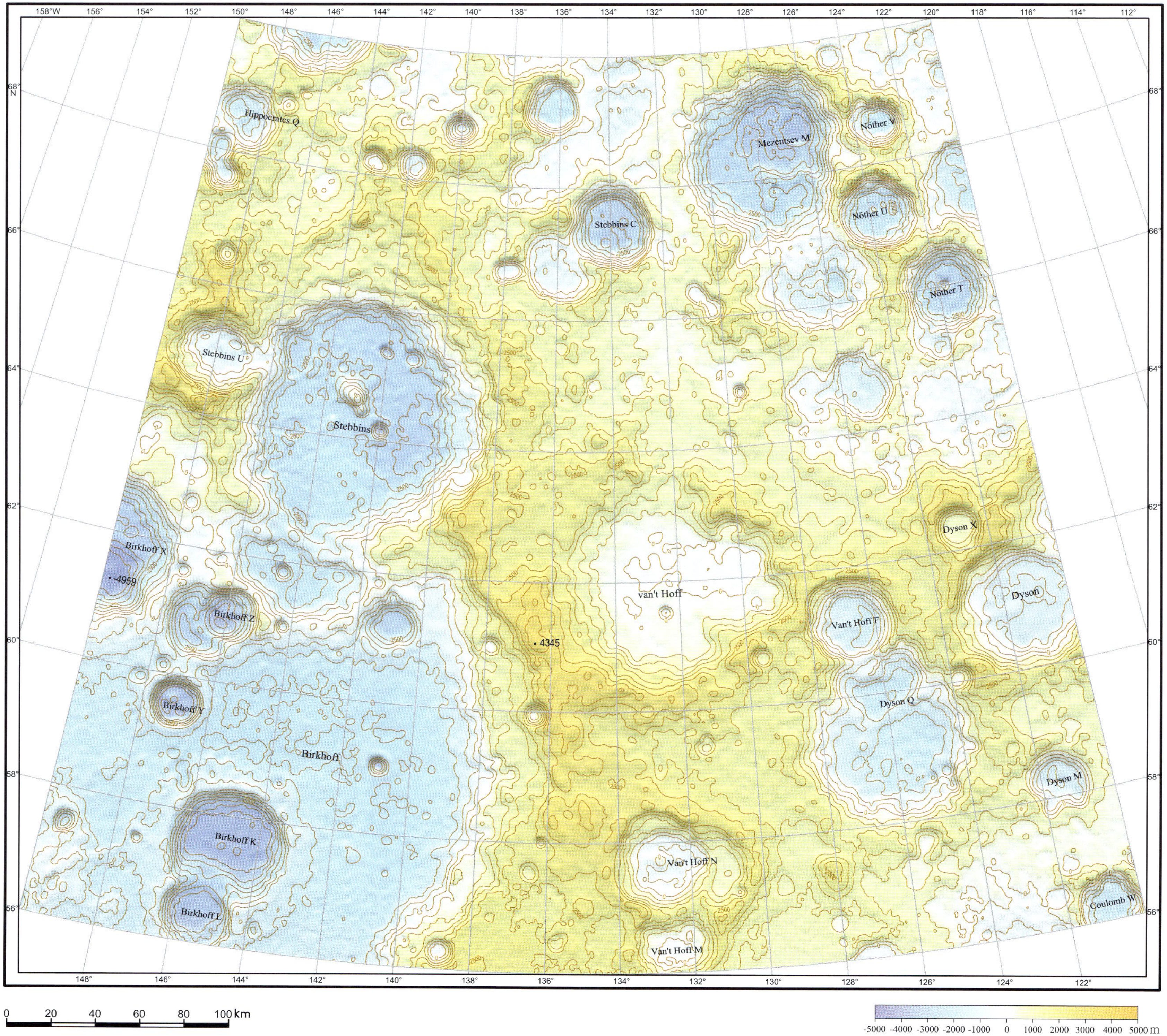

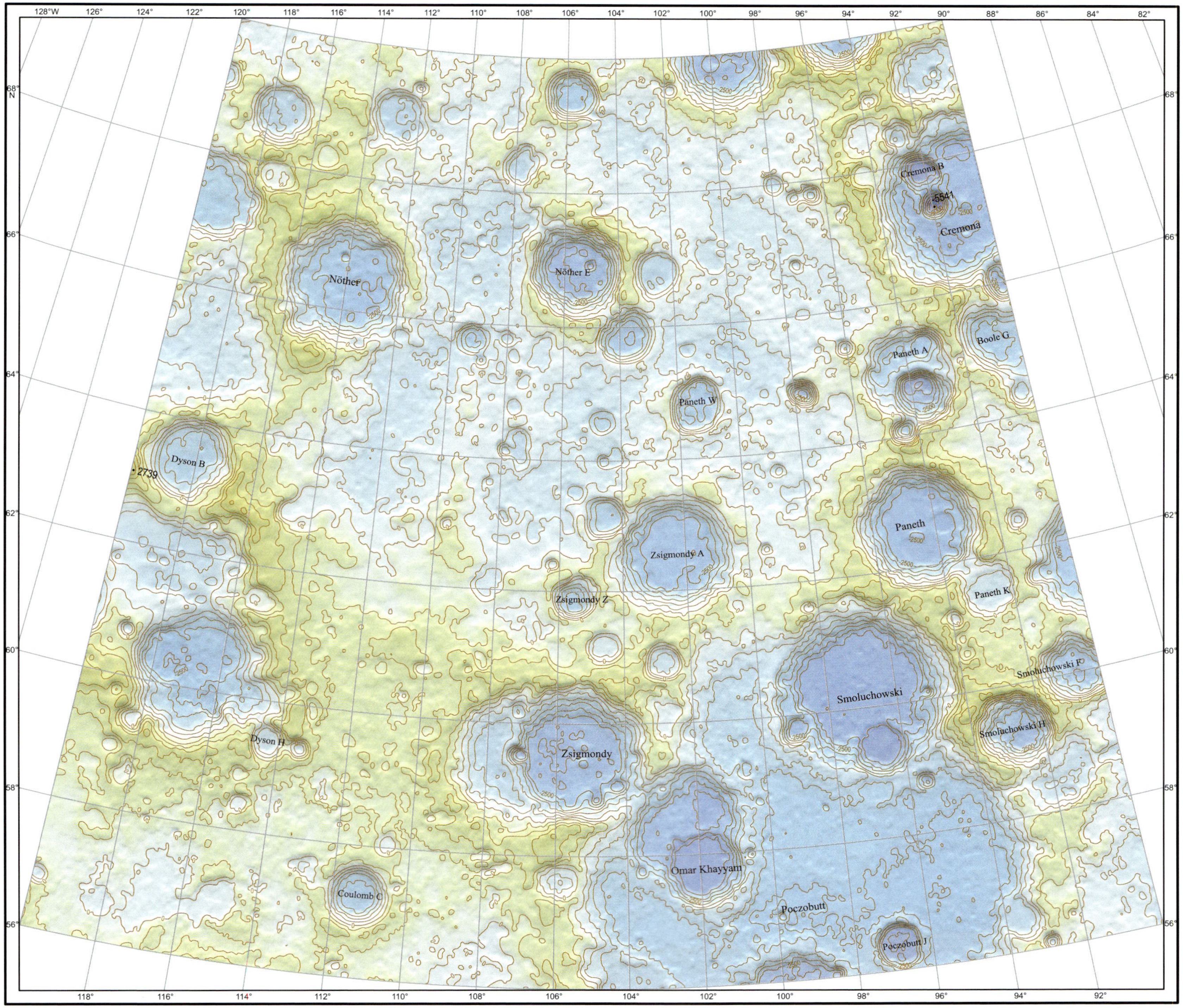

Image Code C003

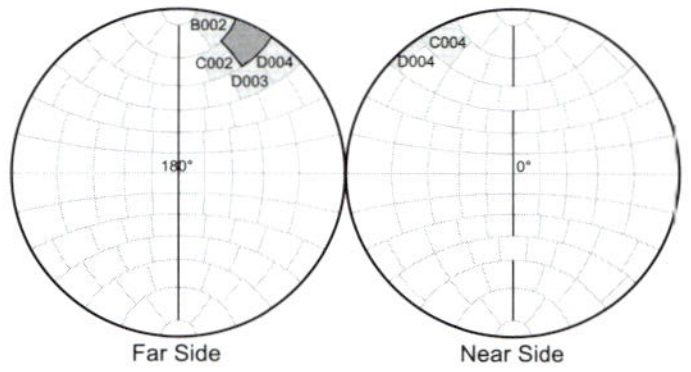

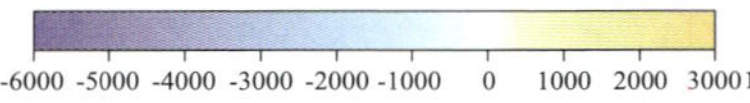

Image Code C004

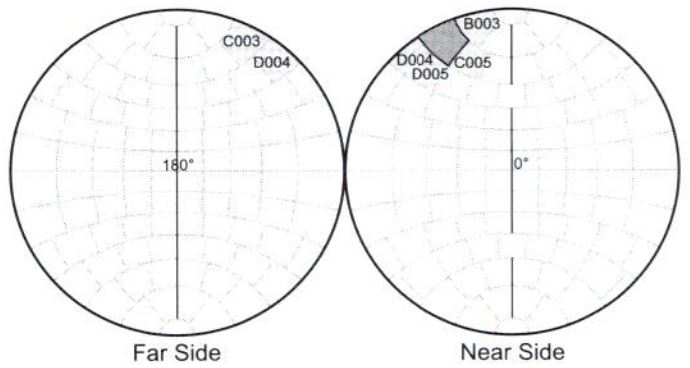

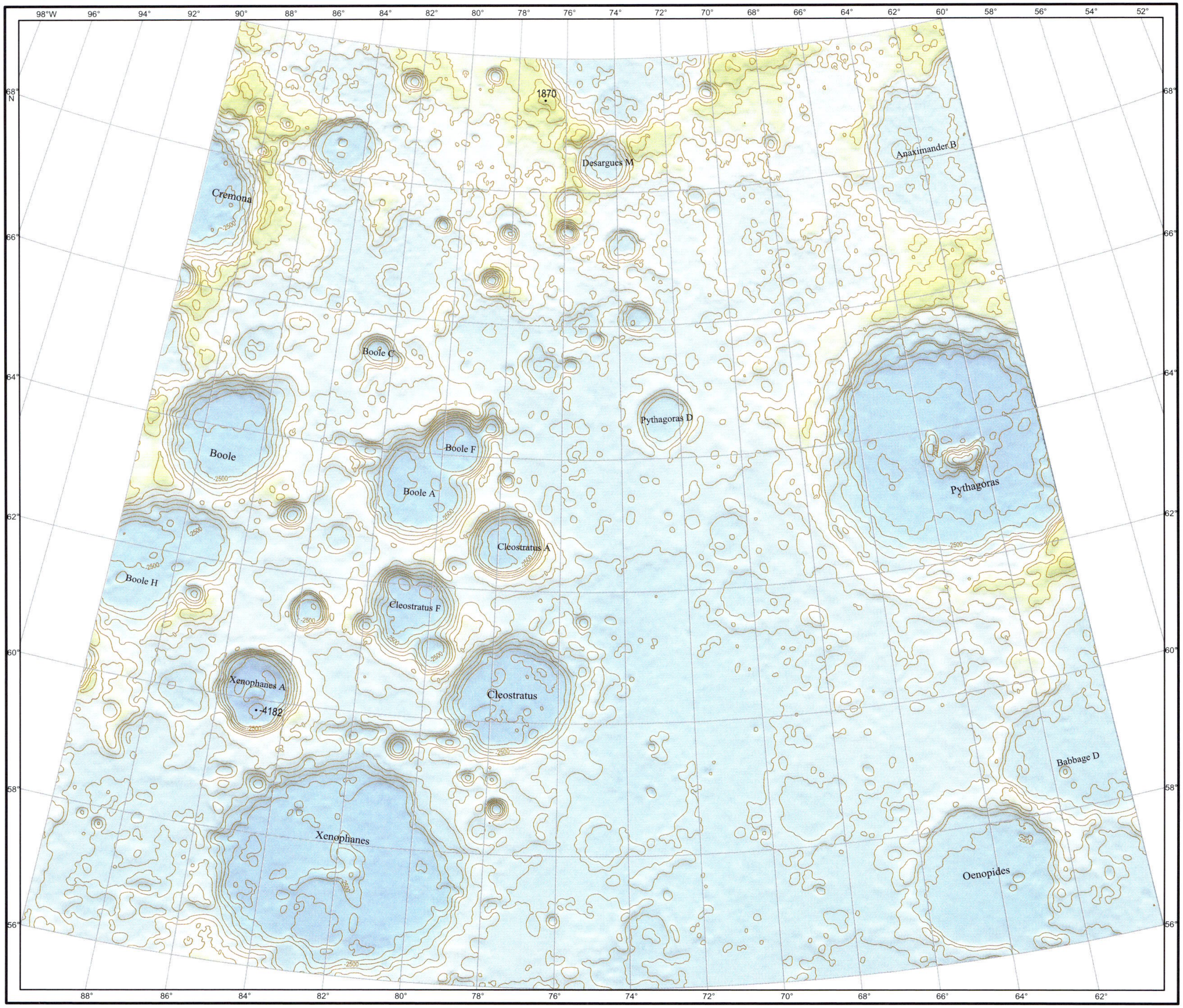

0 20 40 60 80 100 km

-5000 -4000 -3000 -2000 -1000 0 1000 2000 m

Image Code C005

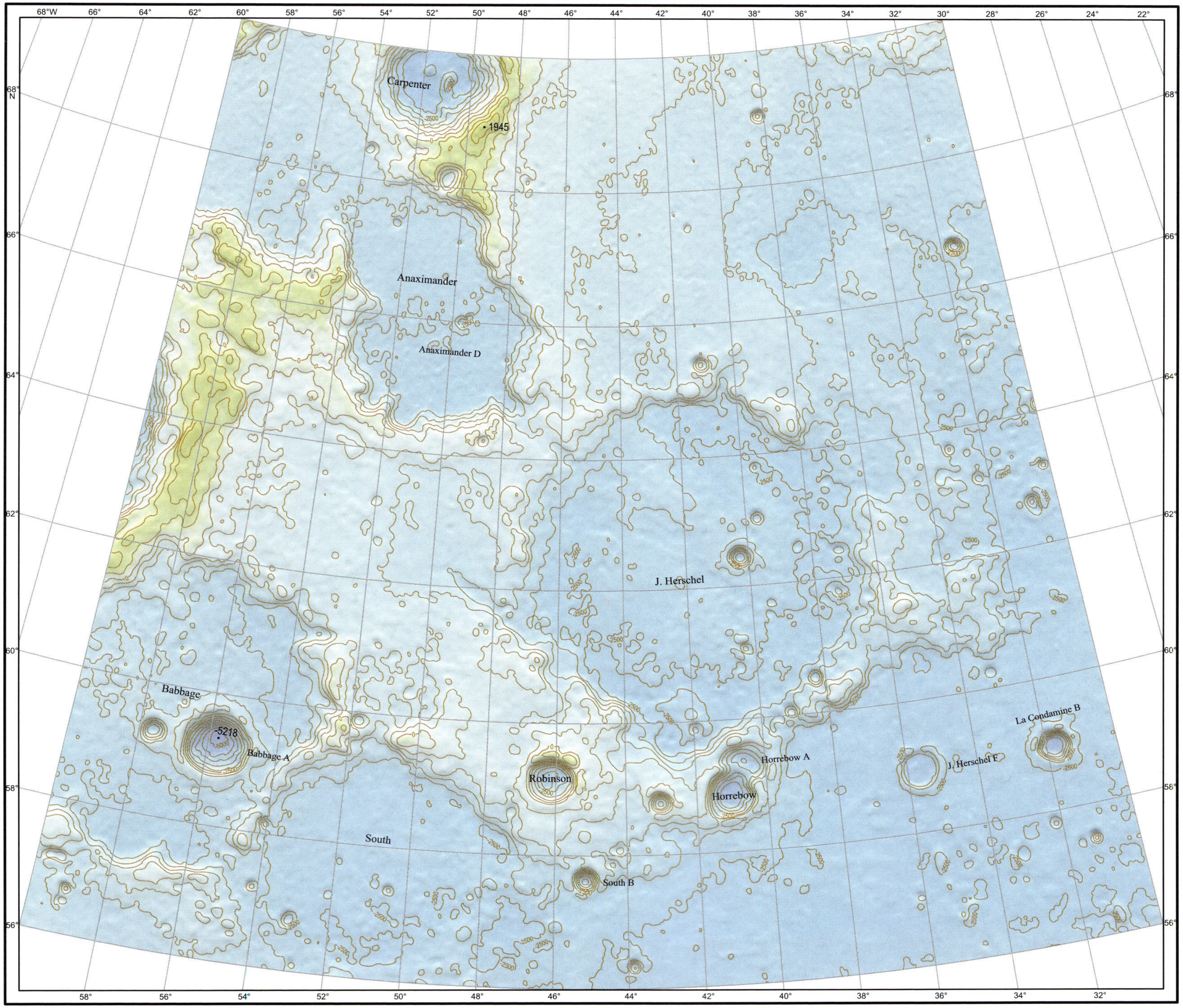

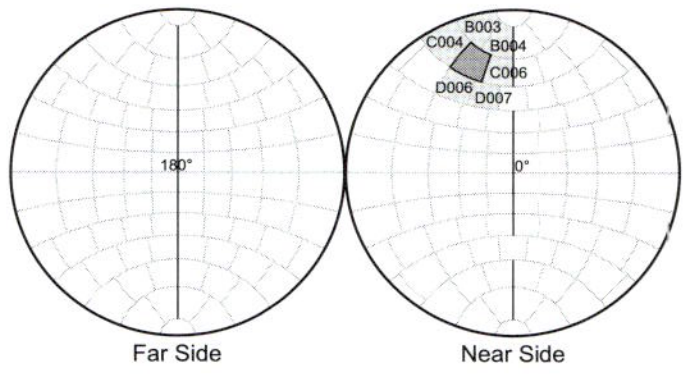

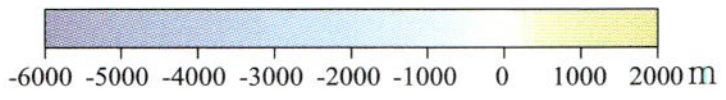

Image Code C006

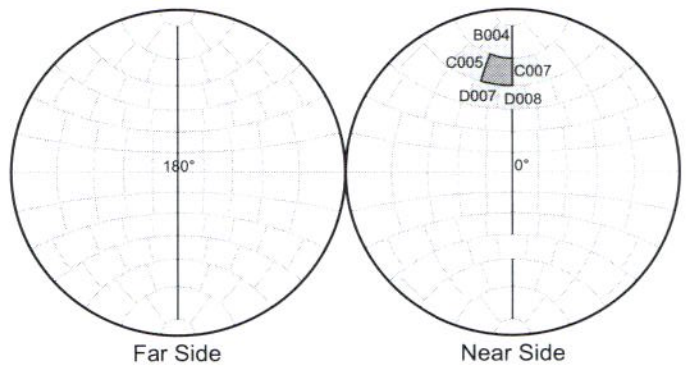

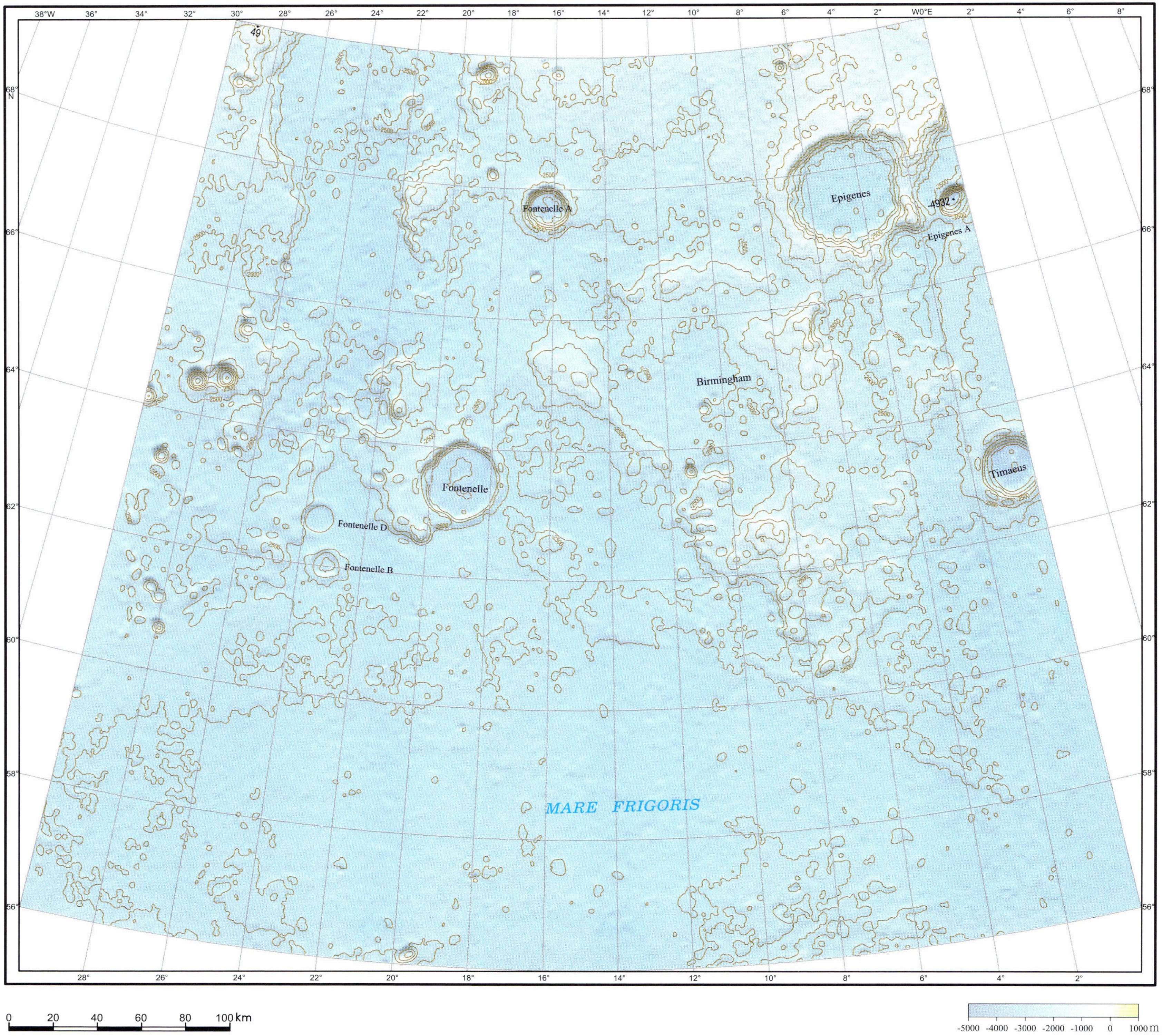

Image Code C007

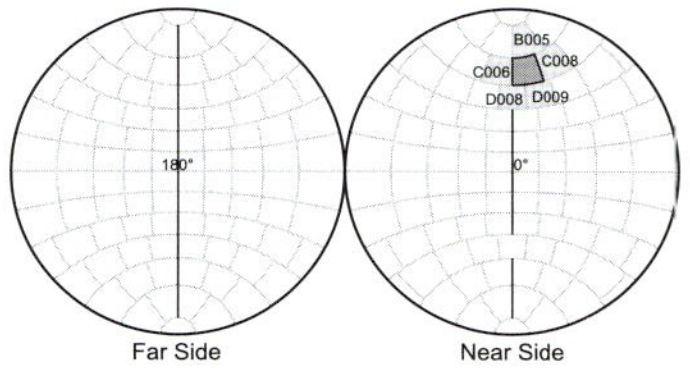

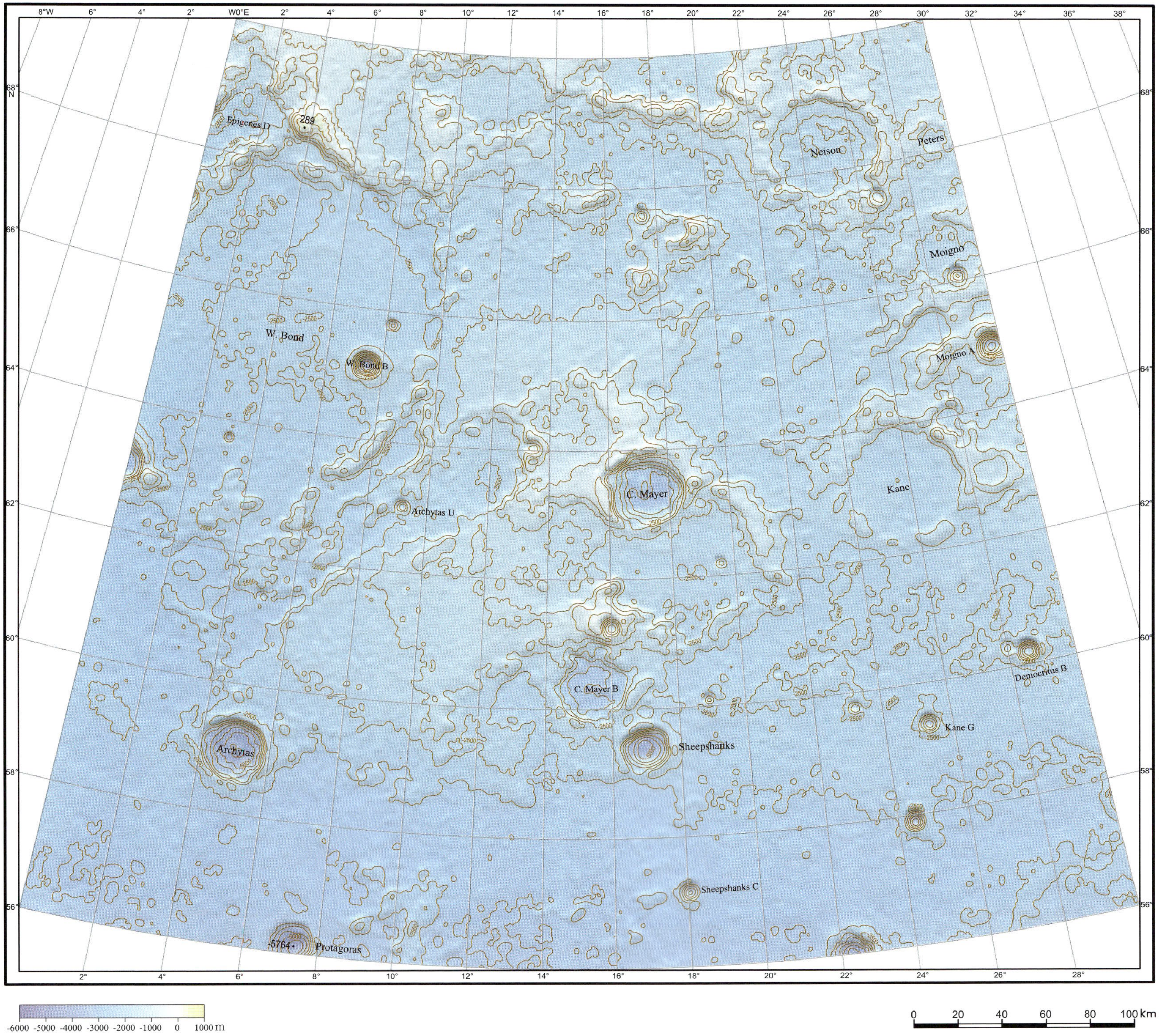

Image Code **C008**

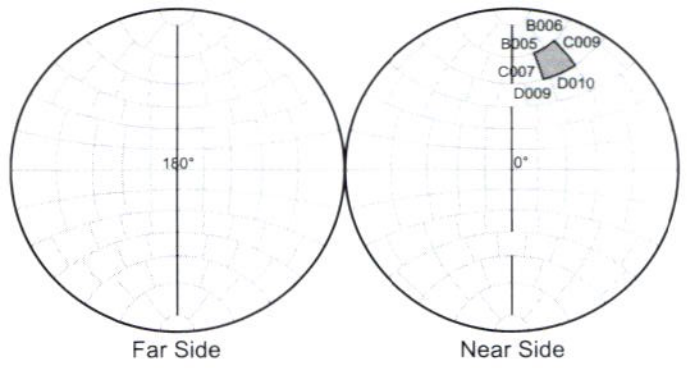

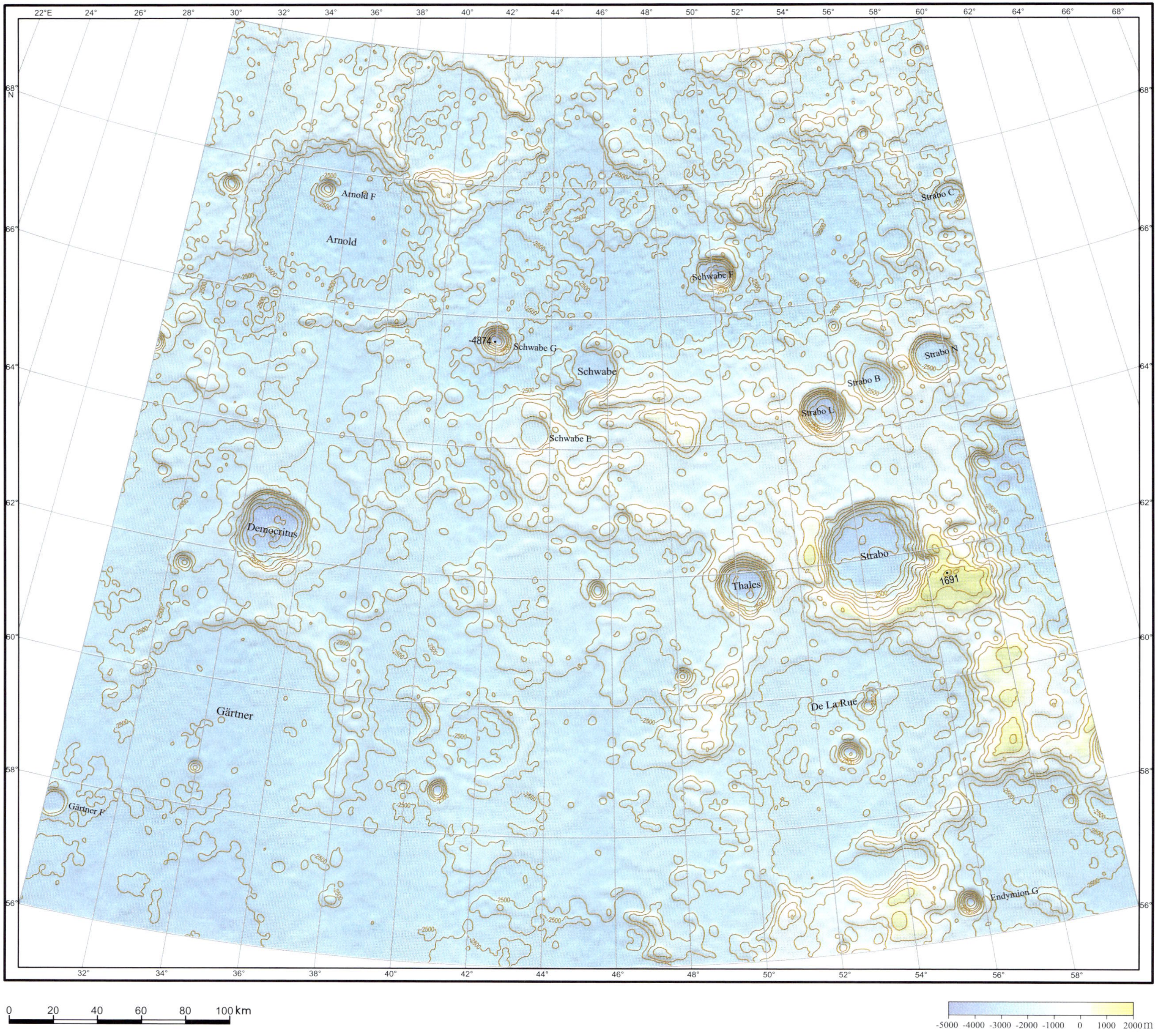

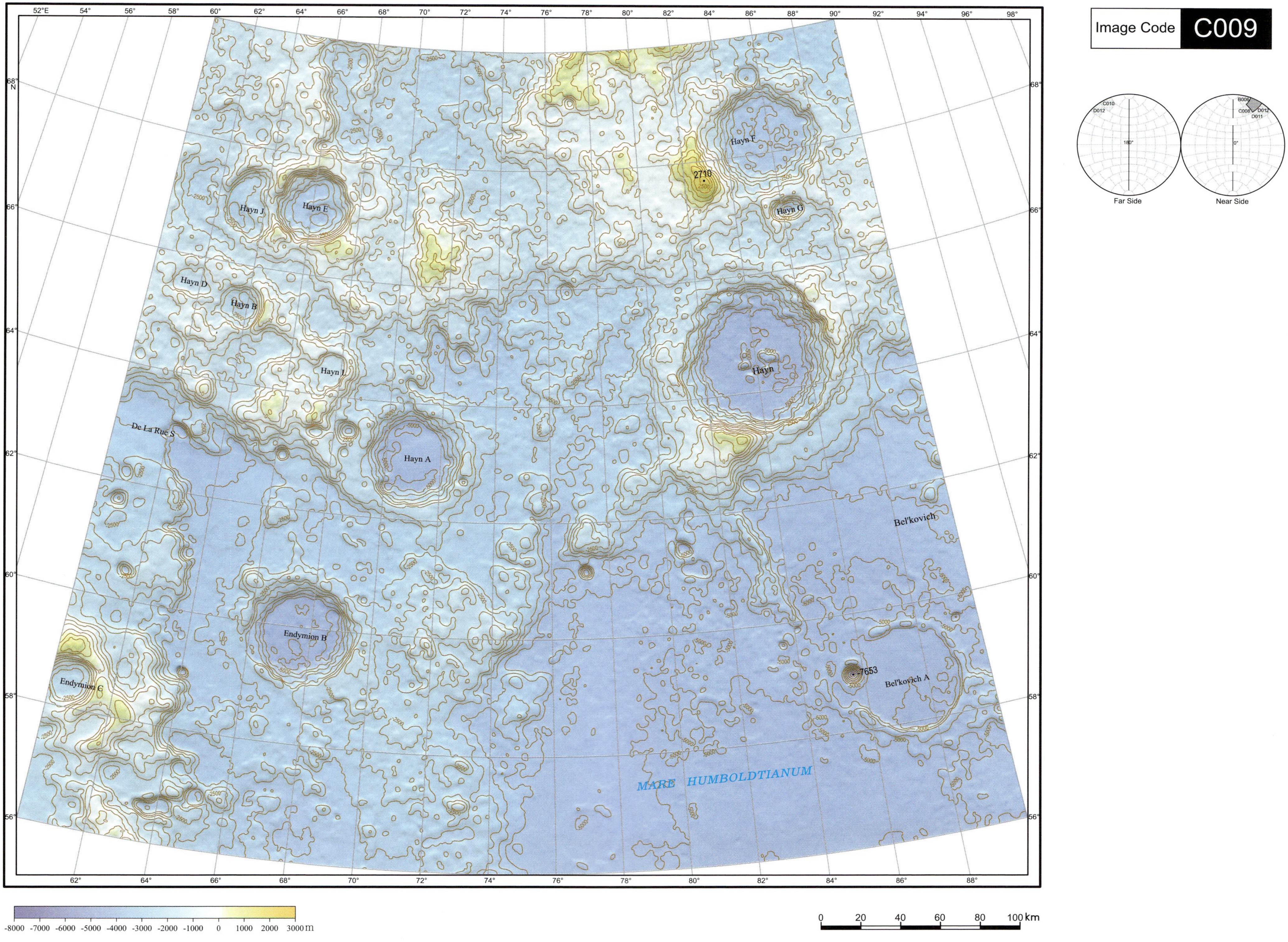
Image Code C009
Hayn F
2710
Hayn J
Hayn E
Hayn G
Hayn D
Hayn B
Hayn
Hayn L
De La Rue S
Hayn A
Bel'kovich
Endymion B
Endymion C
7653
Bel'kovich A
MARE HUMBOLDTIANUM
Far Side
Near Side
-8000 -7000 -6000 -5000 -4000 -3000 -2000 -1000 0 1000 2000 3000 m
0 20 40 60 80 100 km

Image Code C010

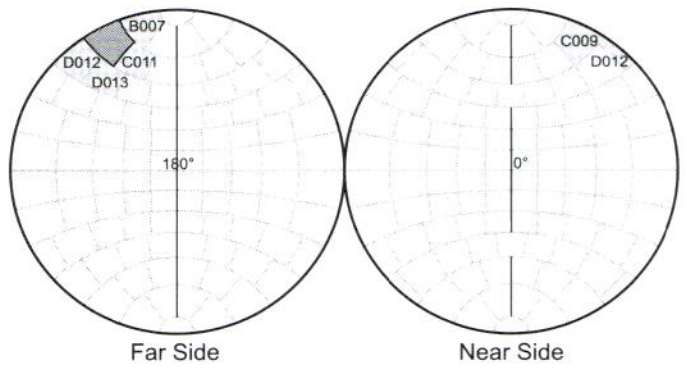

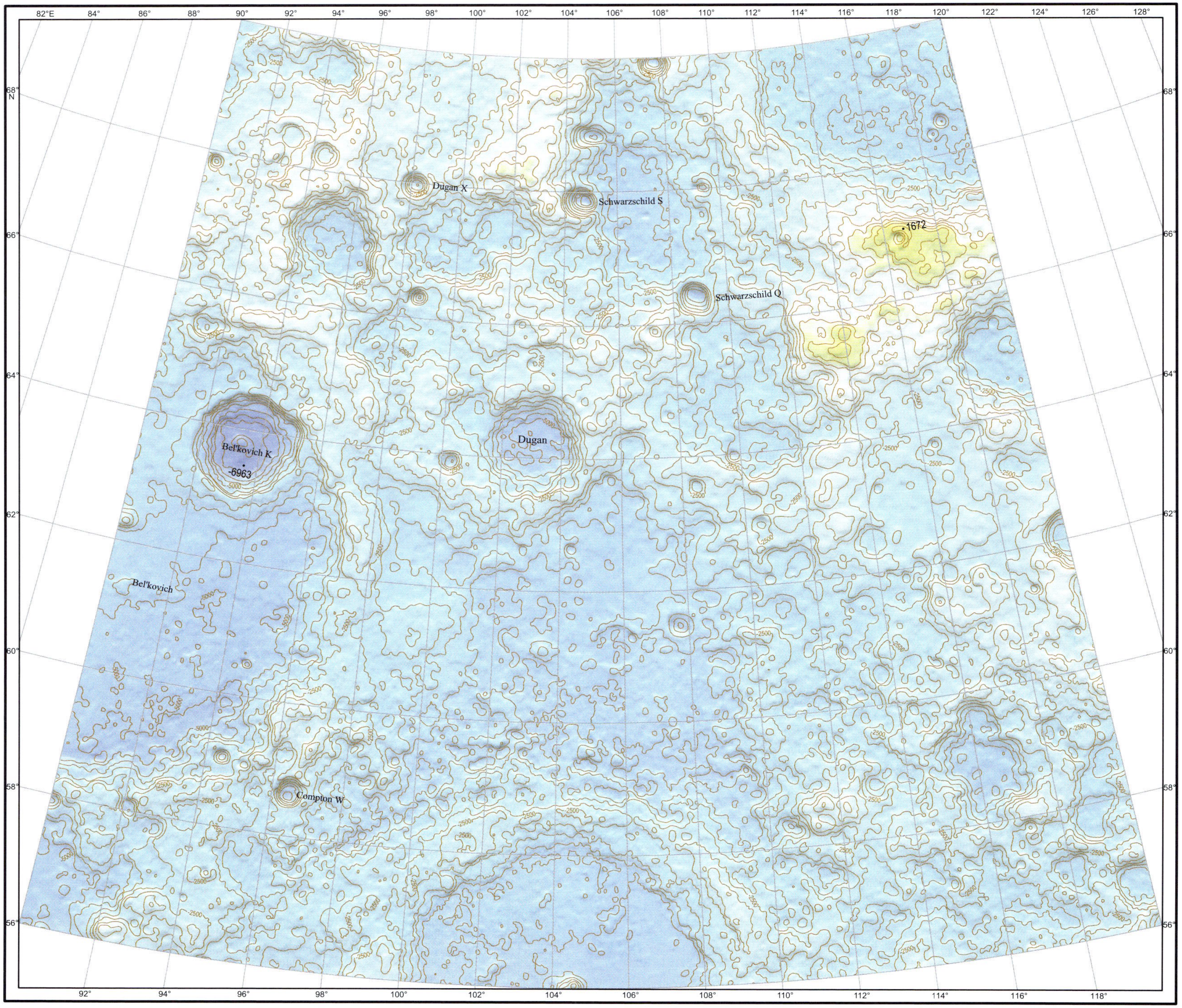

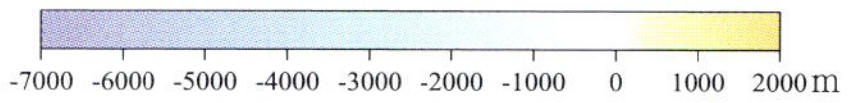

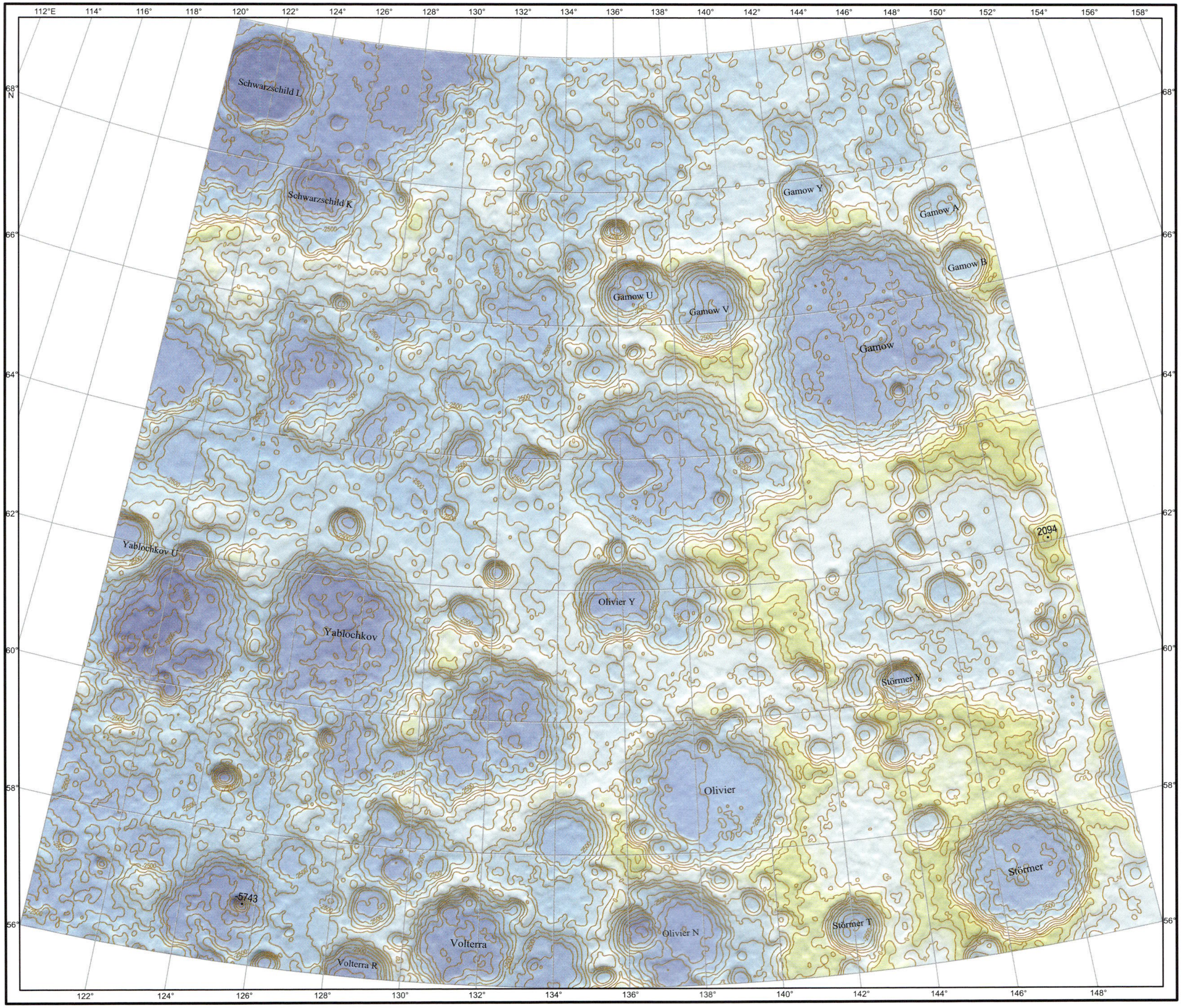

Image Code C011

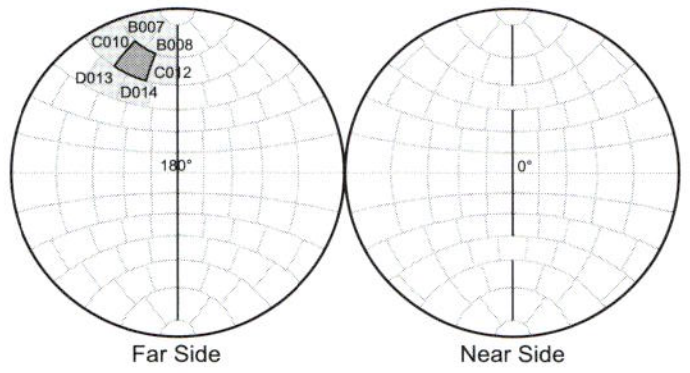

-6000 -5000 -4000 -3000 -2000 -1000 0 1000 2000 3000 m

0 20 40 60 80 100 km

Image Code C012

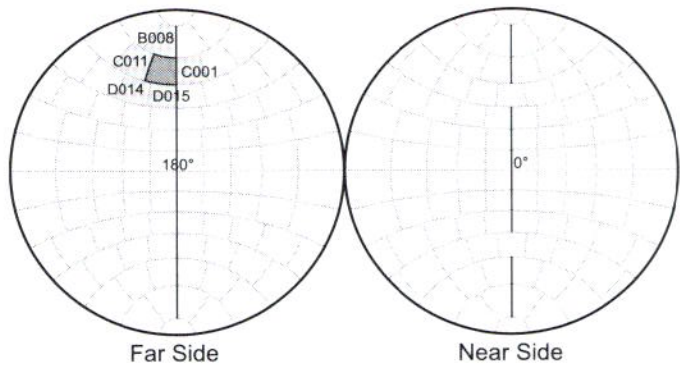

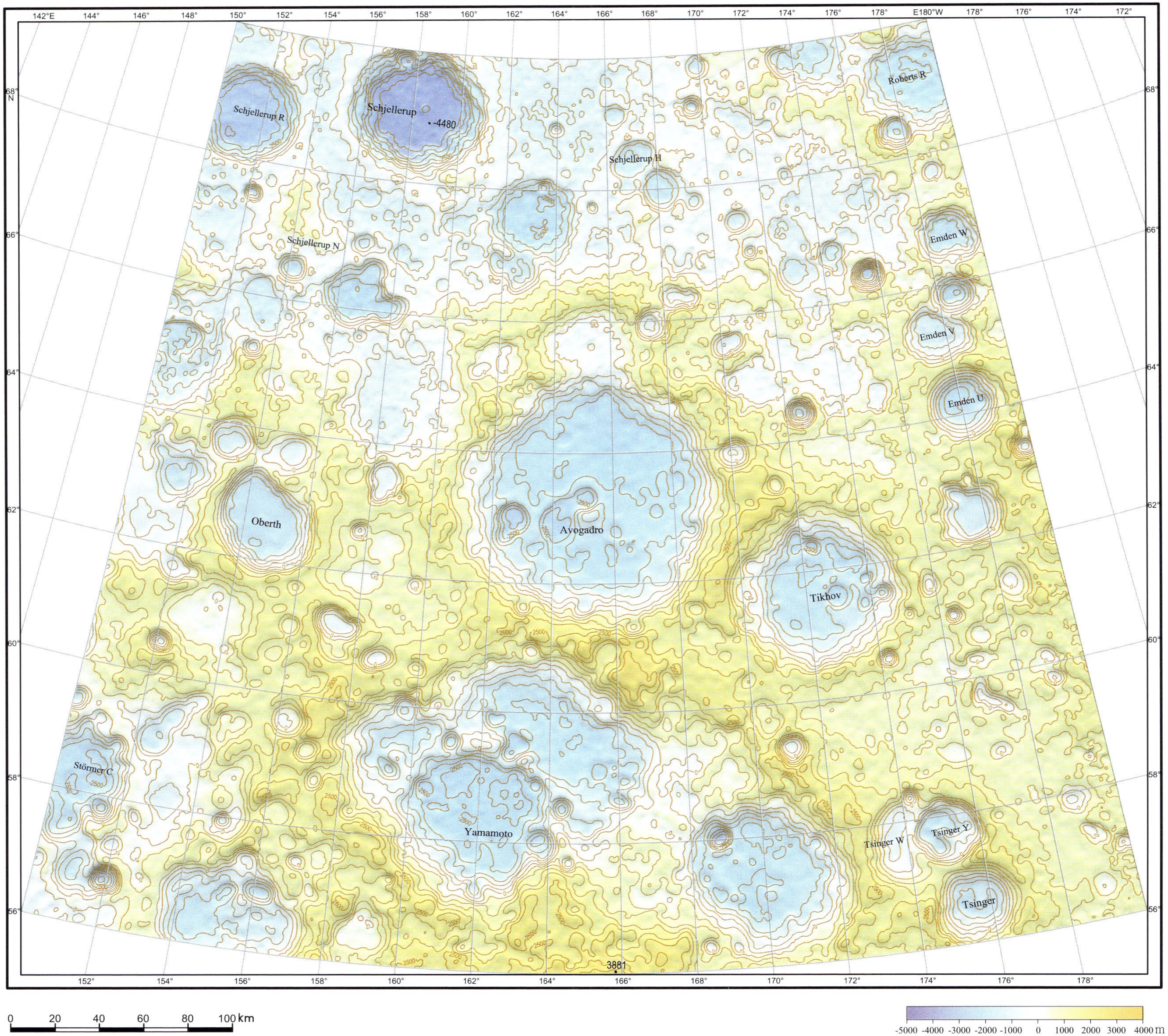

Image Code D001

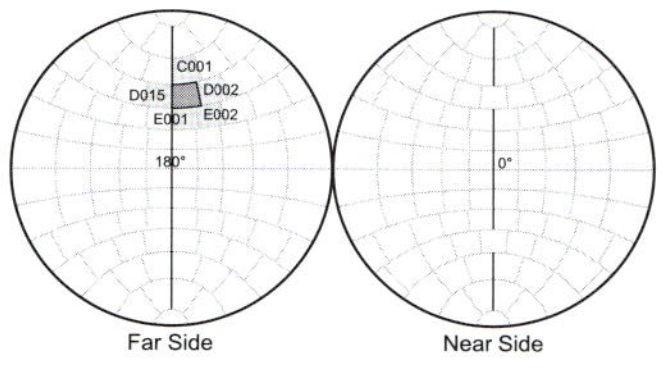

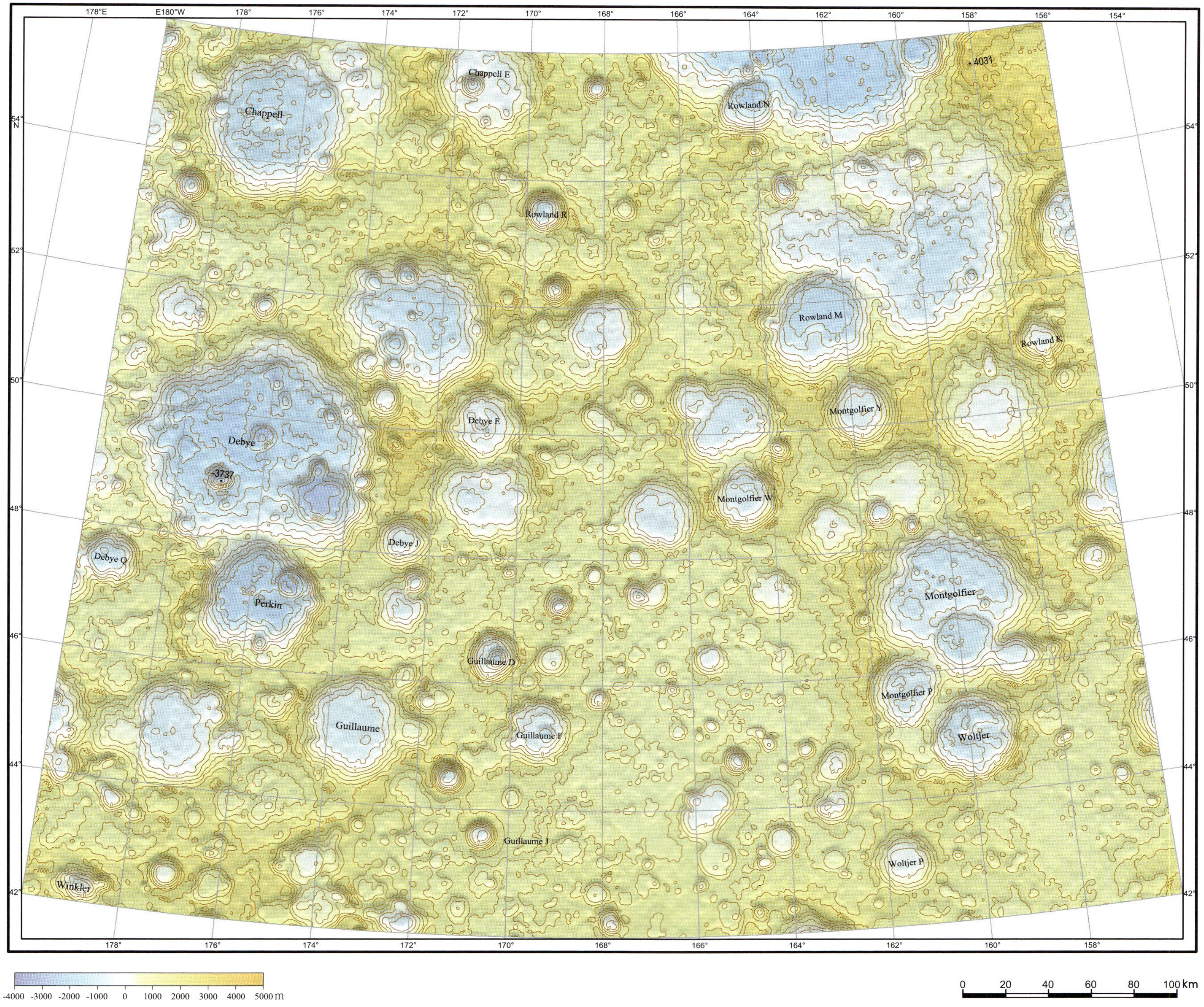

Image Code D002

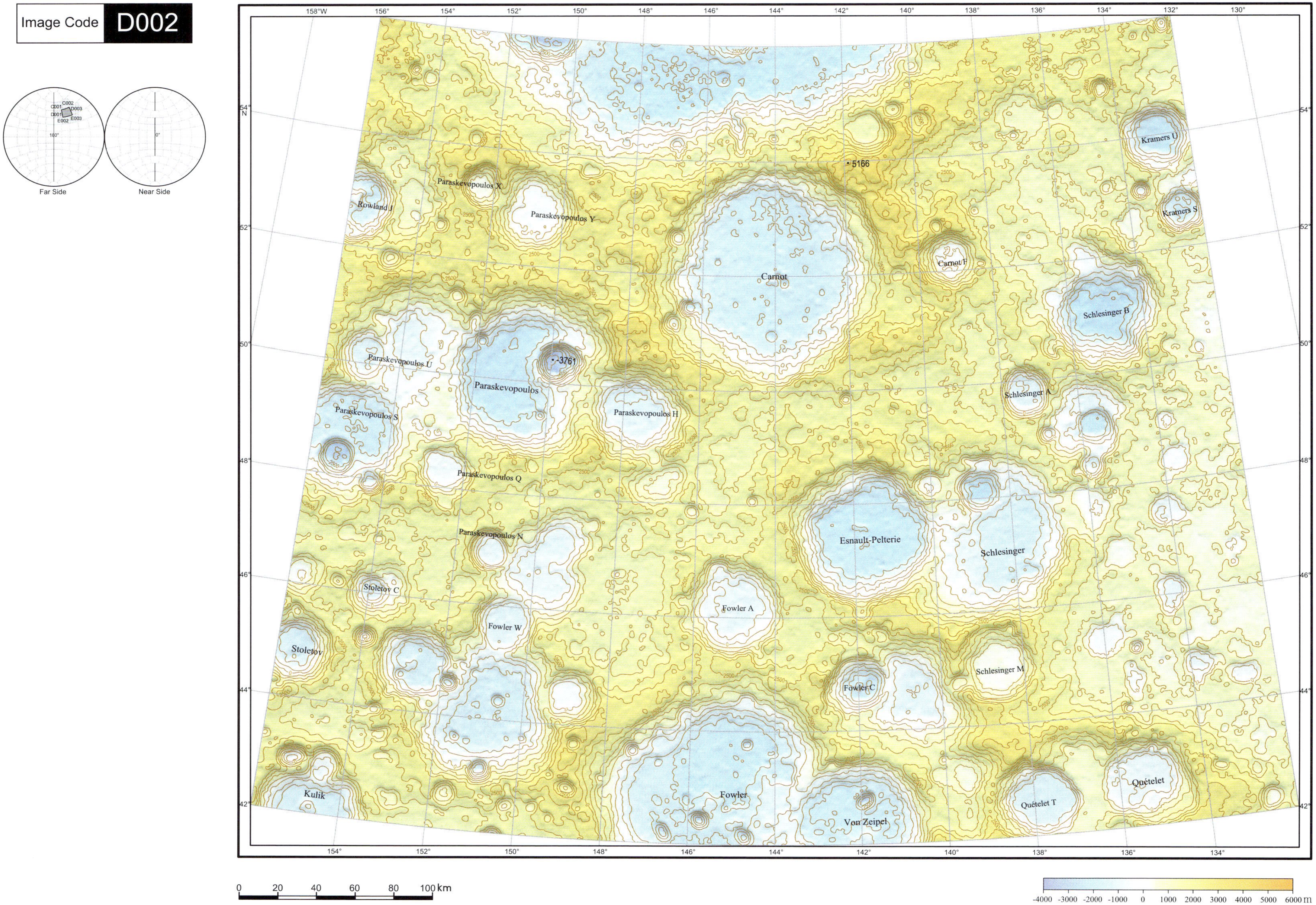

Image Code D003

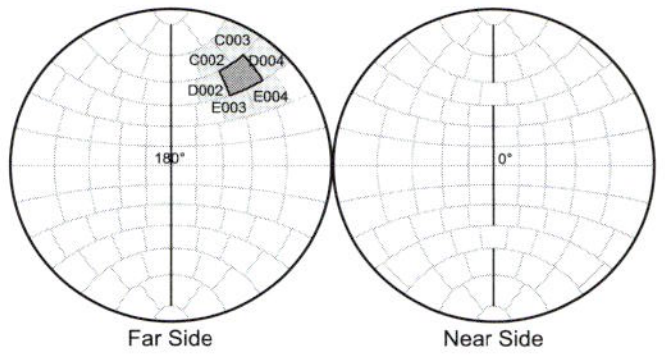

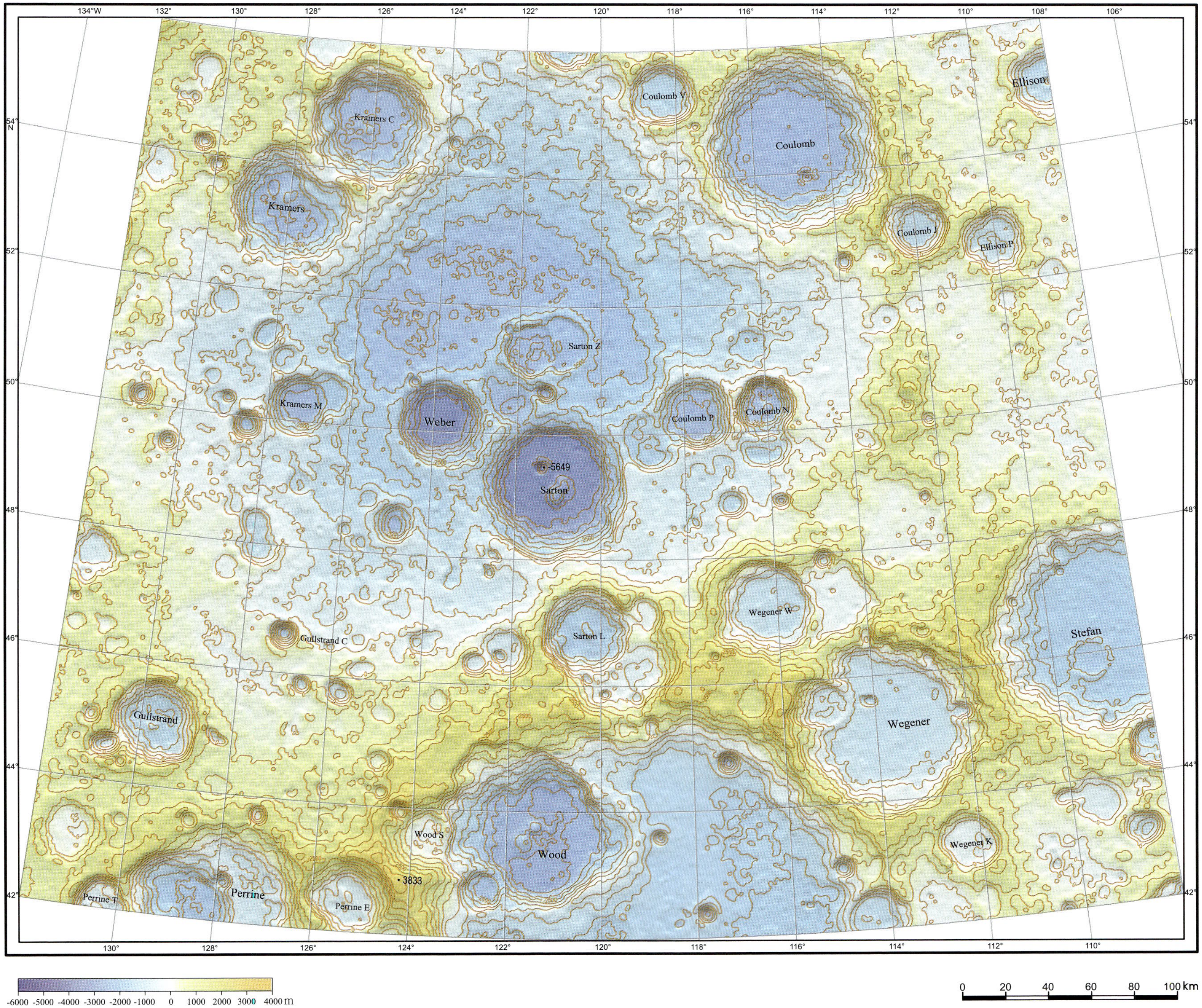

Image Code D004

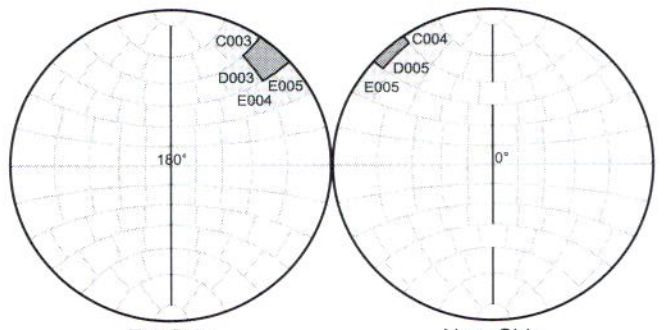

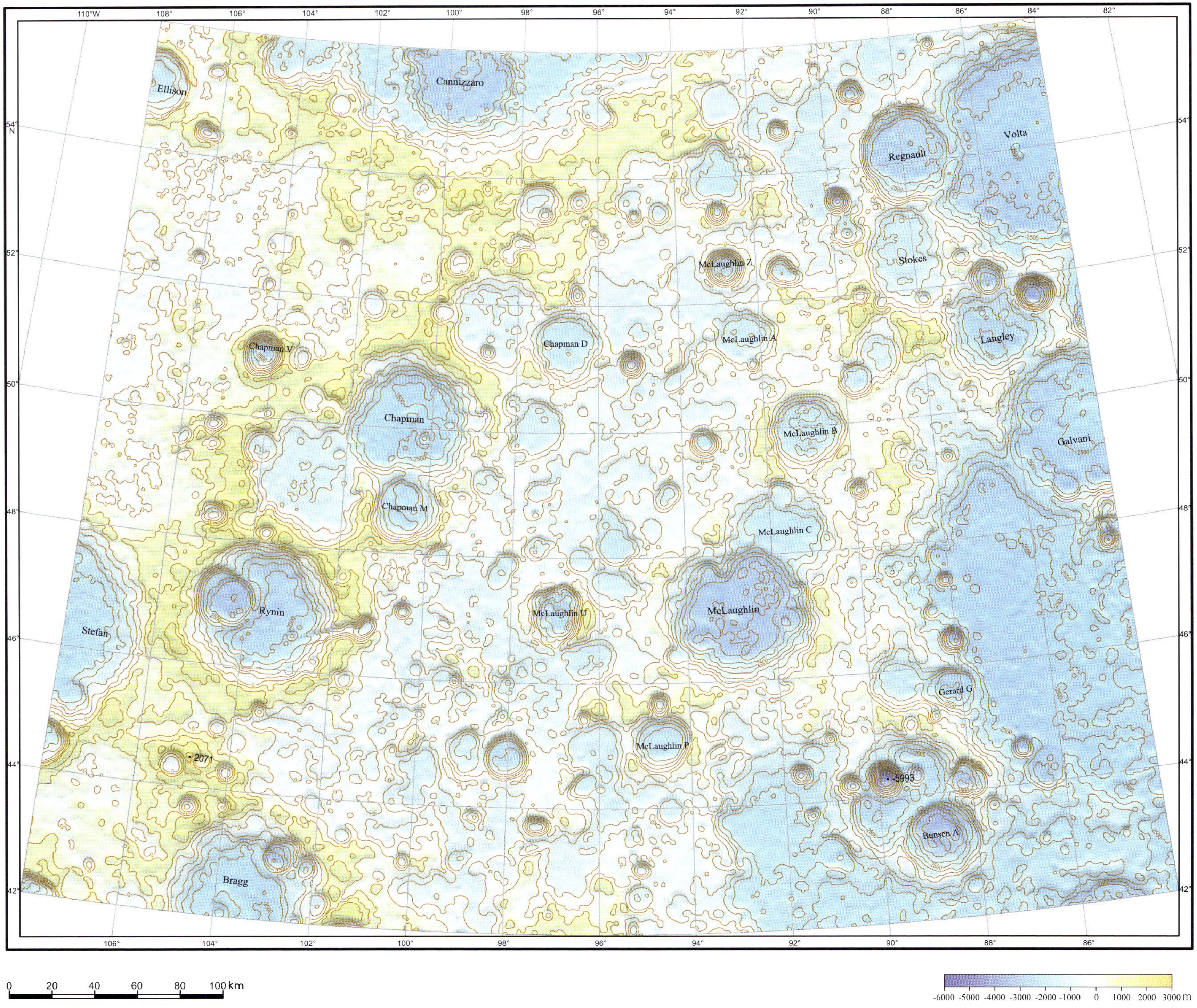

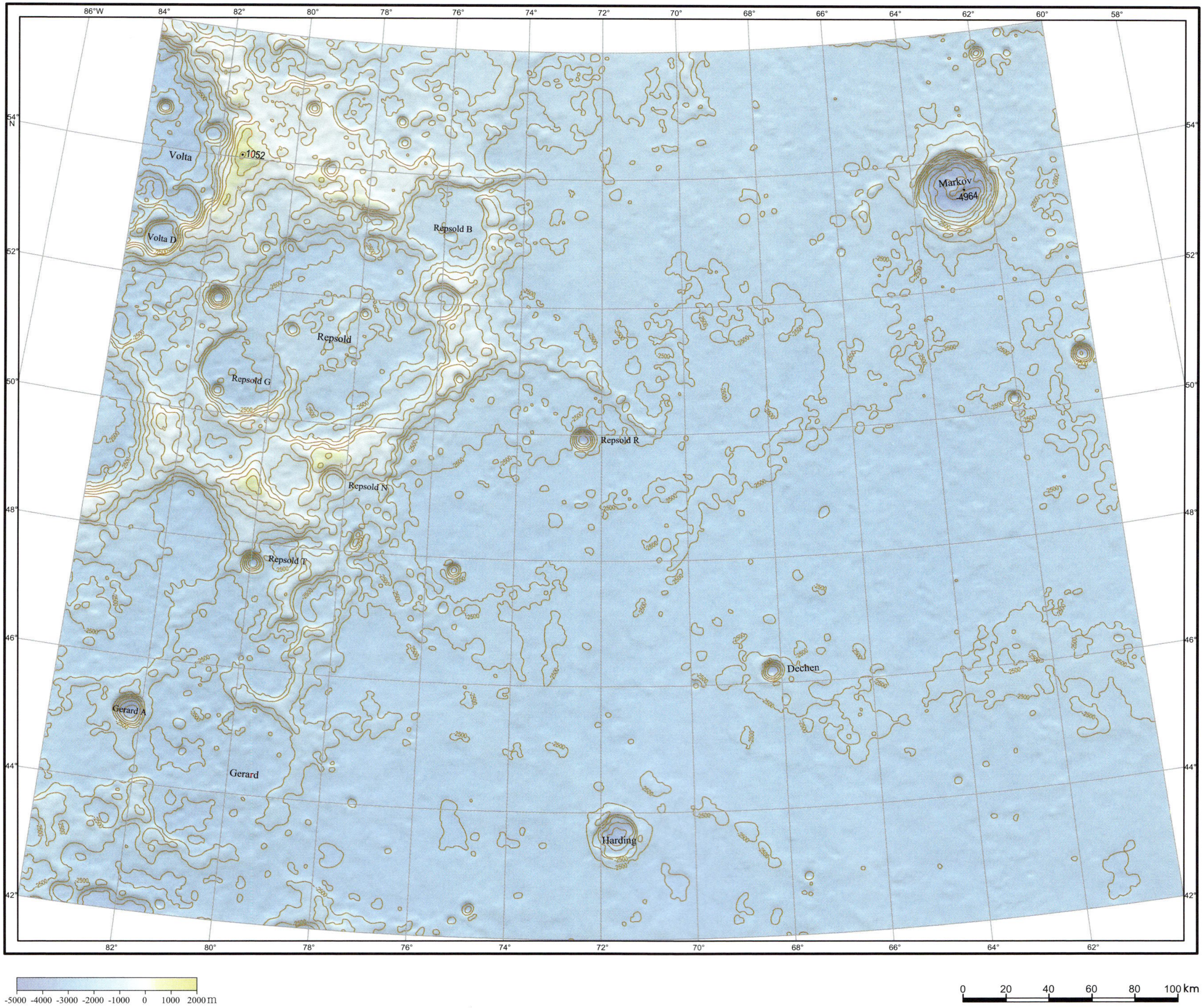

Image Code D005

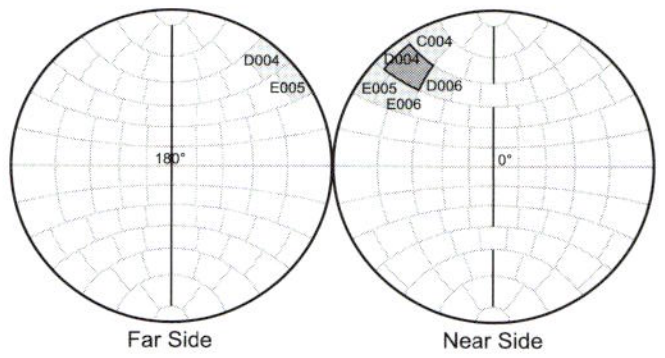

Image Code D006

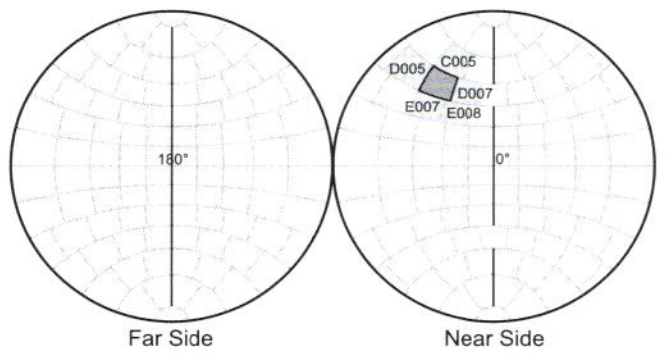

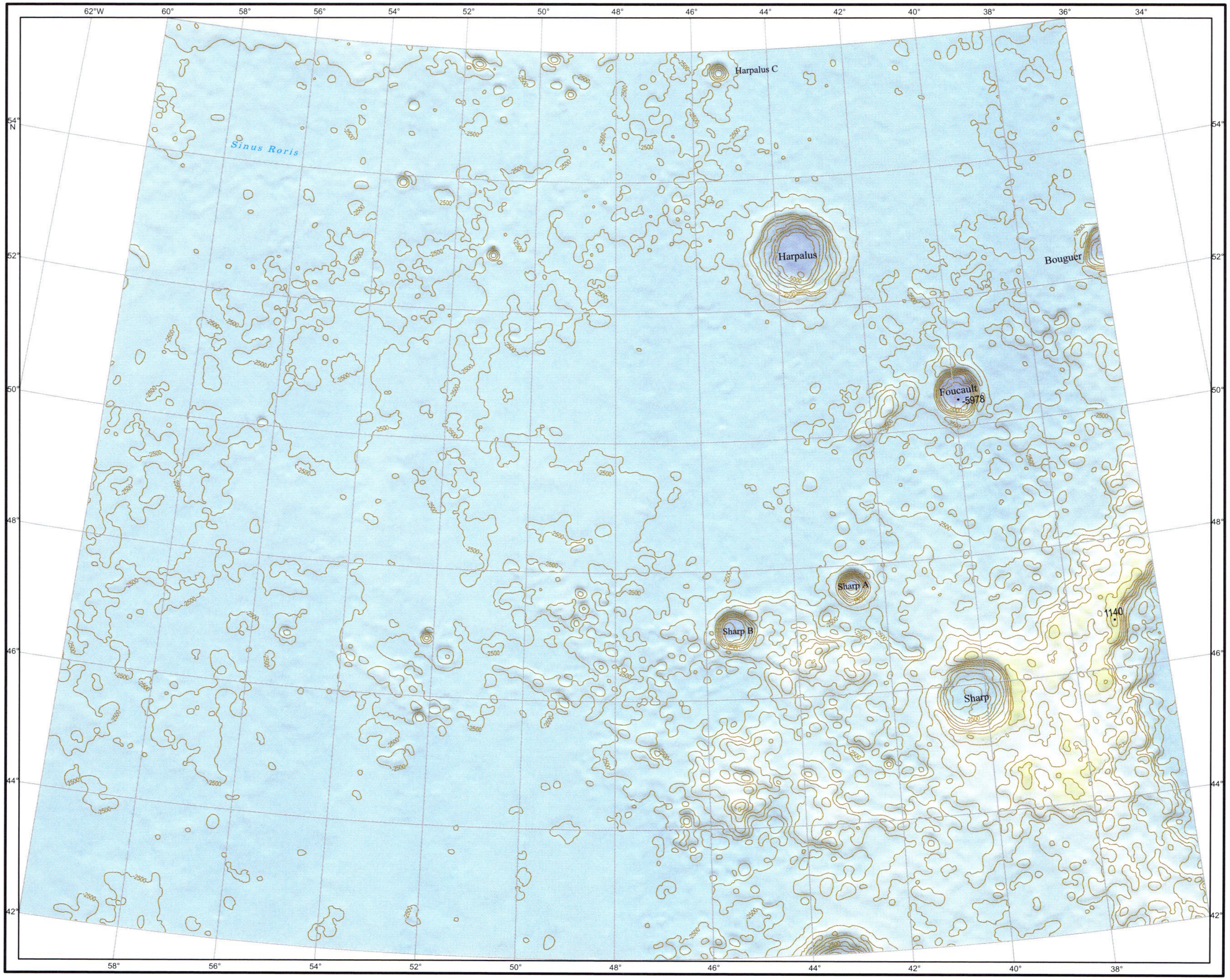

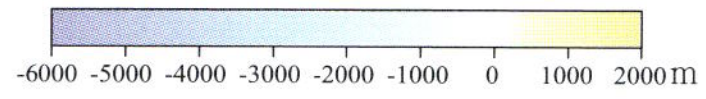

Image Code D007

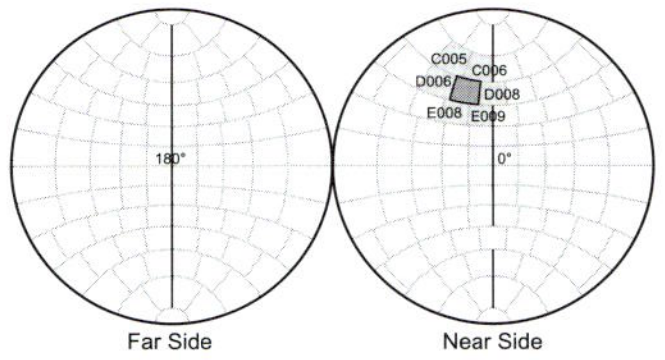

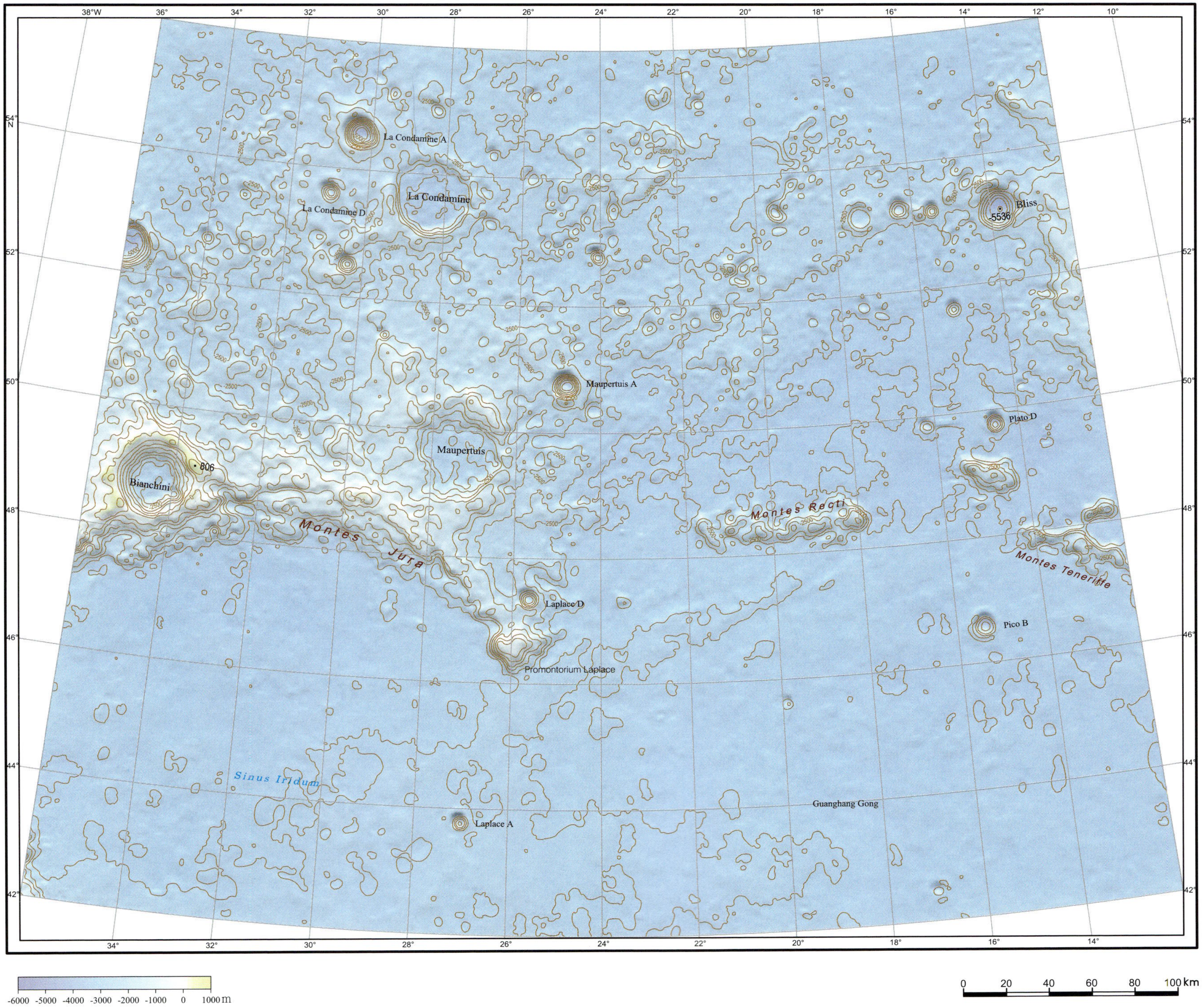

Image Code D008

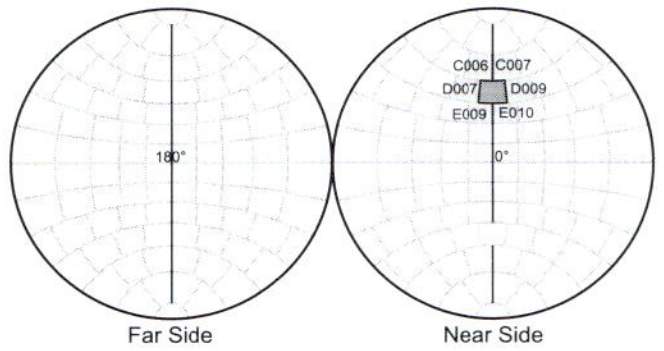

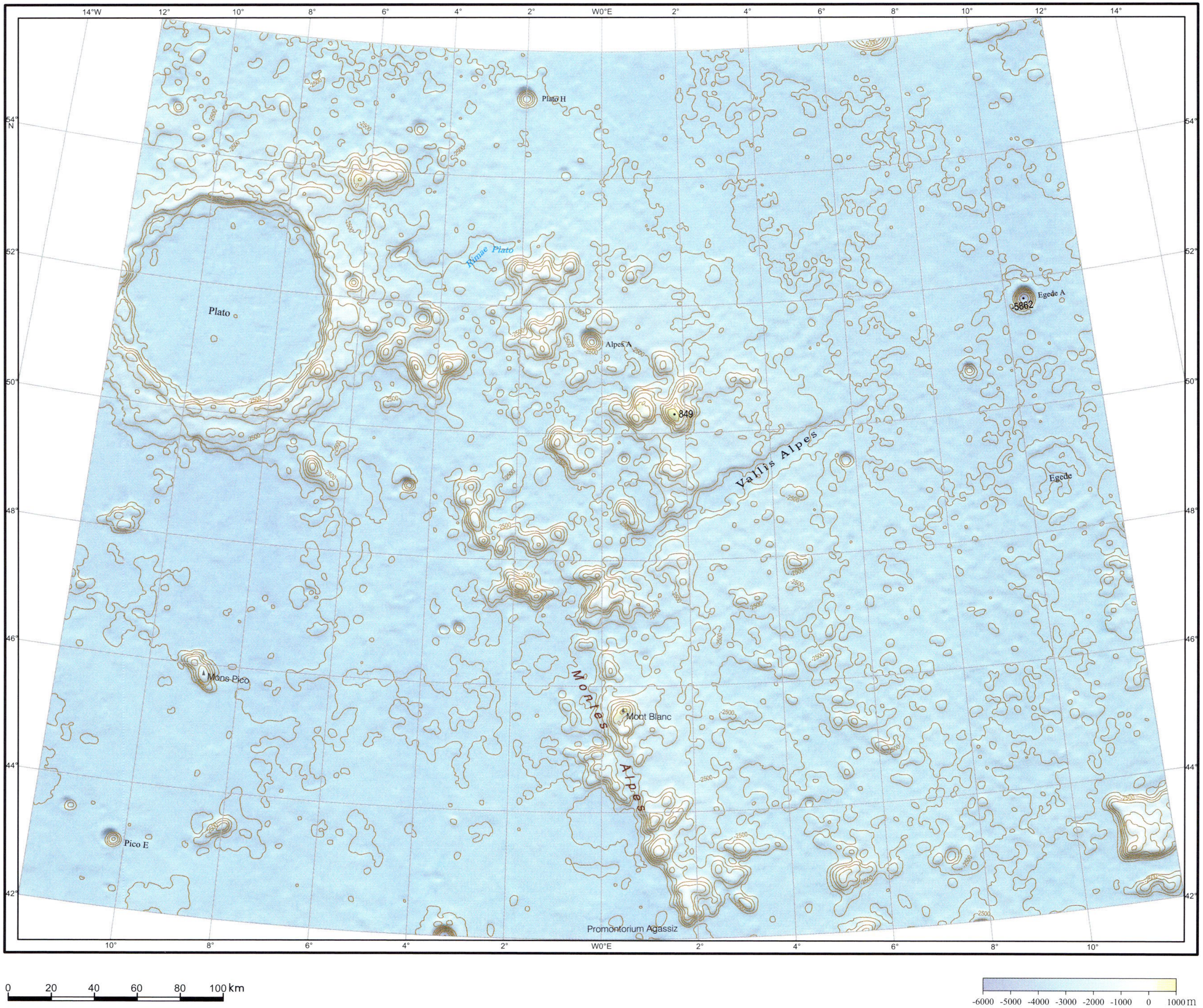

Image Code D009

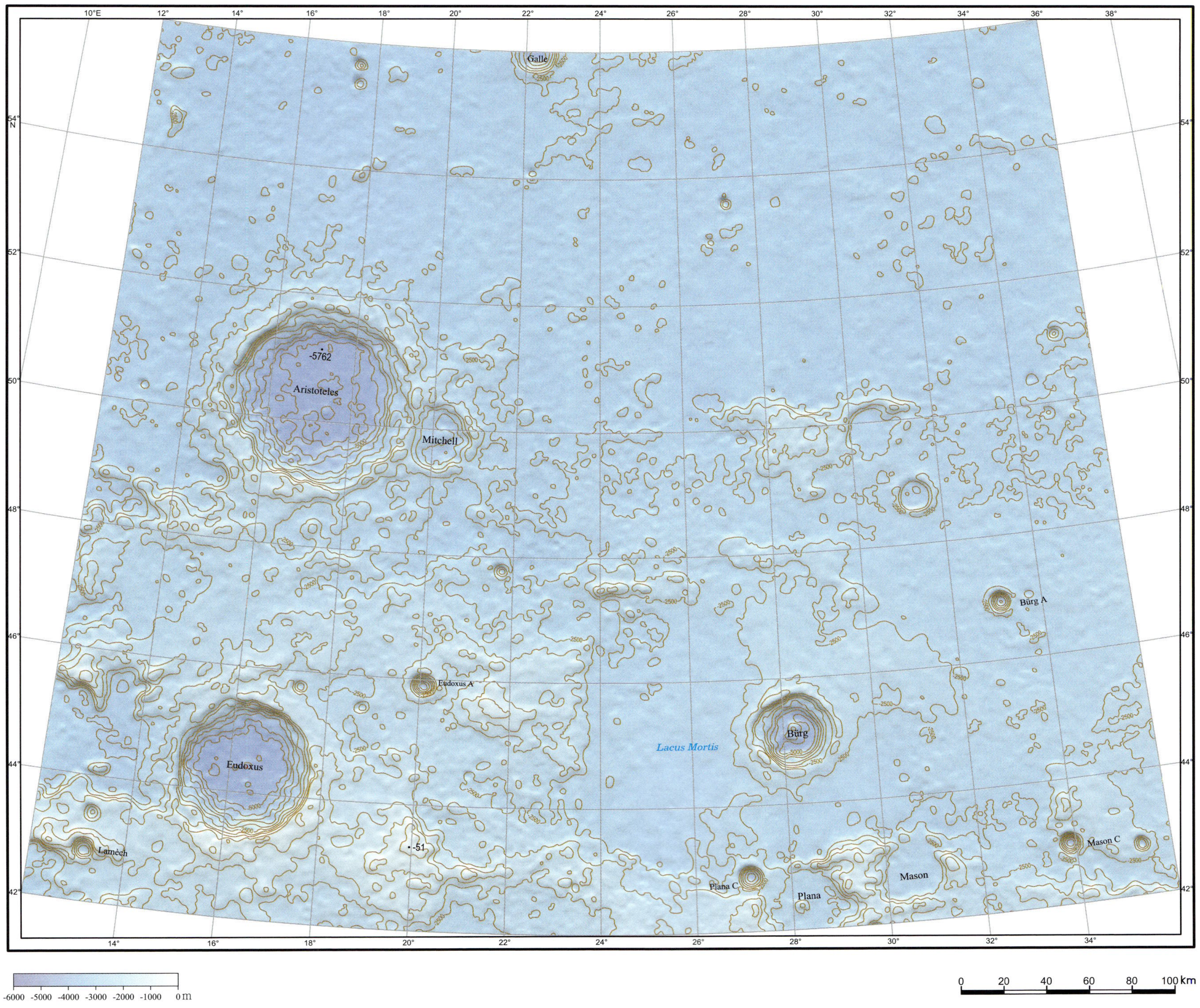

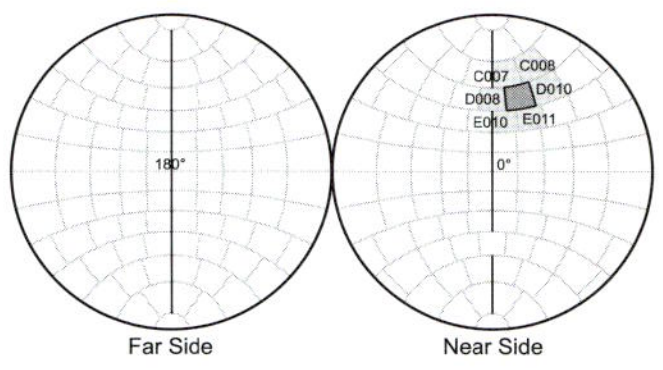

Image Code D010

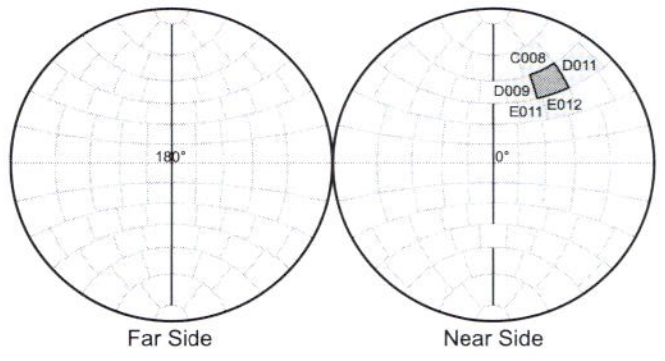

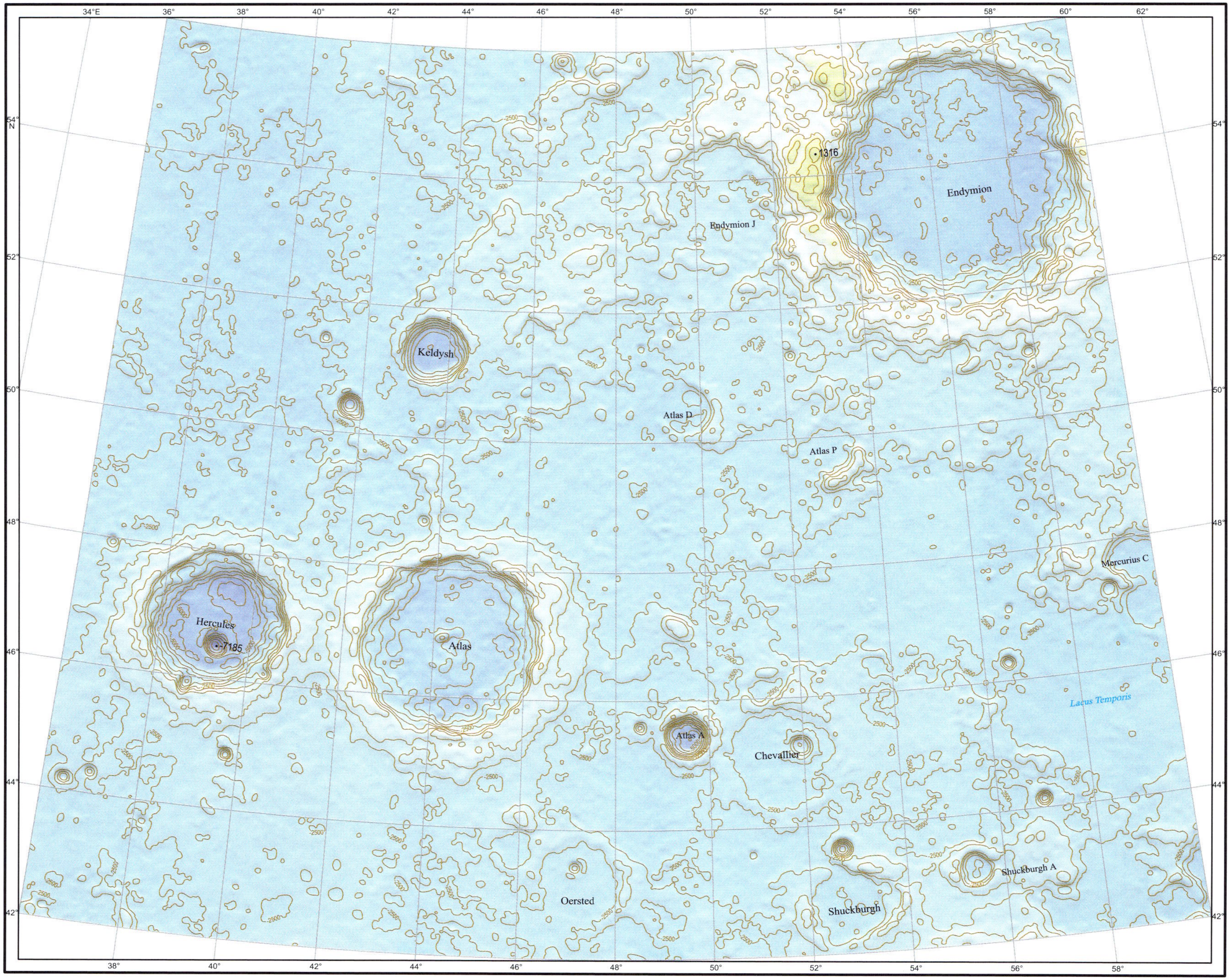

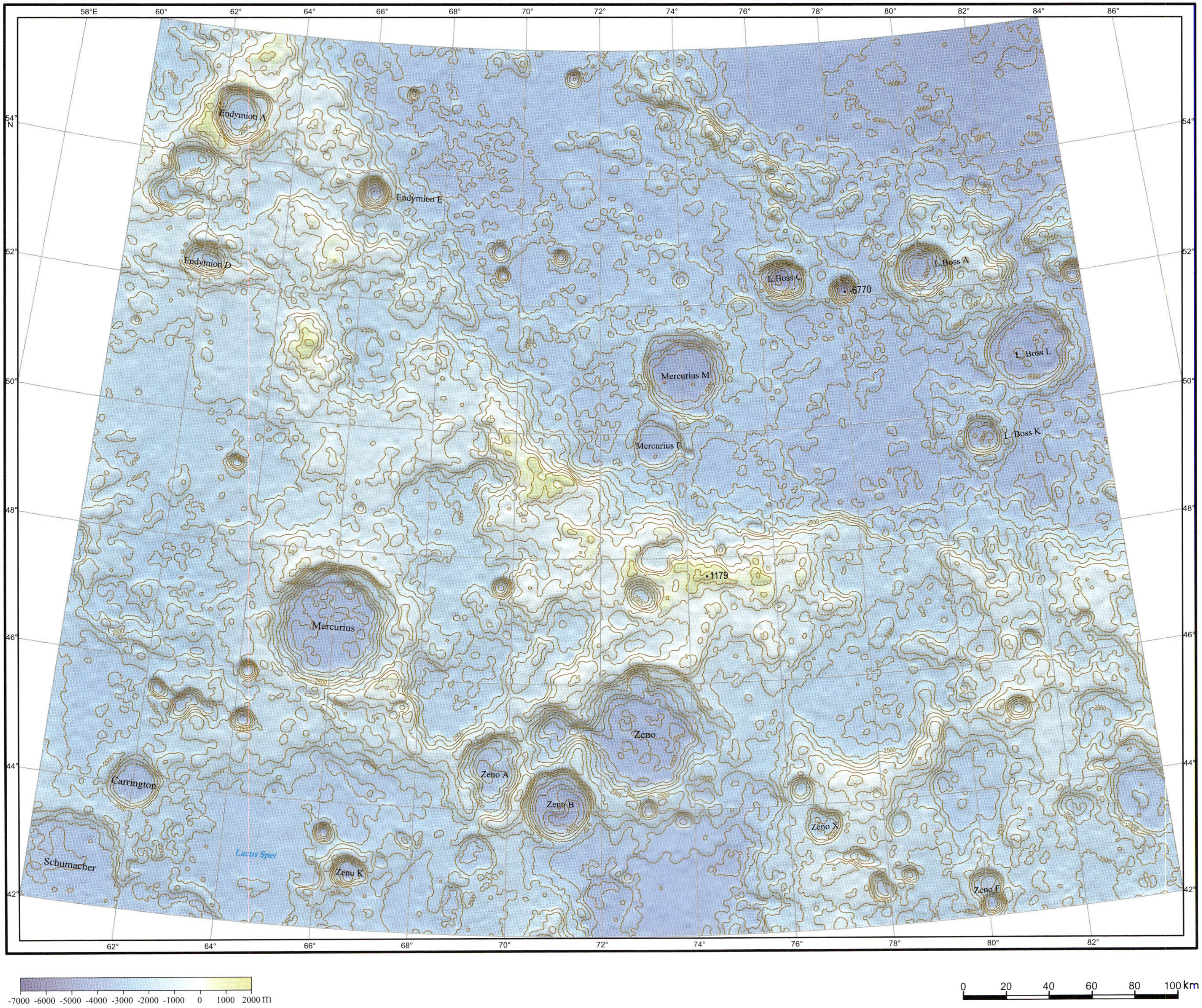

Image Code D011

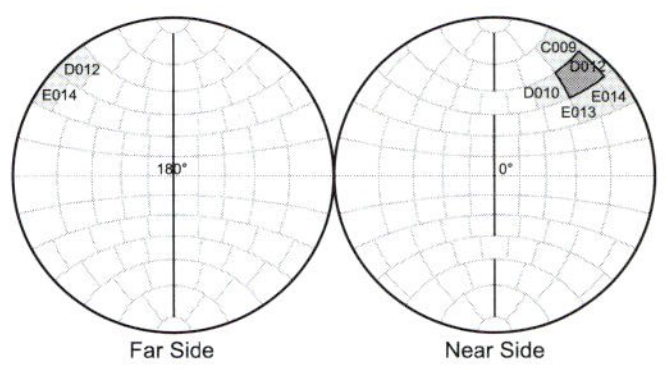

Image Code D012

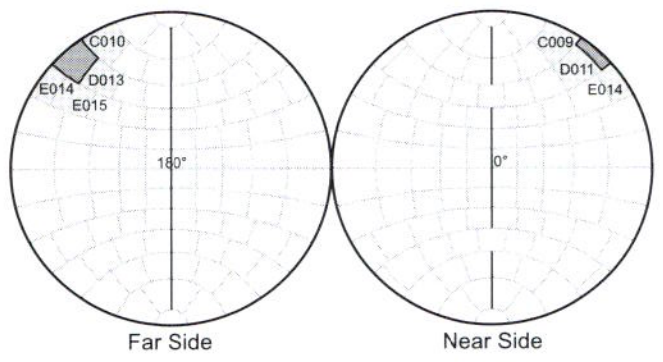

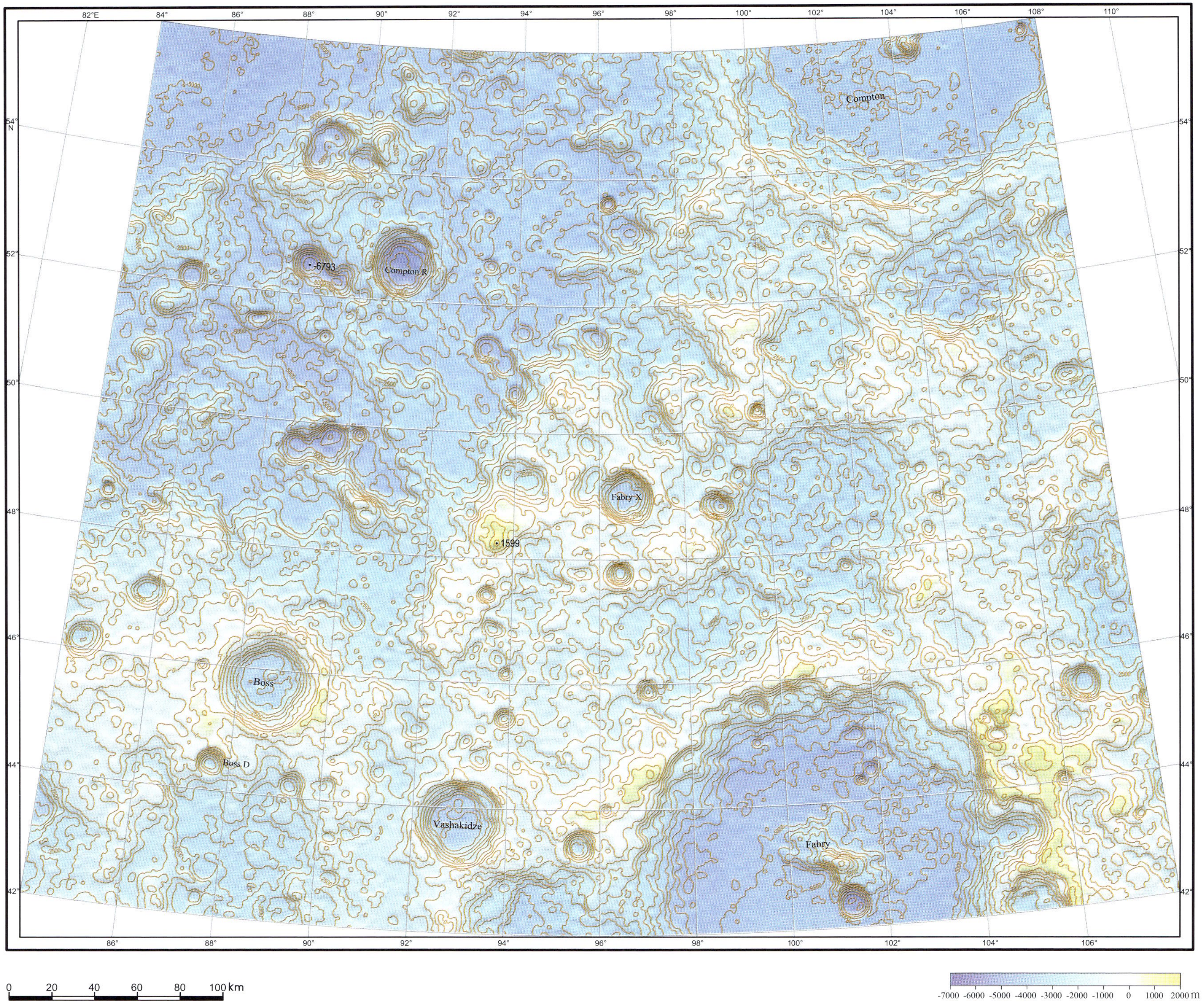

Image Code **D013**

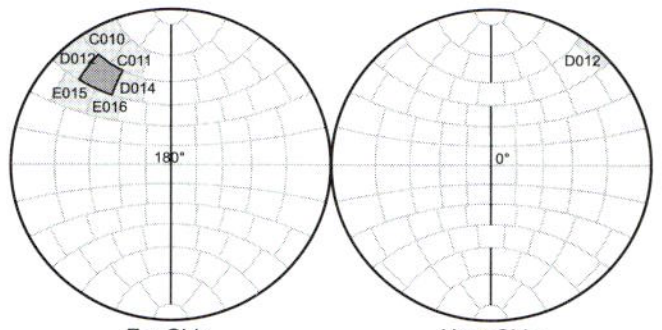

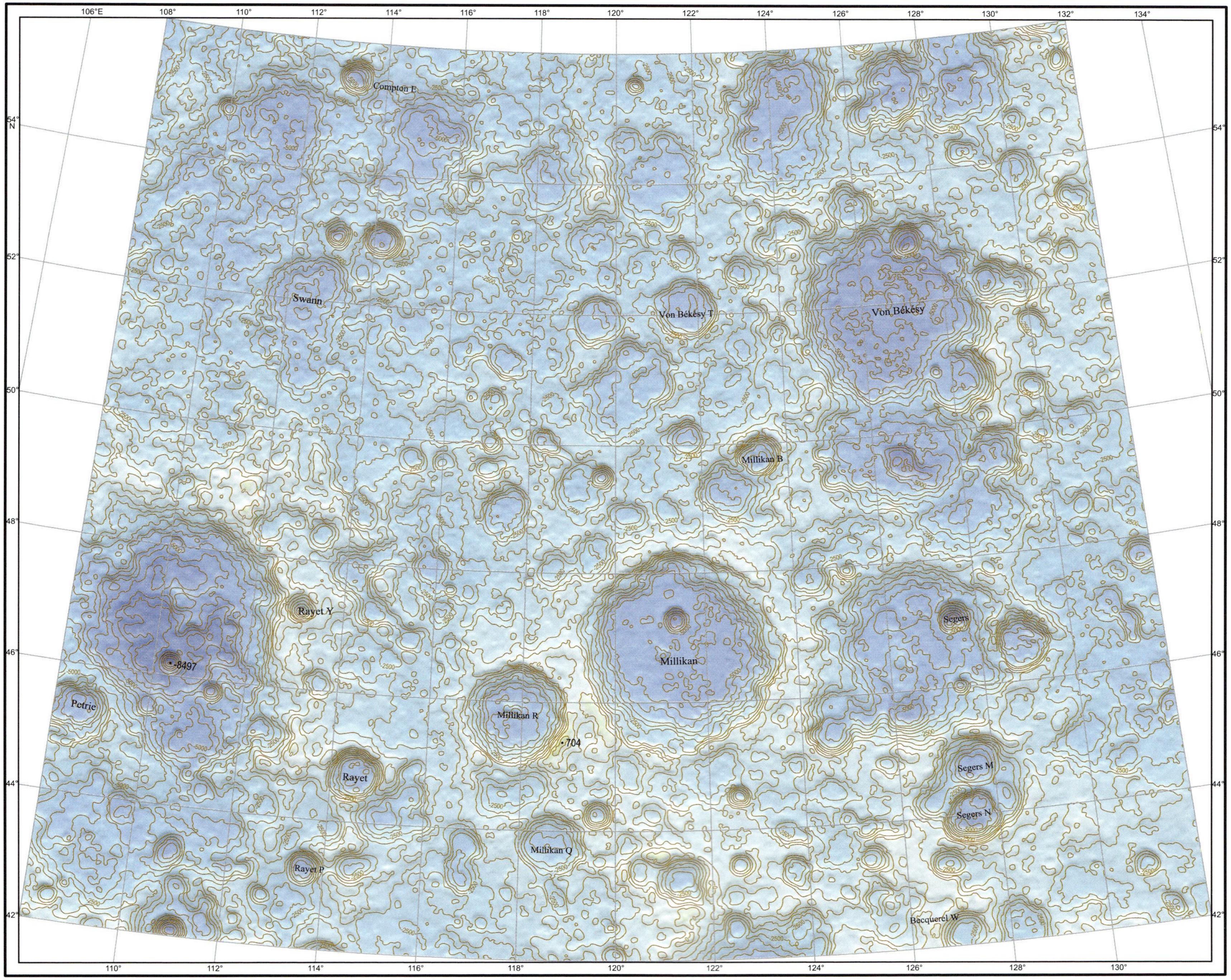

Image Code D014

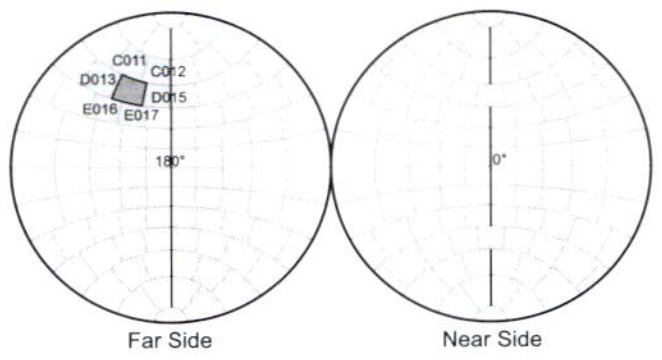

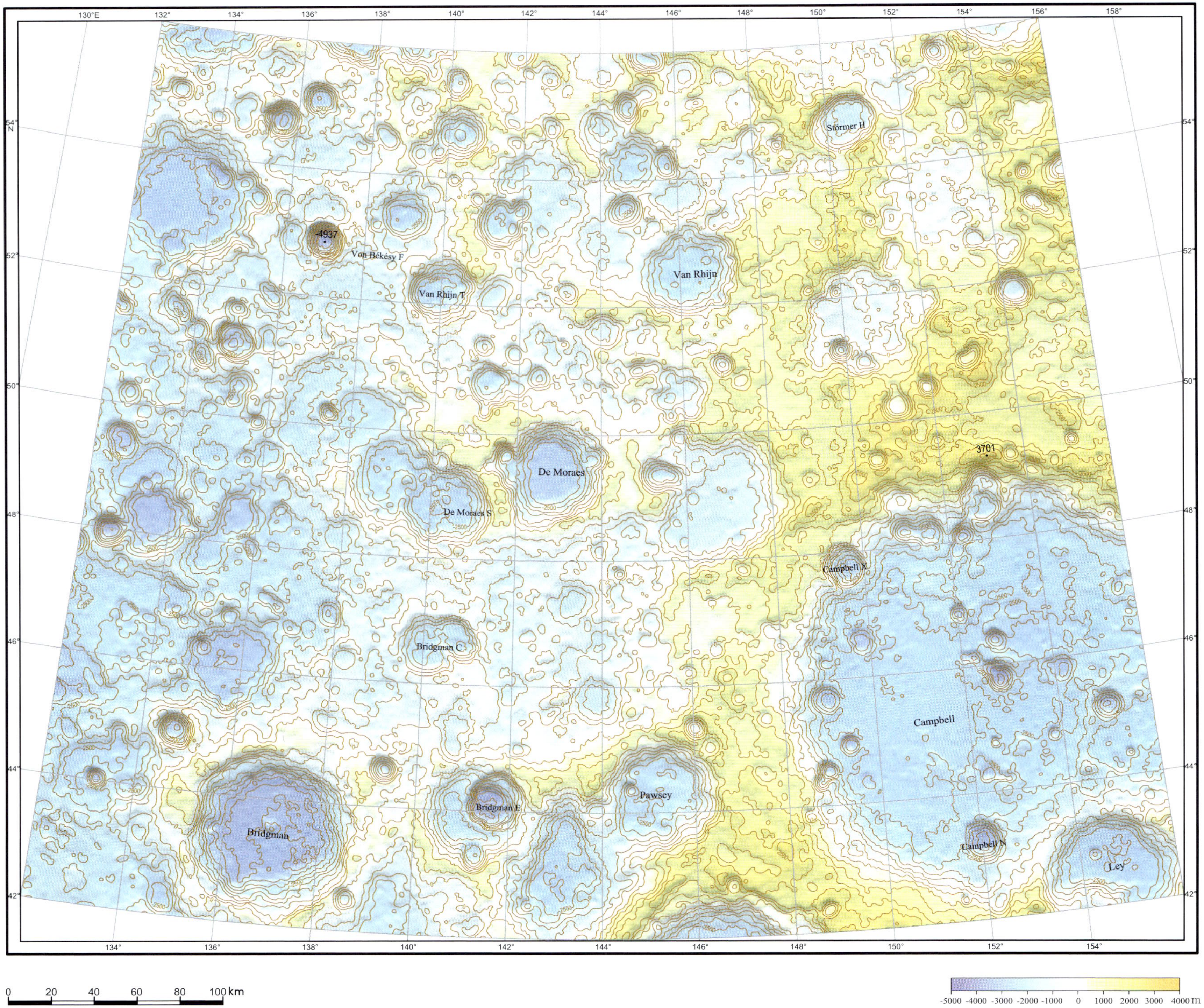

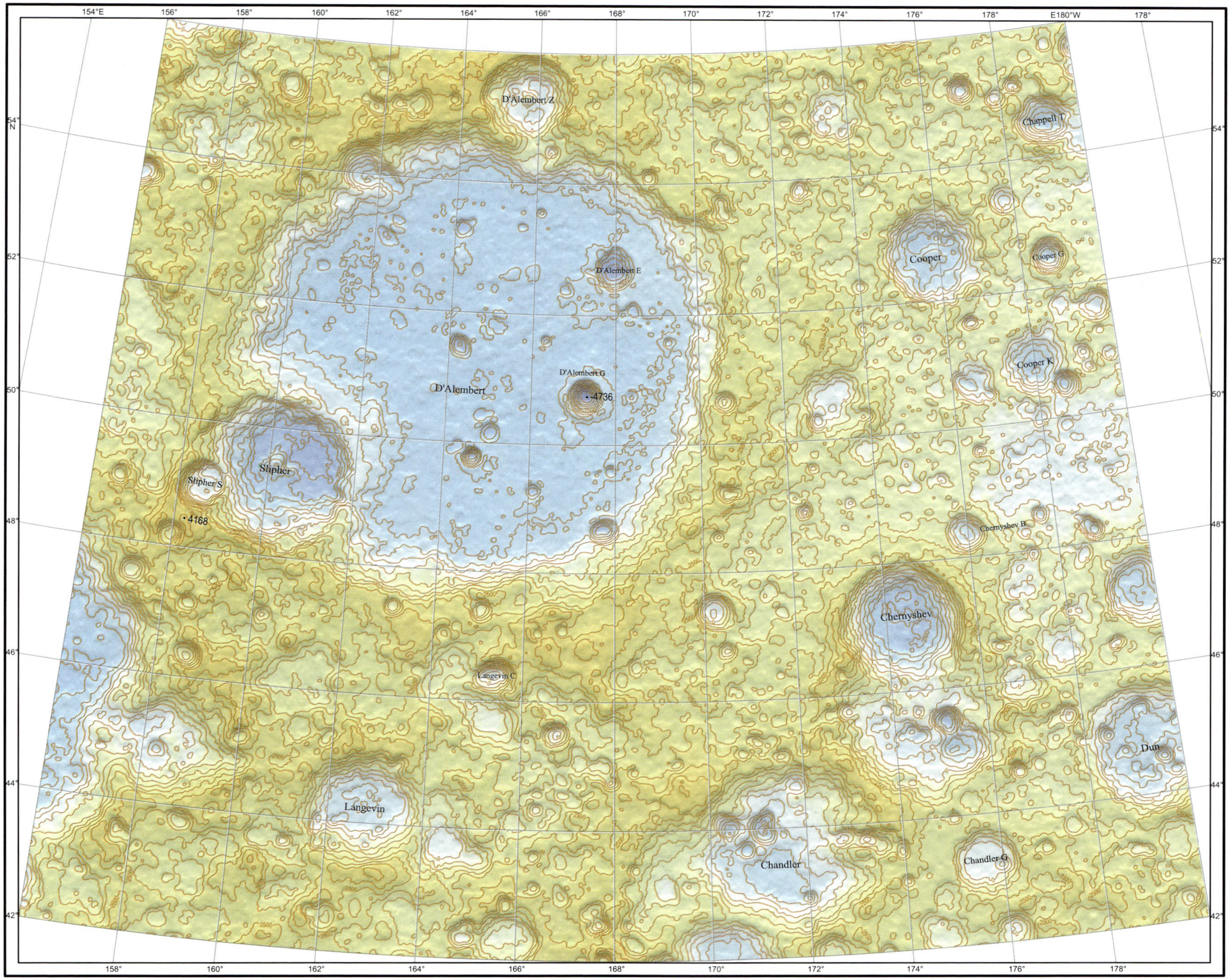

Image Code **D015**

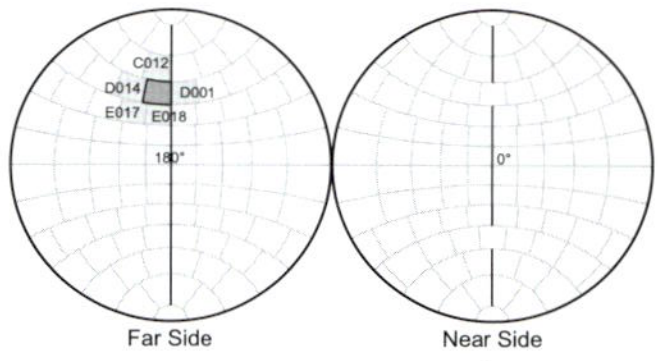

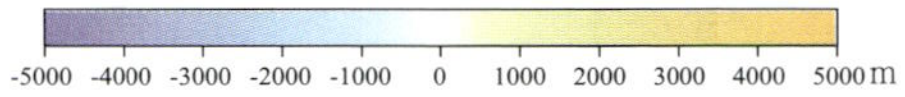

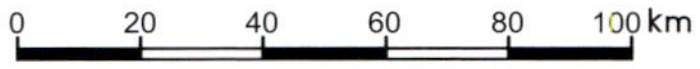

Image Code **E001**

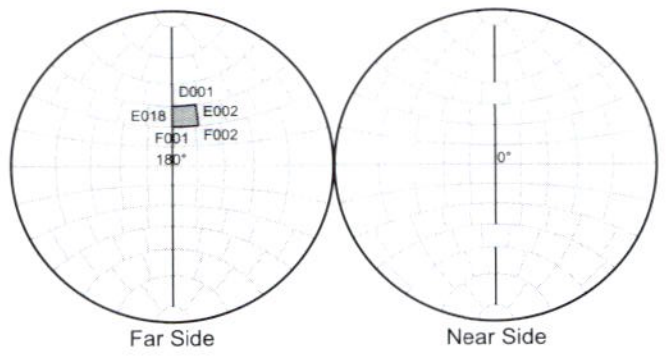

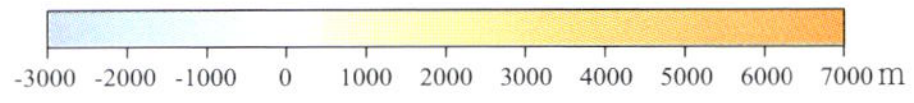

Image Code E002

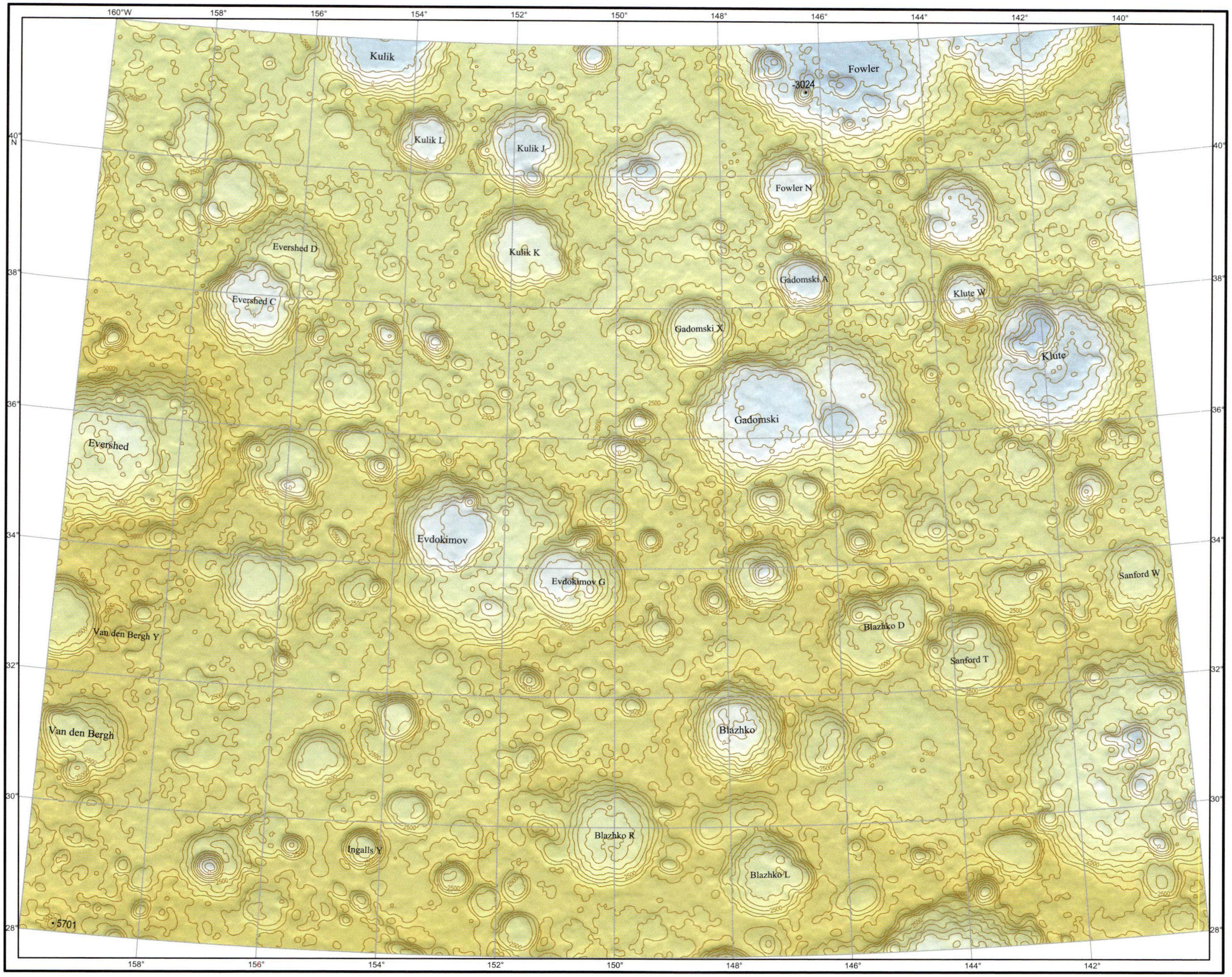

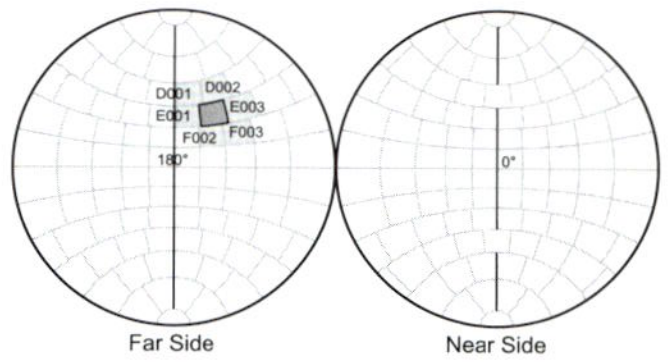

-4000 -3000 -2000 -1000 0 1000 2000 3000 4000 5000 6000 m

0 20 40 60 80 100 km

Image Code **E003**

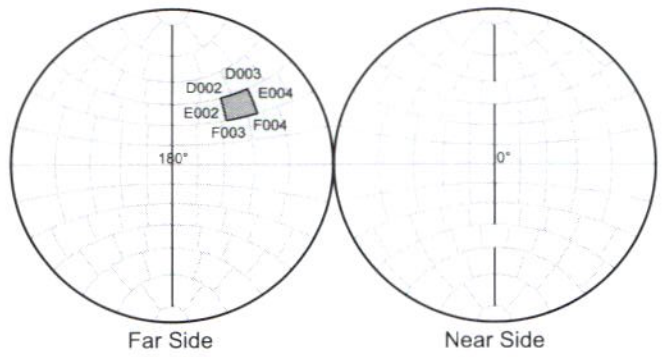

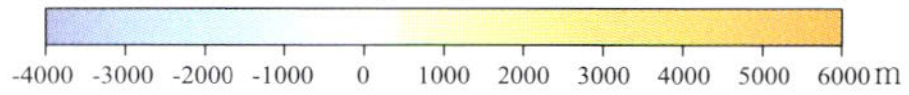

Image Code **E004**

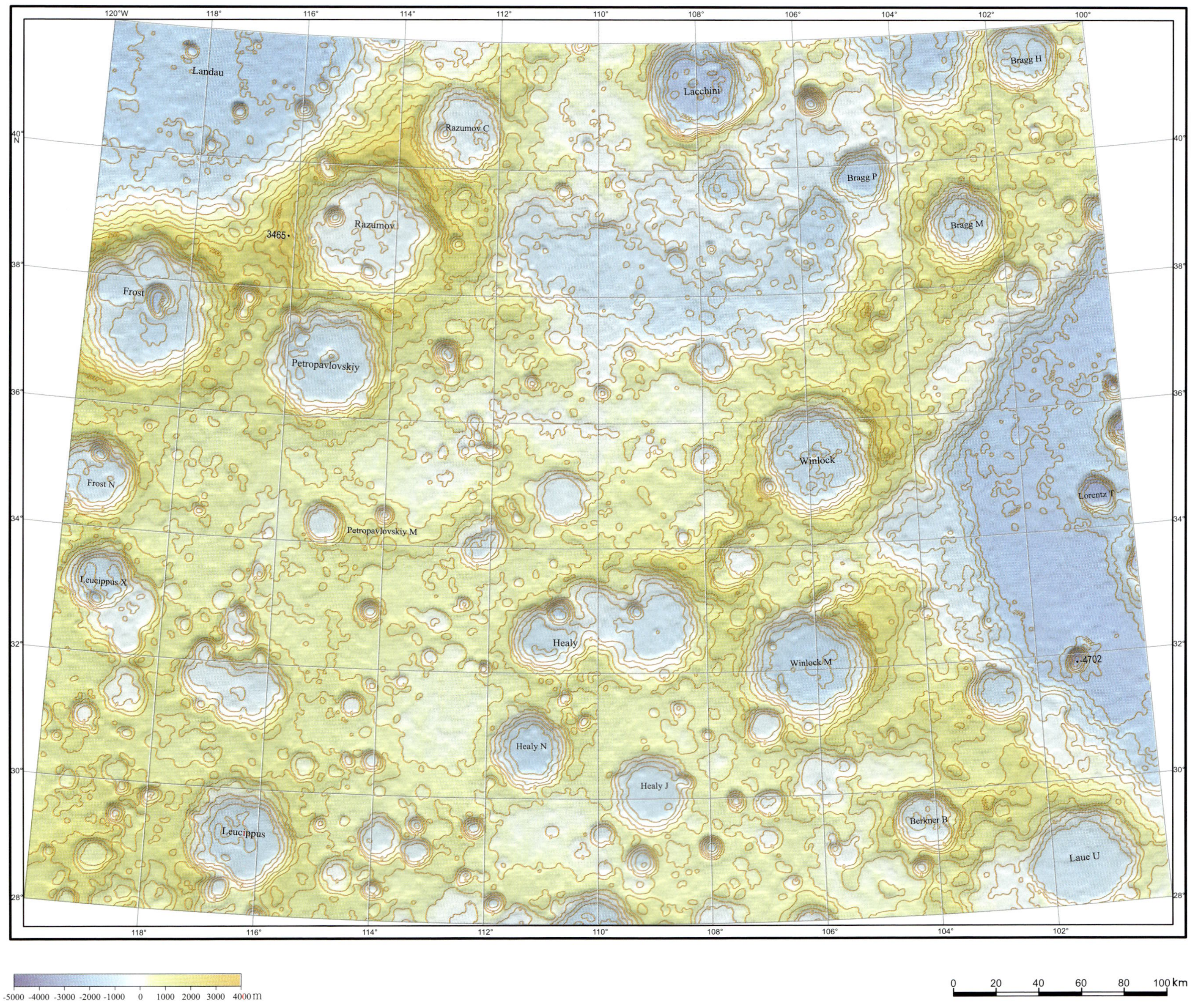

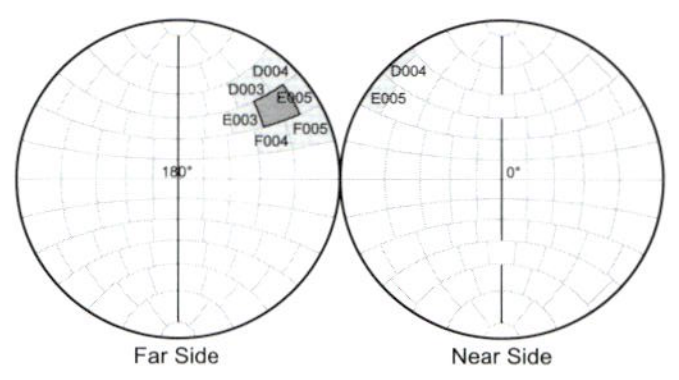

Image Code E005

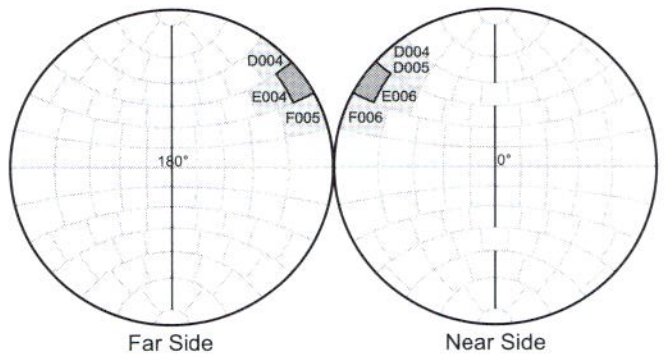

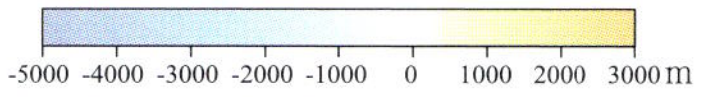

Image Code E006

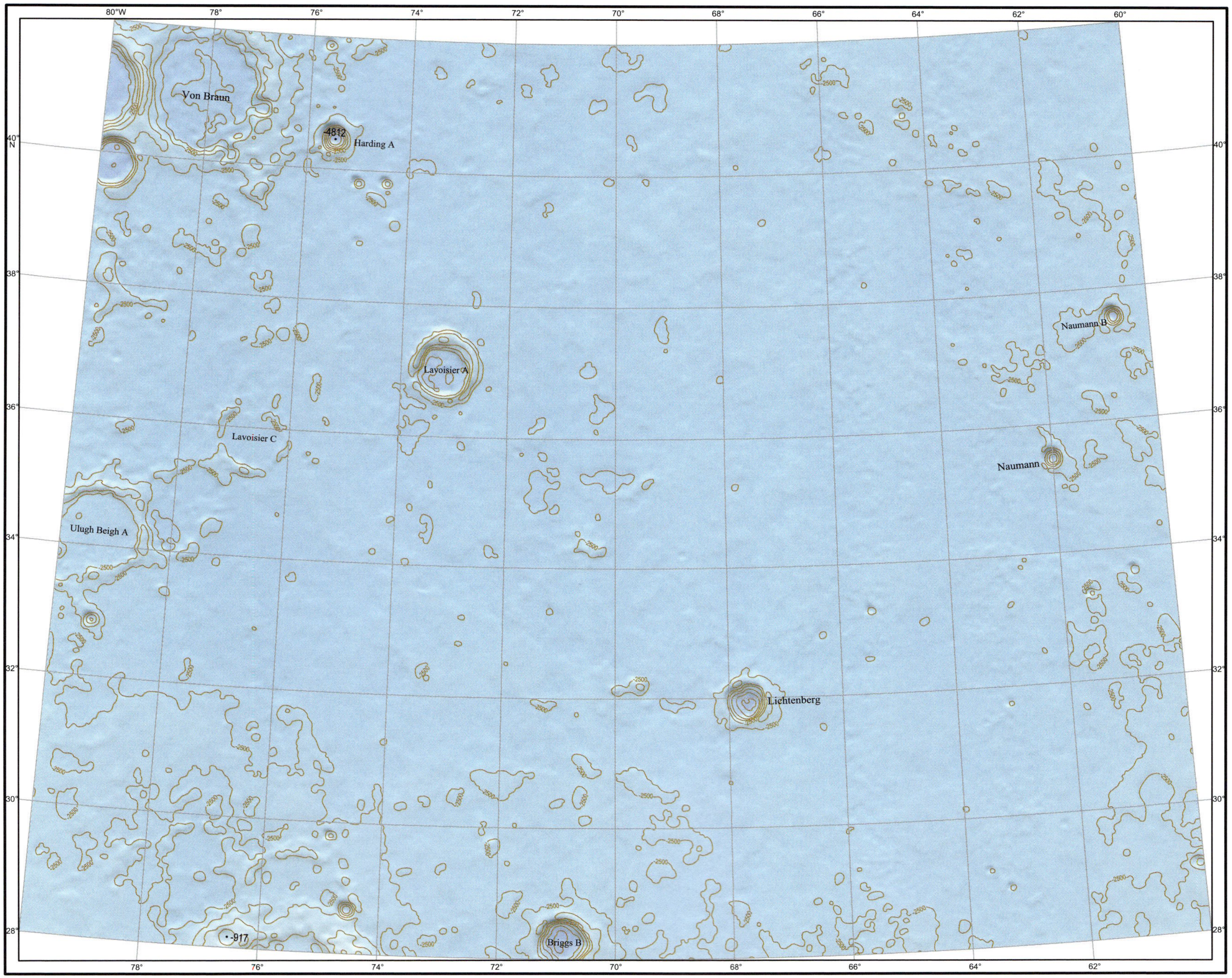

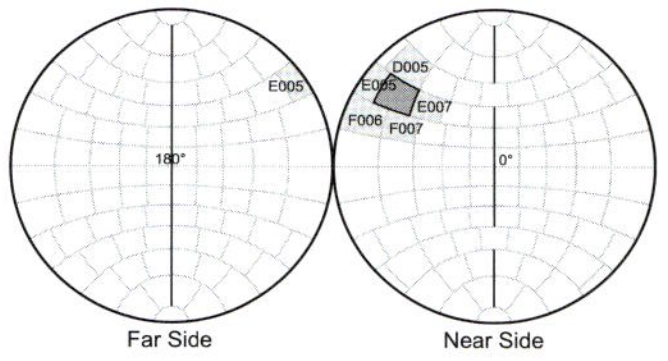

Image Code **E007**

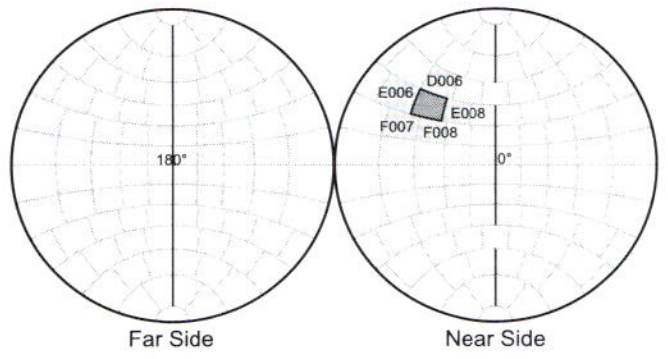

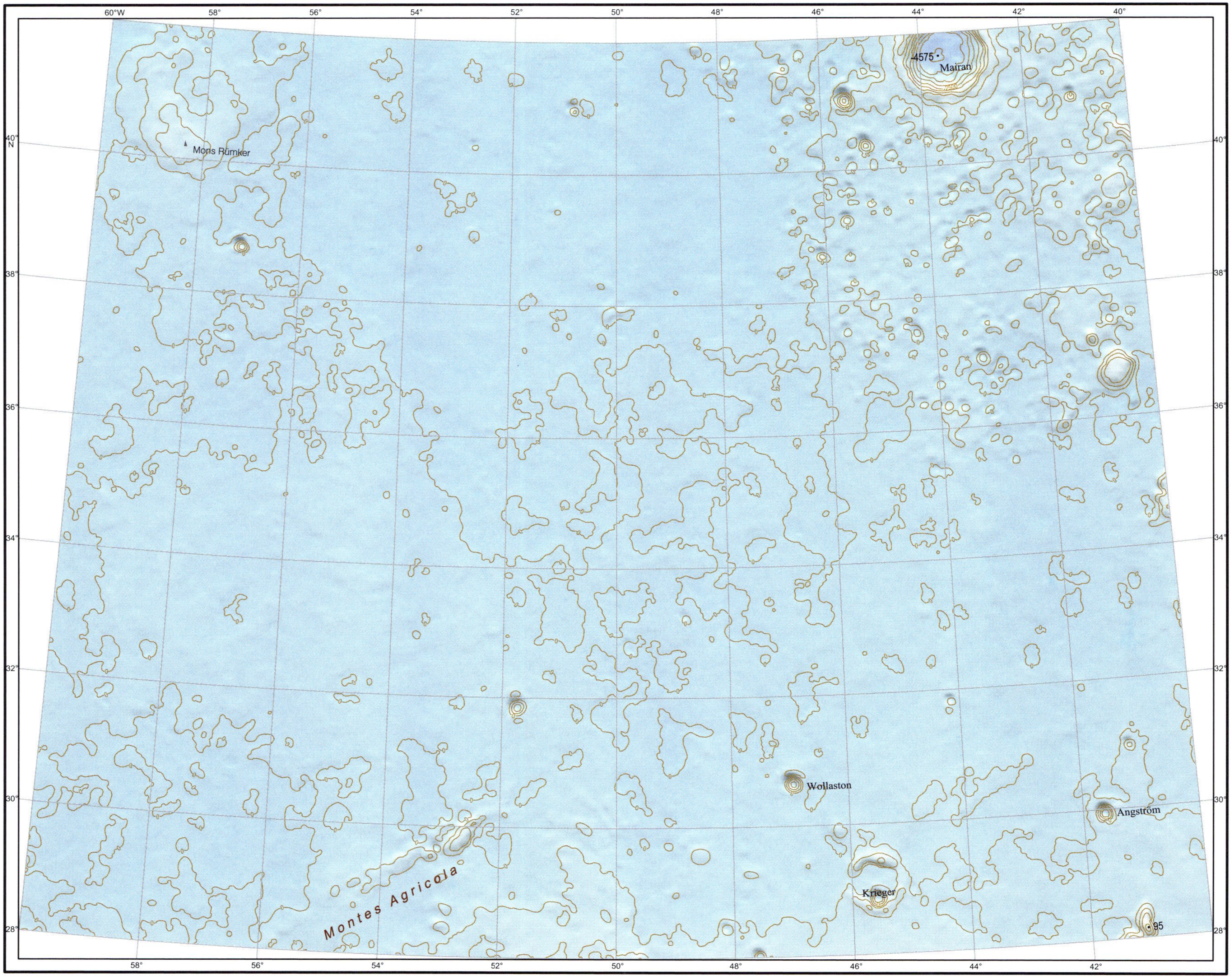

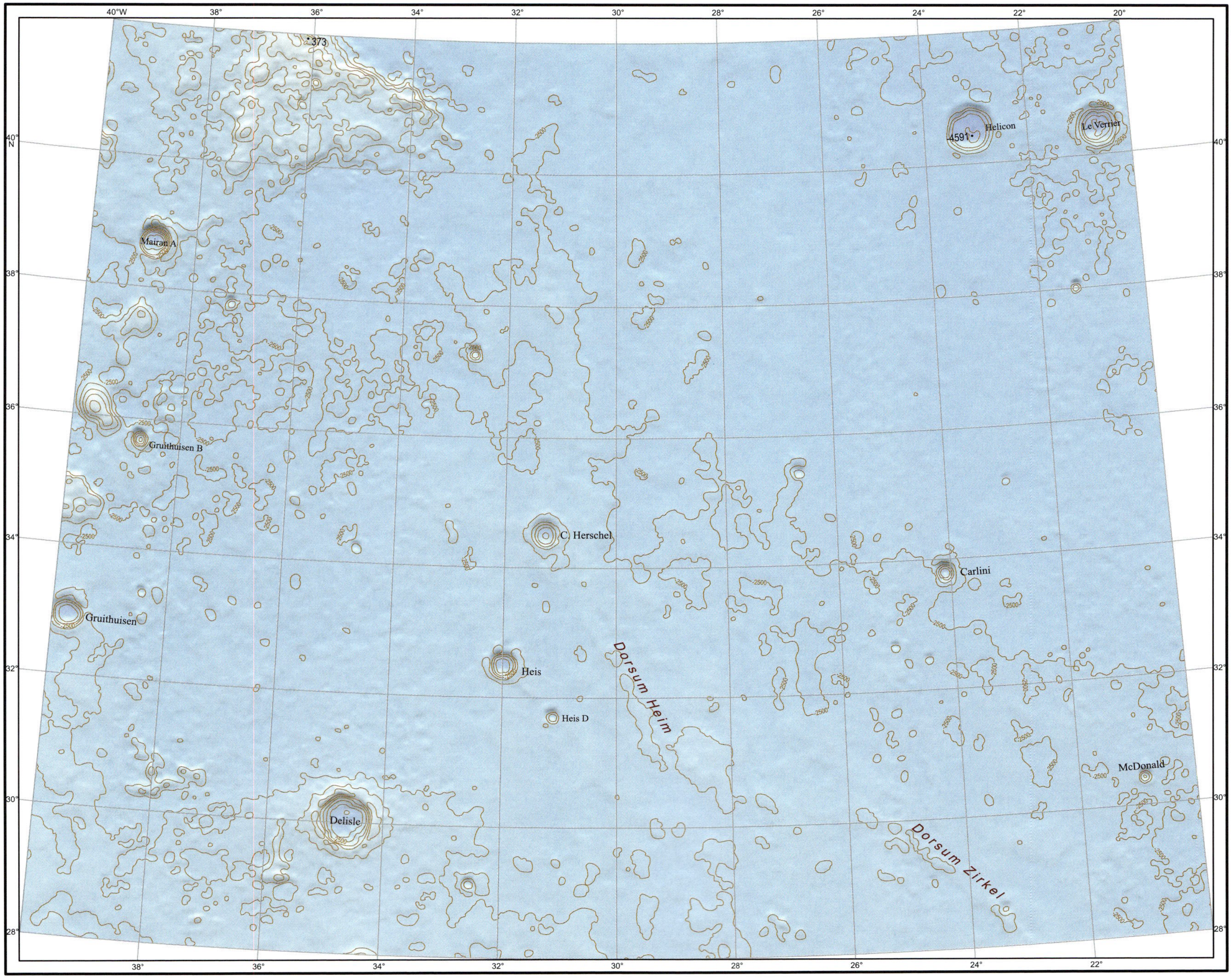

Image Code E008

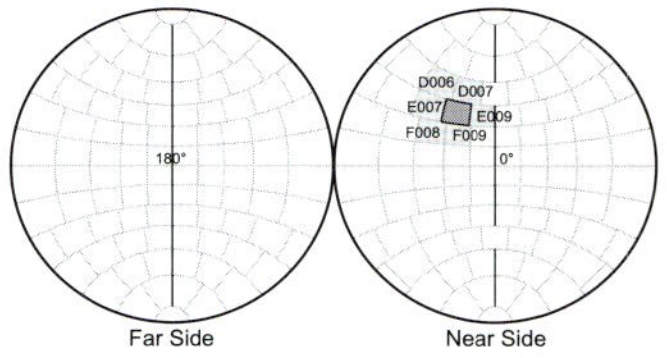

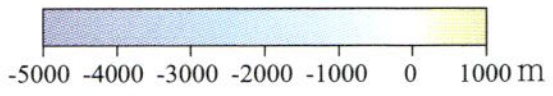

Image Code E009

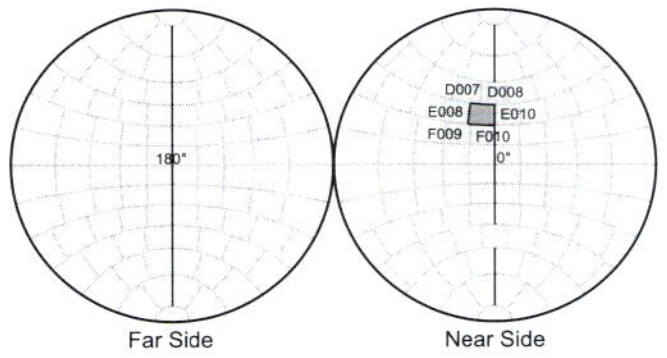

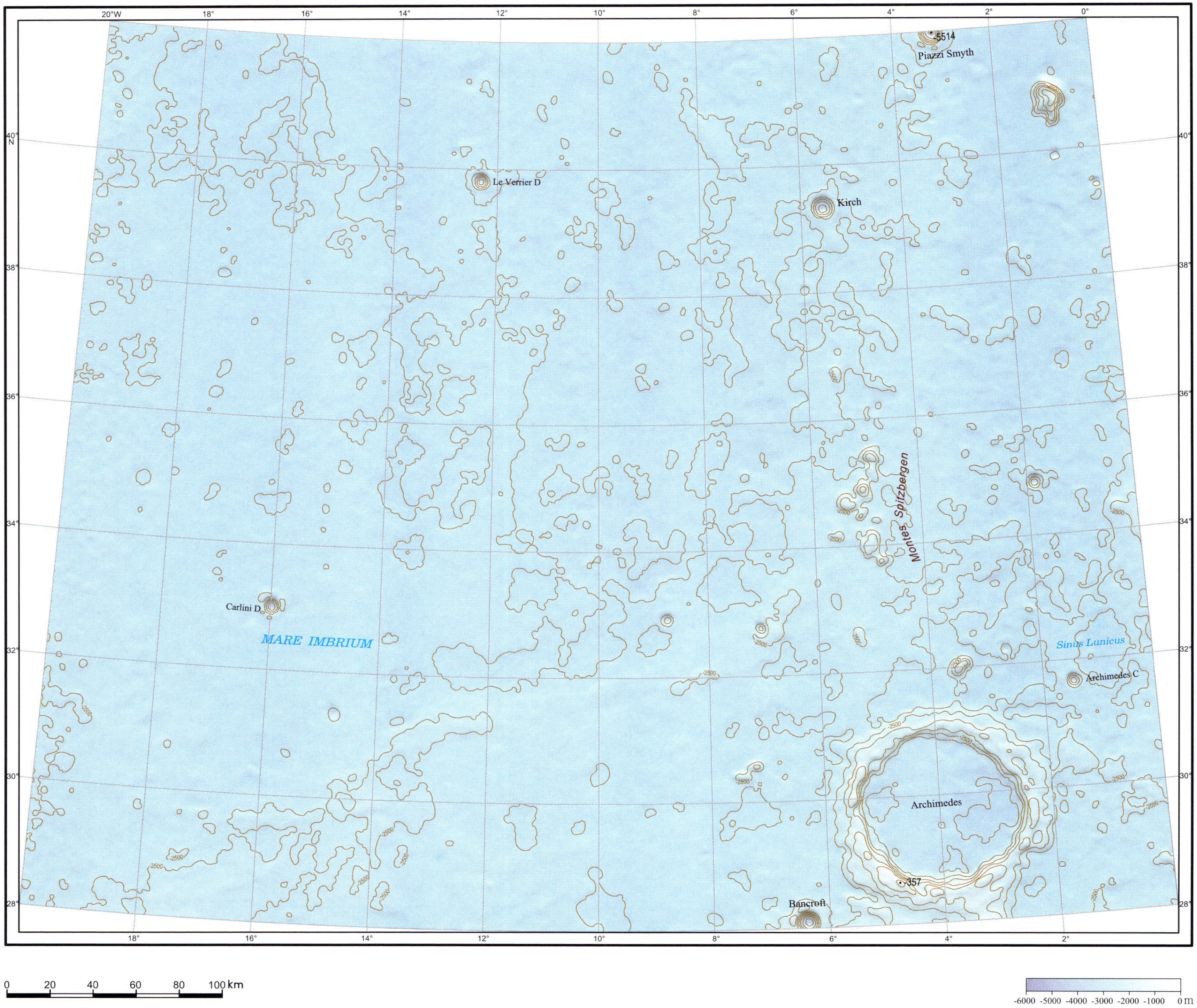

Image Code E010

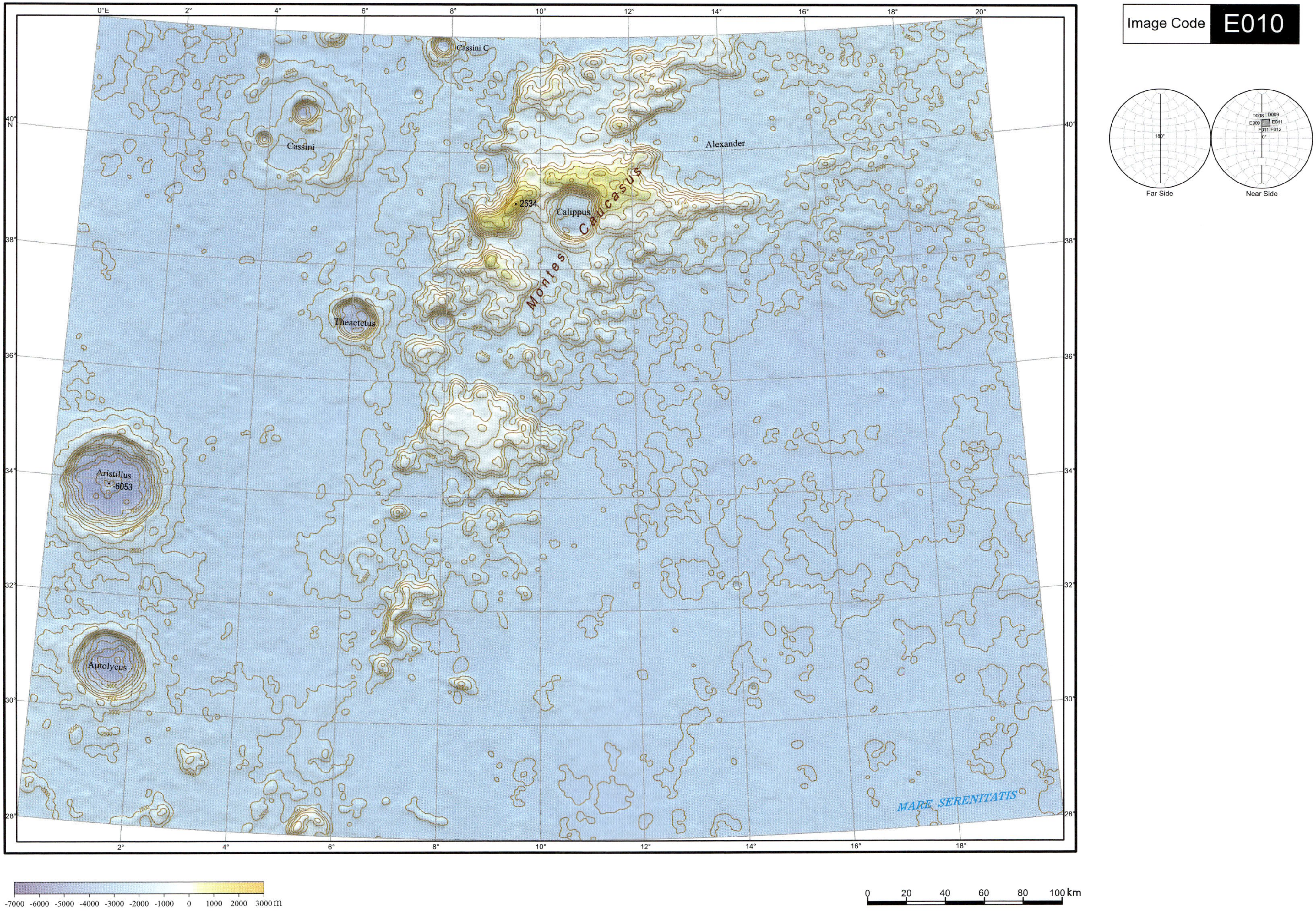

Image Code E011

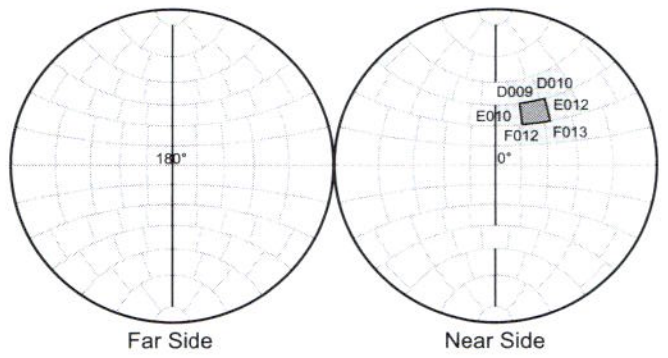

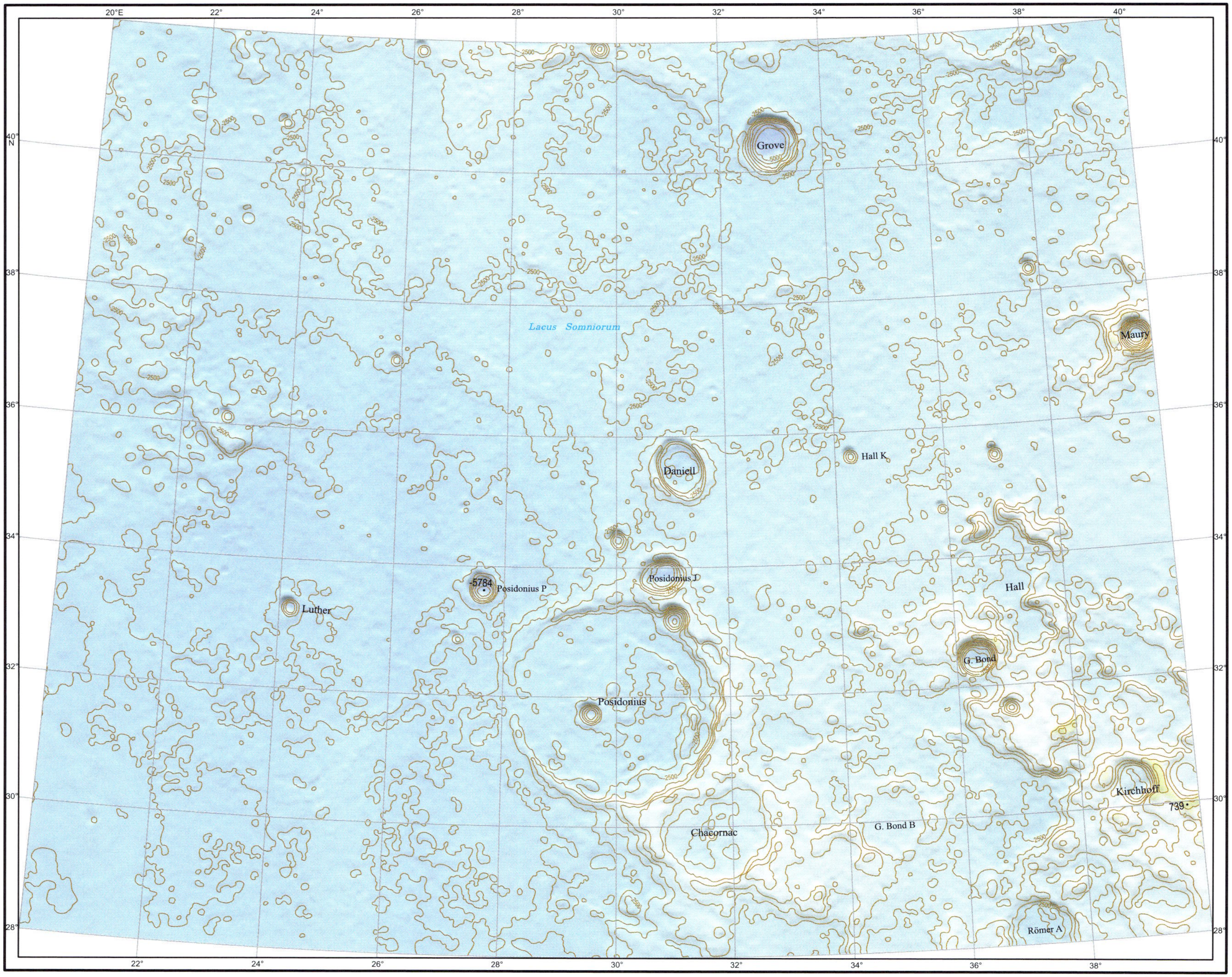

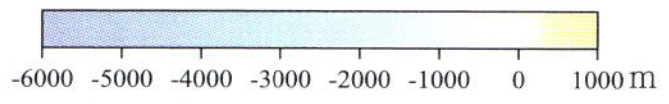

Image Code E012

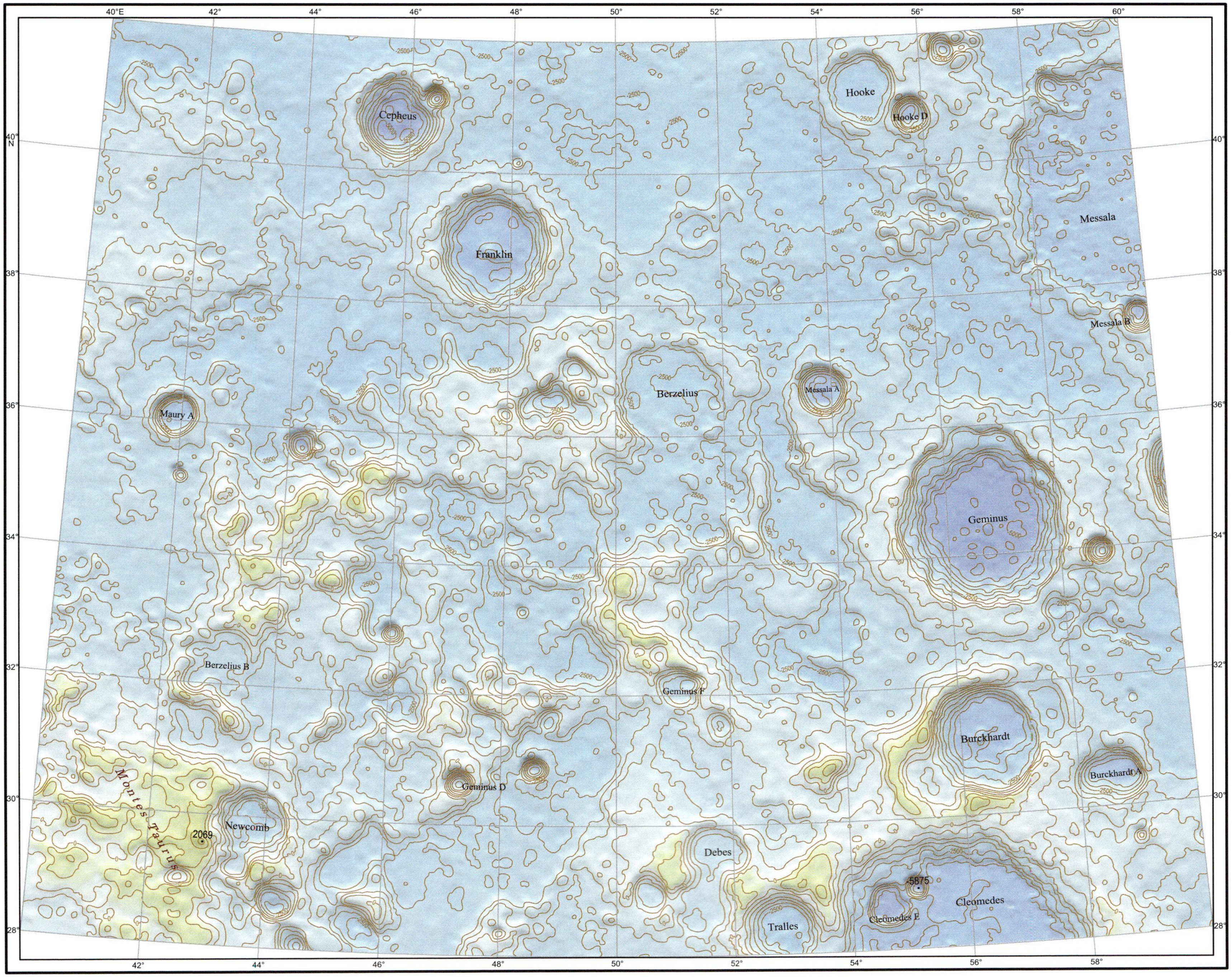

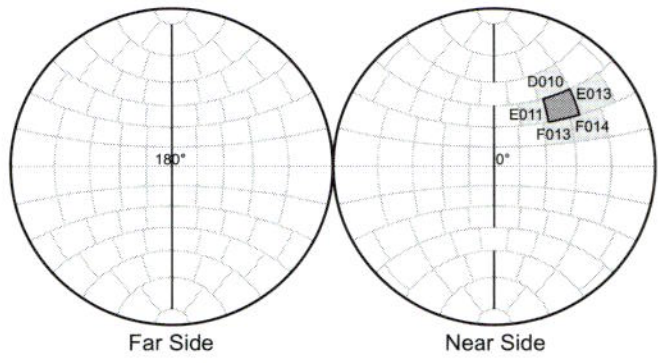

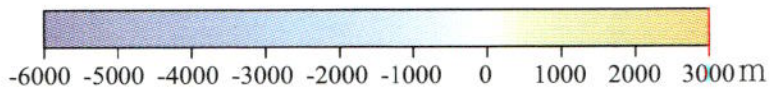

Image Code E013

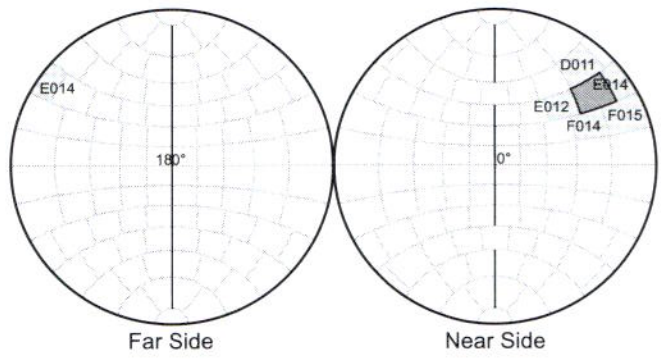

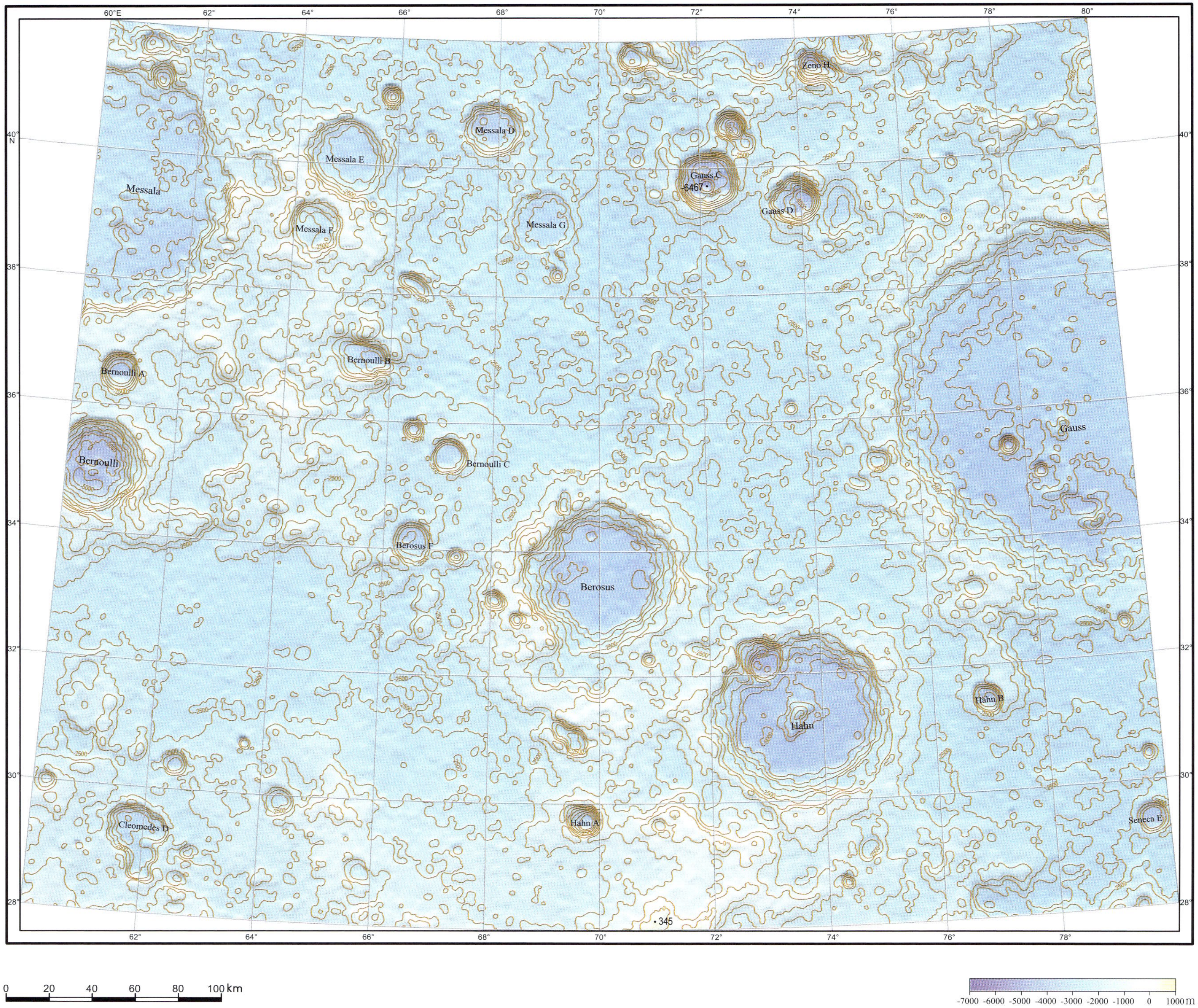

Image Code E014

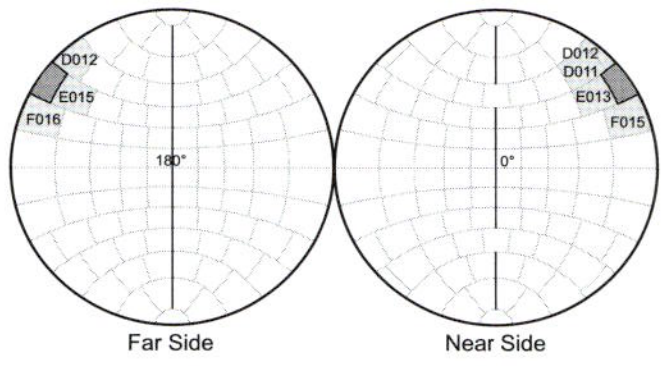

-7000 -6000 -5000 -4000 -3000 -2000 -1000 0 1000 m

0 20 40 60 80 100 km

Image Code E015

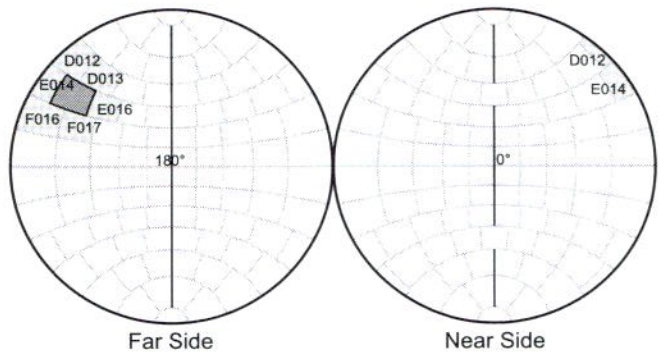

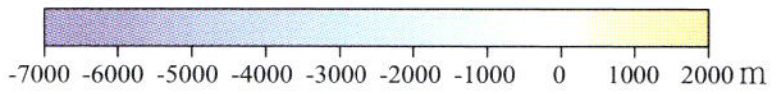

Image Code E016

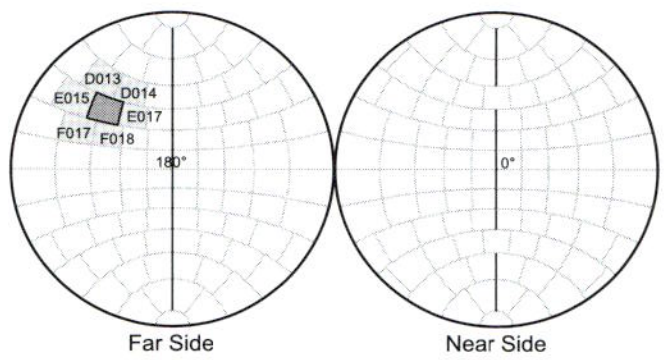

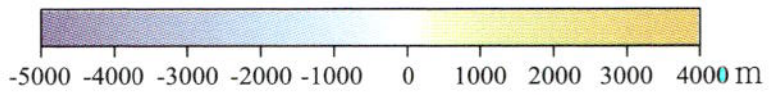

Image Code **E017**

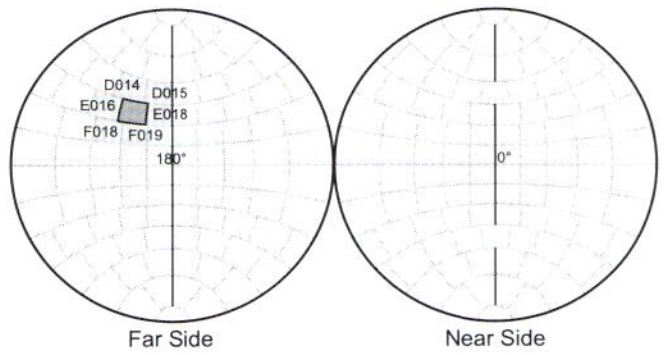

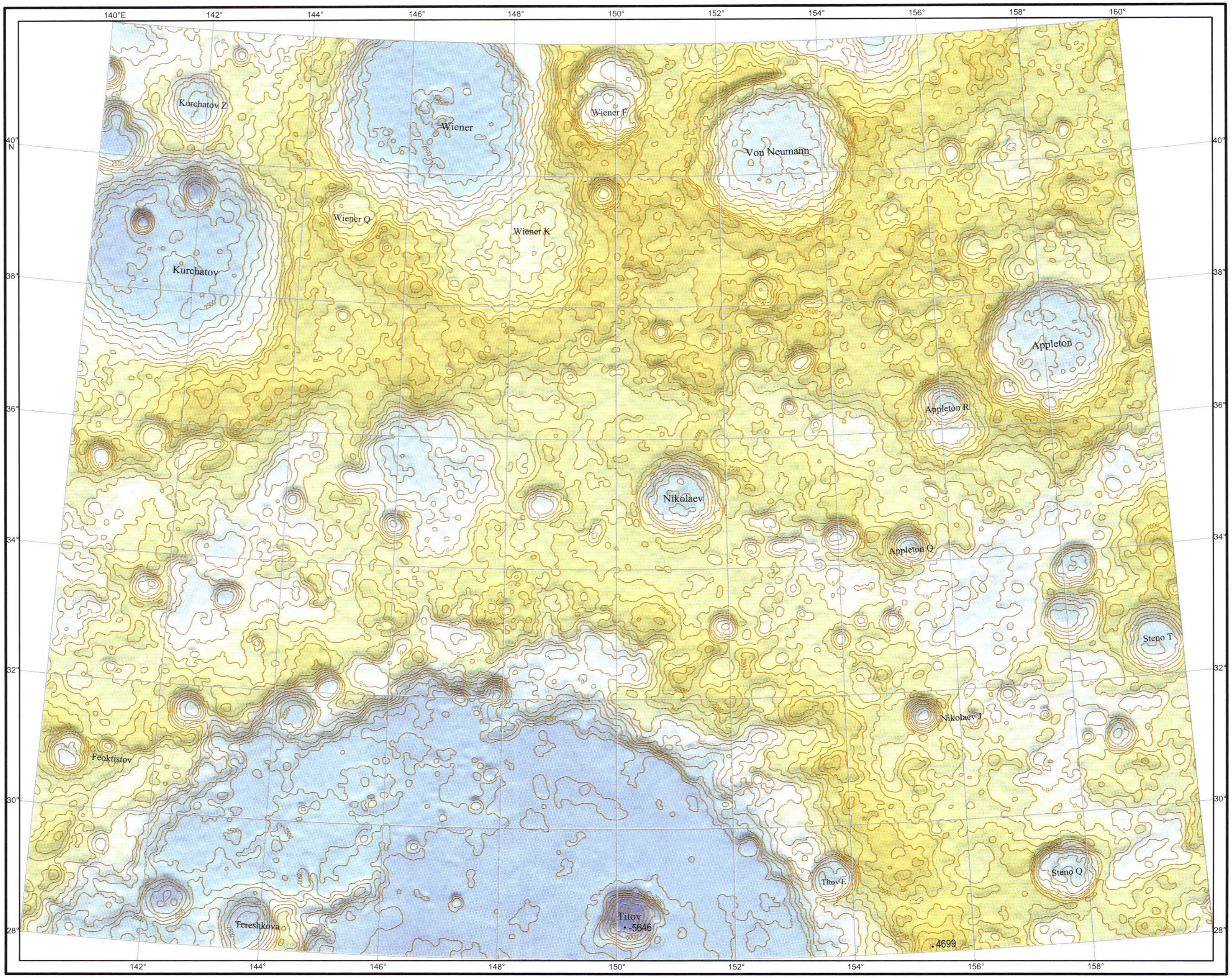

0 20 40 60 80 100 km

-6000 -5000 -4000 -3000 -2000 -1000 0 1000 2000 3000 4000 5000 m

Image Code E018

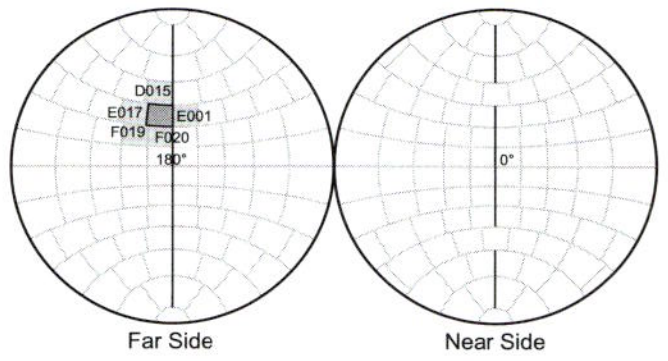

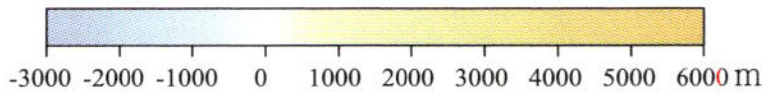

Image Code F001

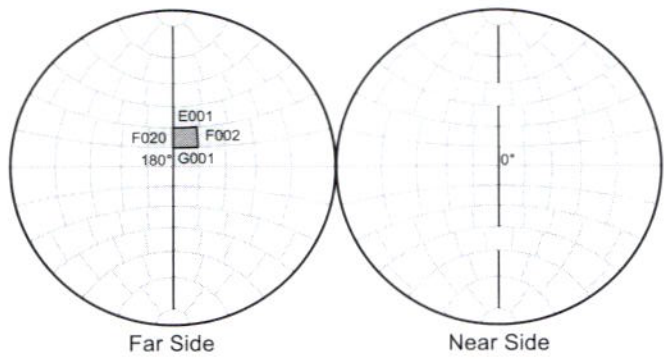

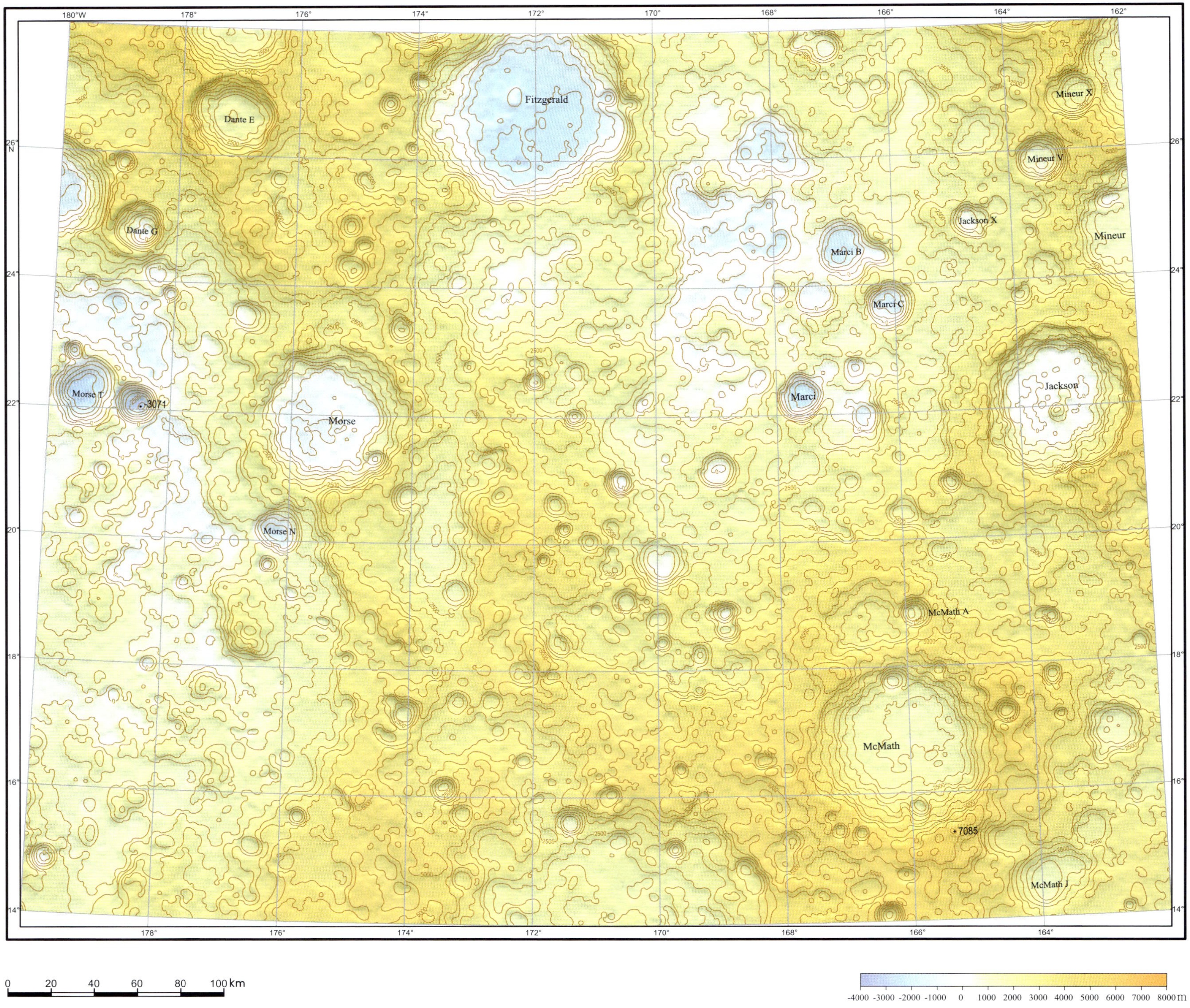

Image Code F002

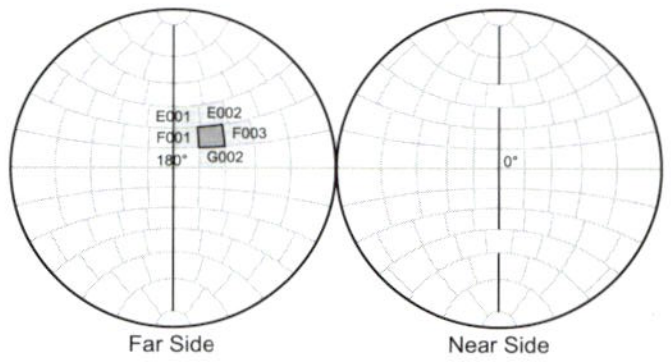

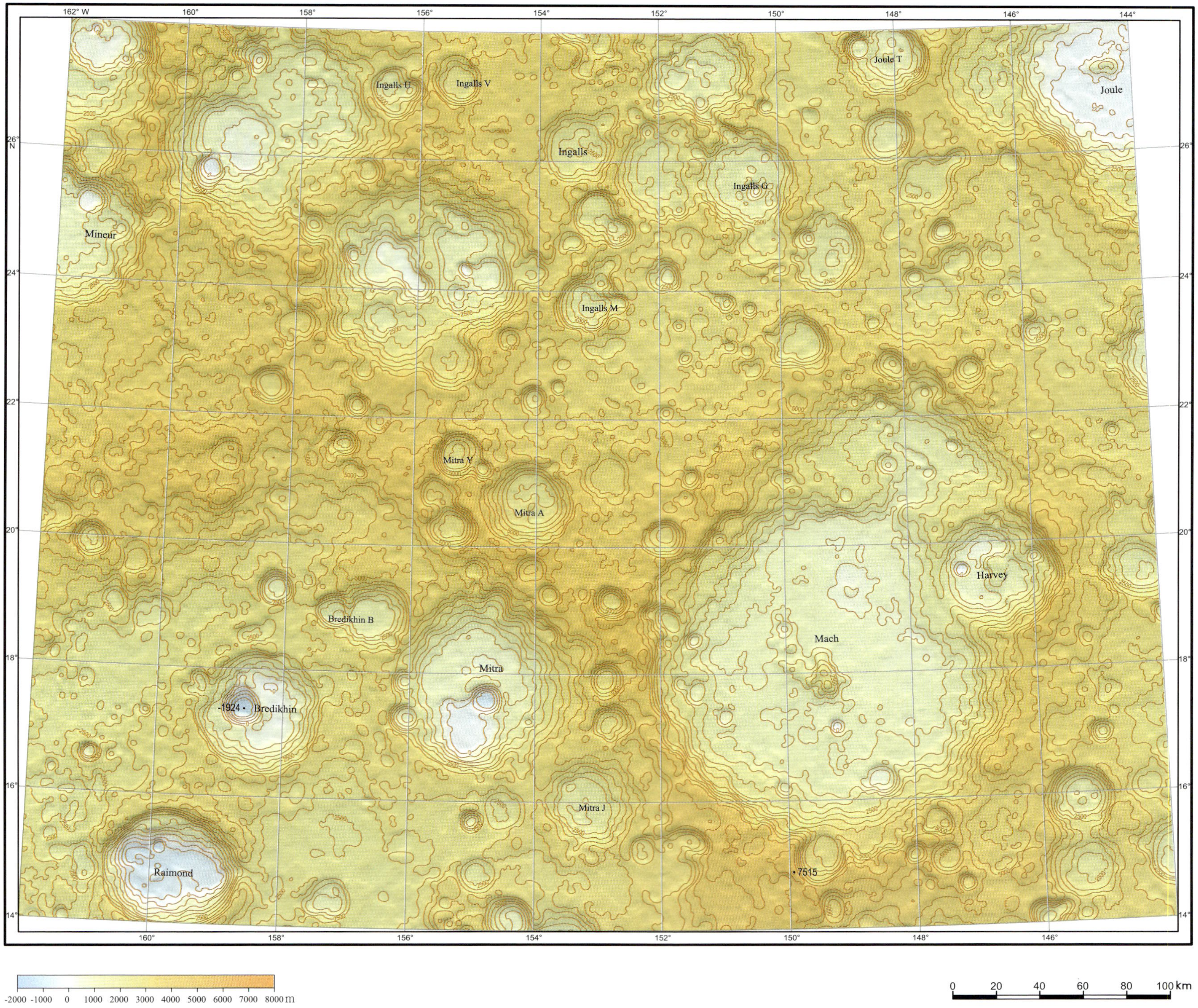

Image Code **F003**

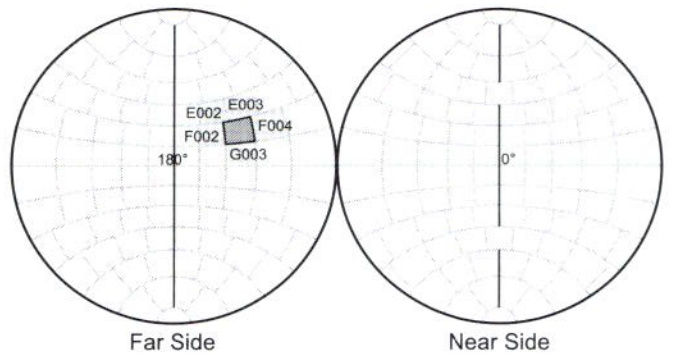

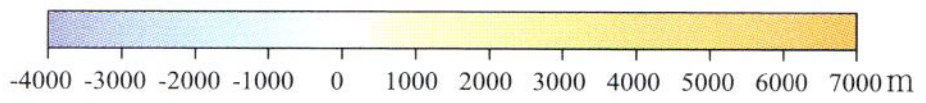

Image Code **F004**

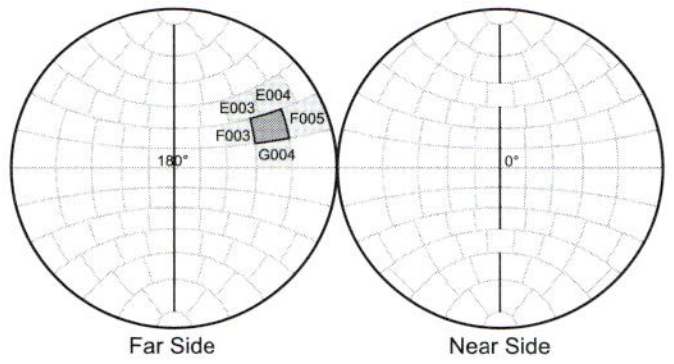

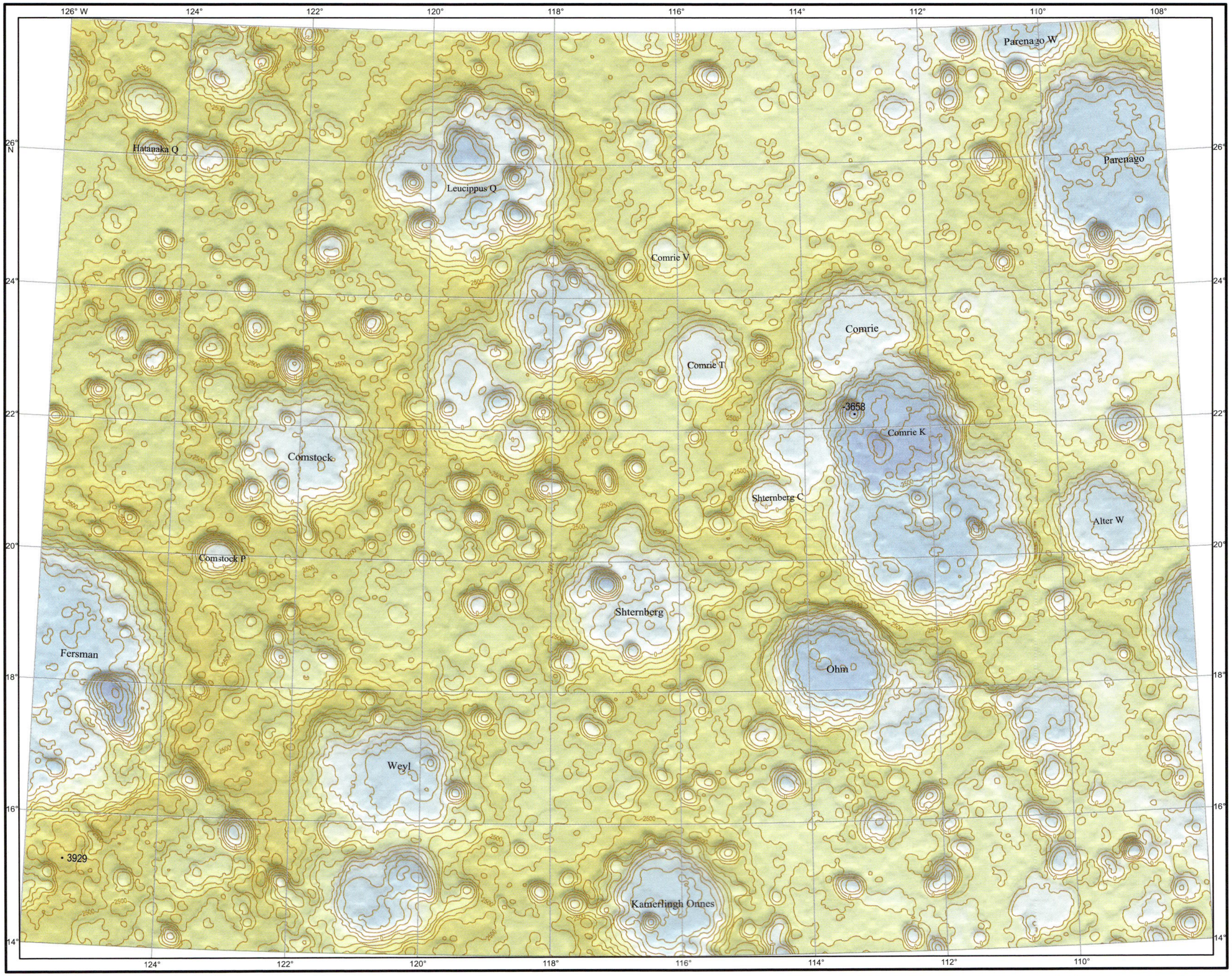

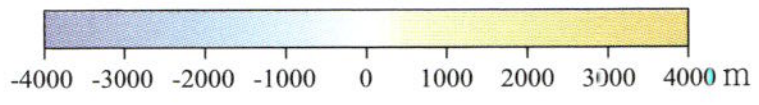

Image Code F005

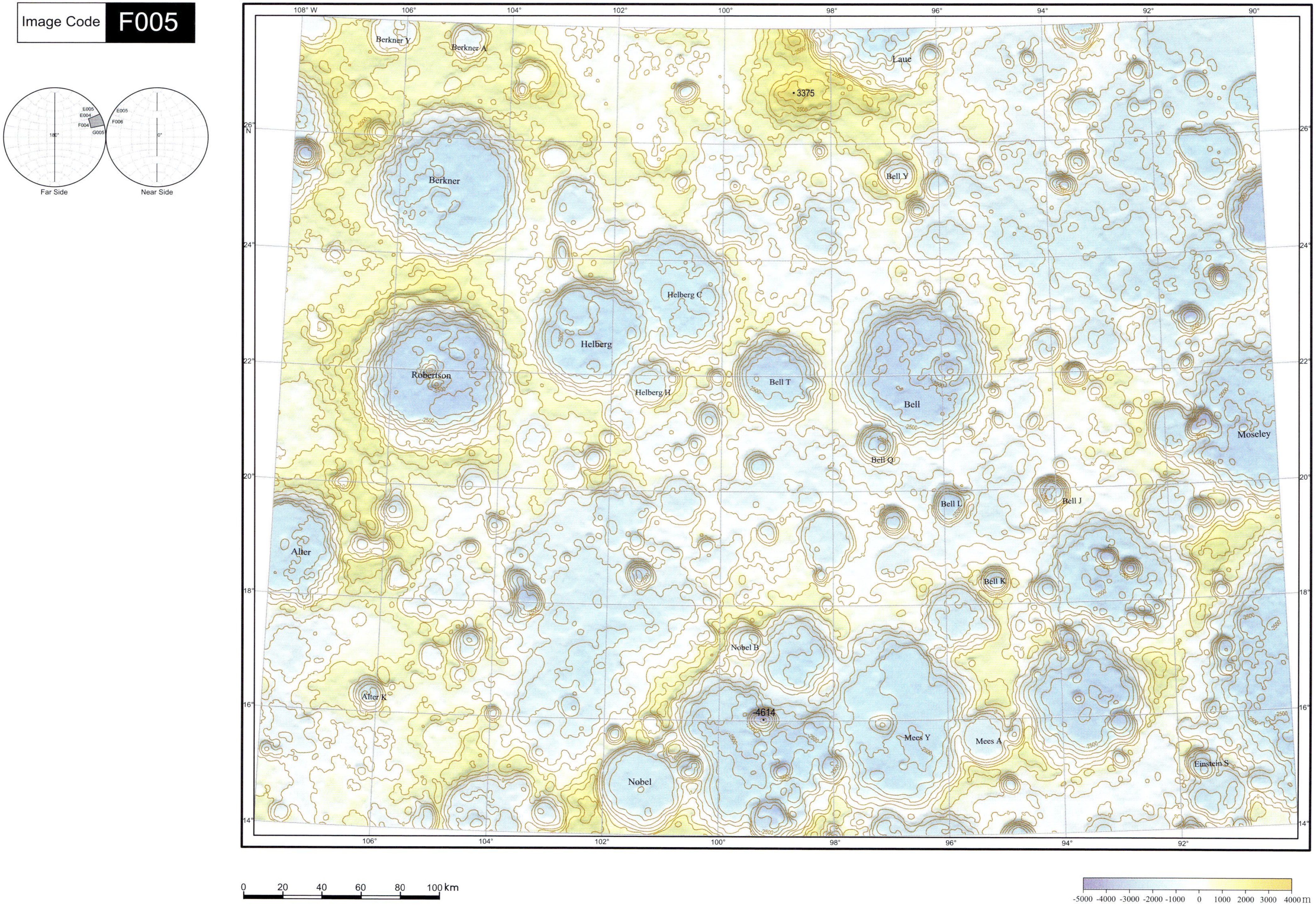

Image Code F006

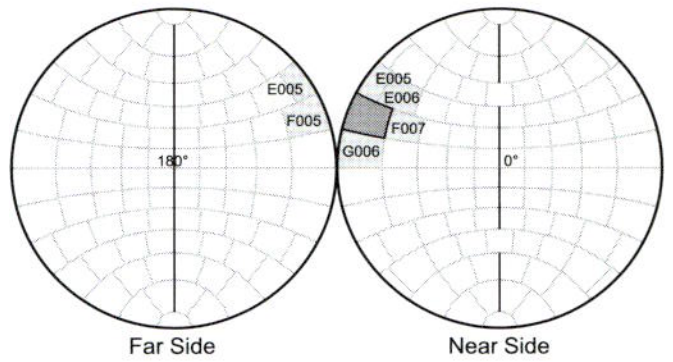

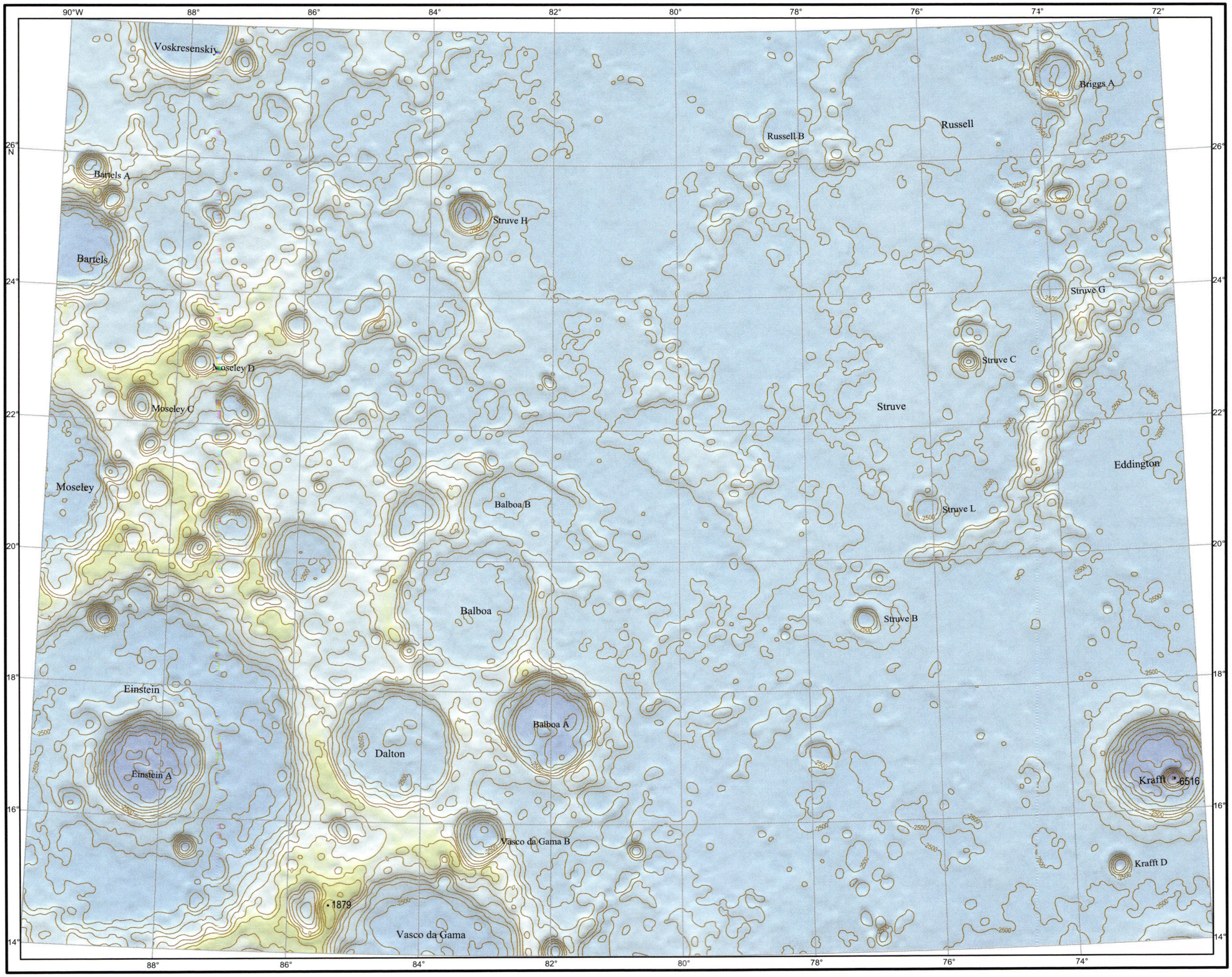

-7000 -6000 -5000 -4000 -3000 -2000 -1000 0 1000 2000 m

0 20 40 60 80 100 km

Image Code F007

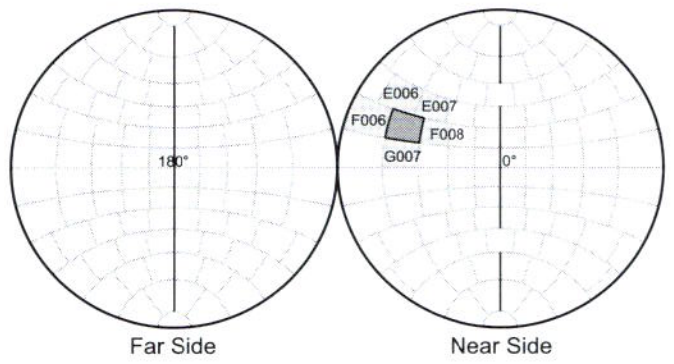

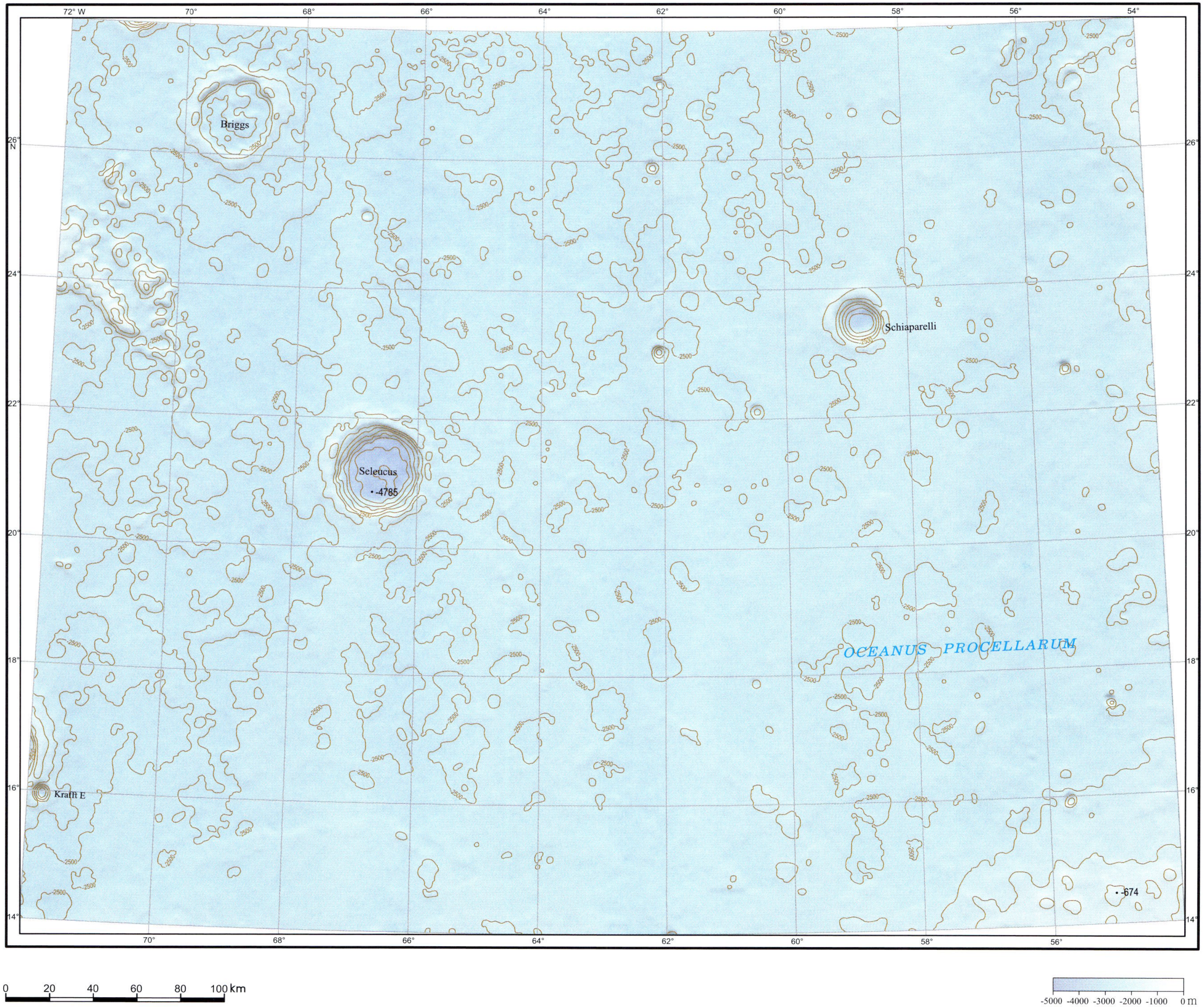

Image Code F008

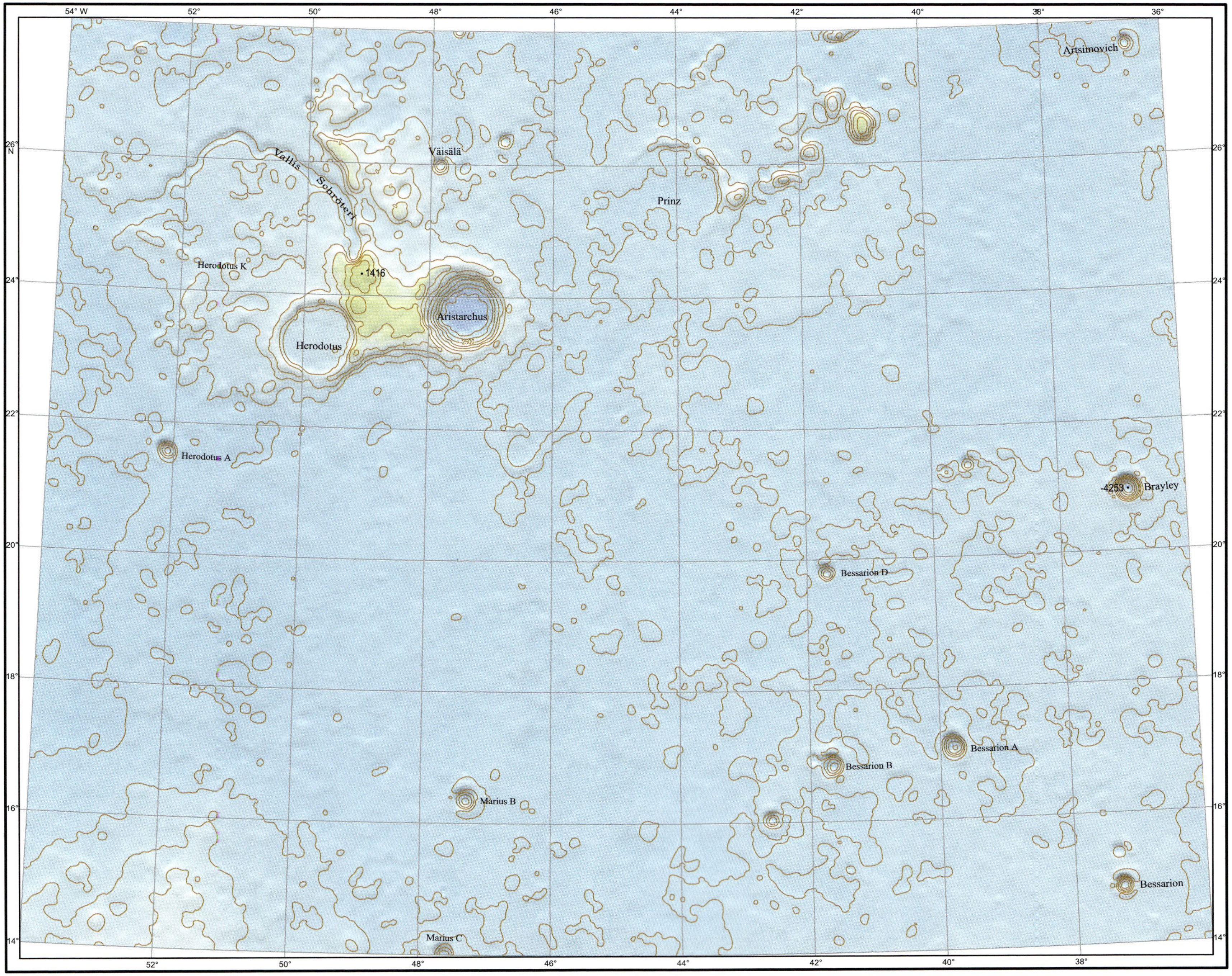

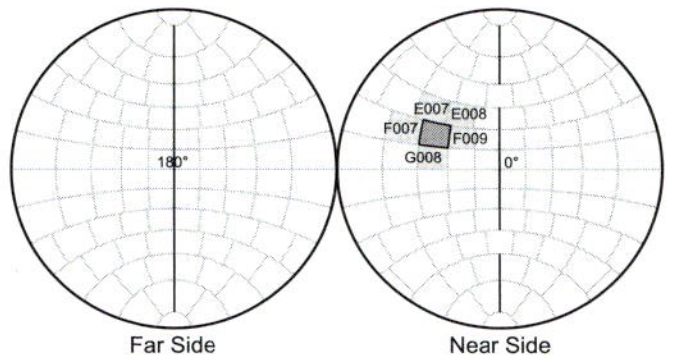

Image Code F009

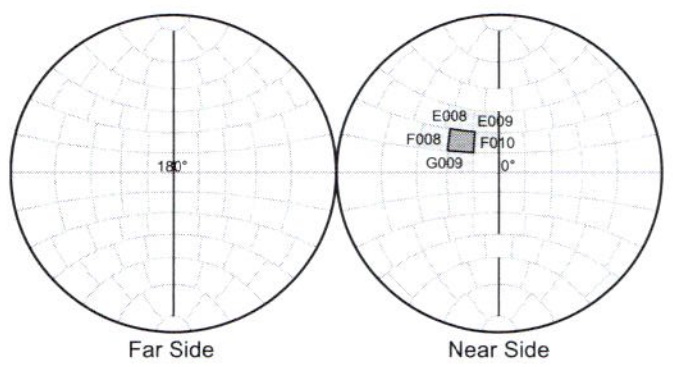

Image Code F010

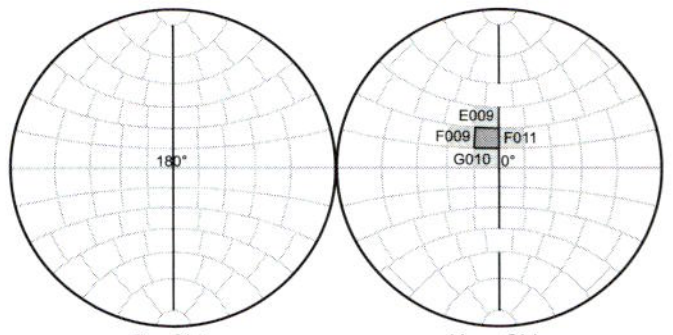

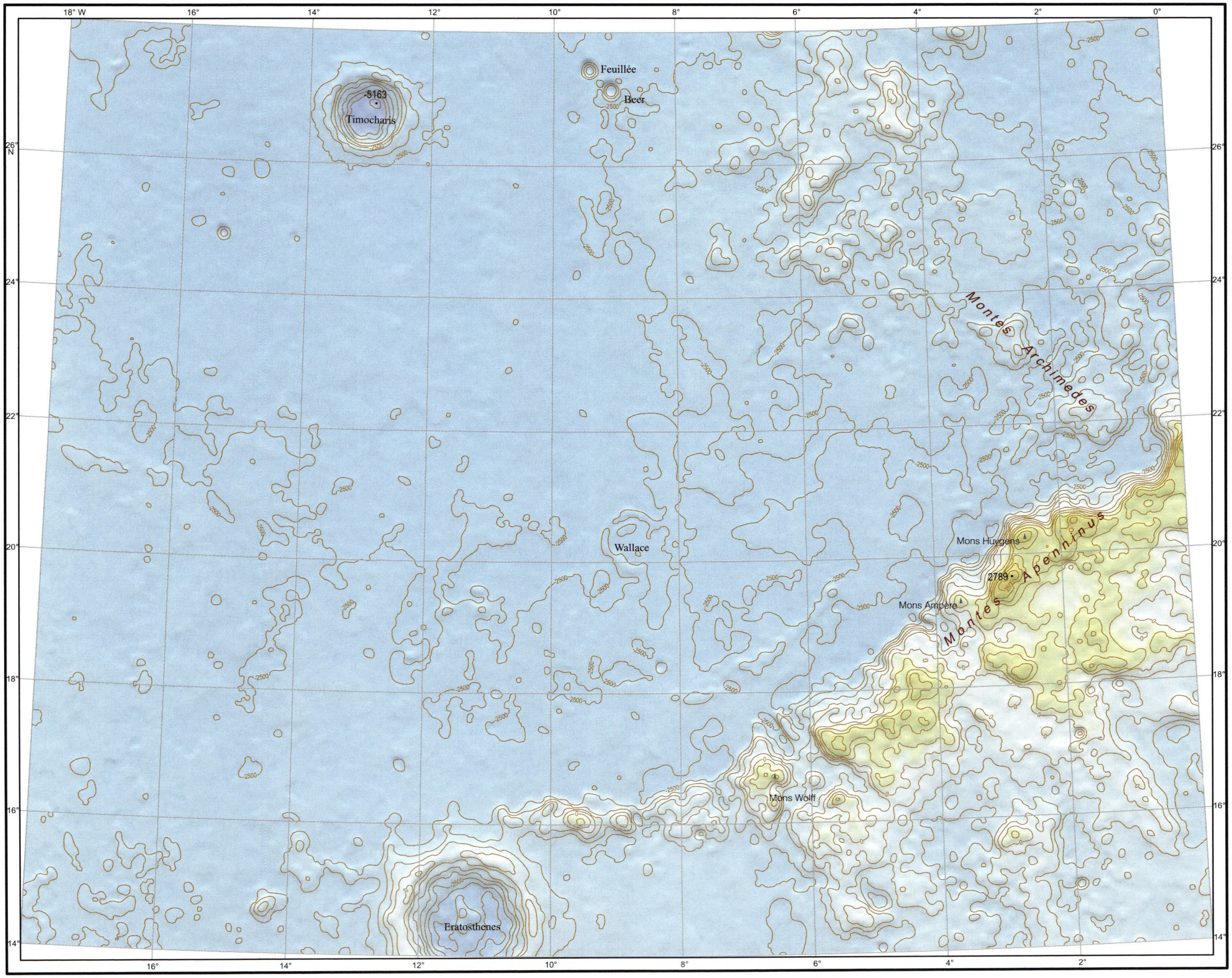

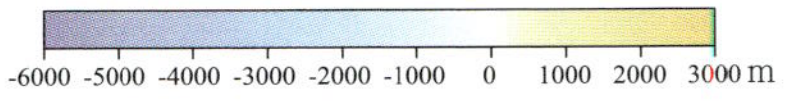

Image Code **F011**

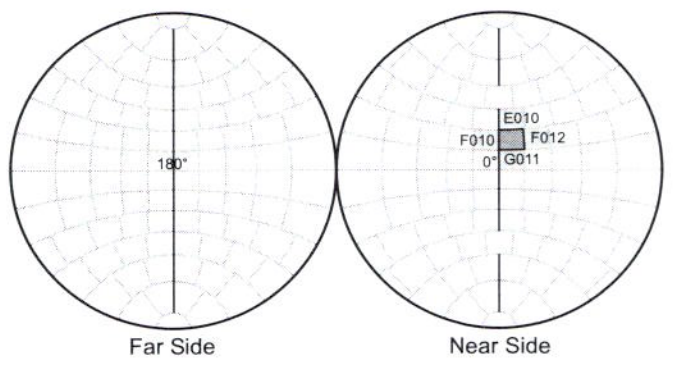

0 20 40 60 80 100 km

-6000 -5000 -4000 -3000 -2000 -1000 0 1000 2000 3000 m

Image Code **F012**

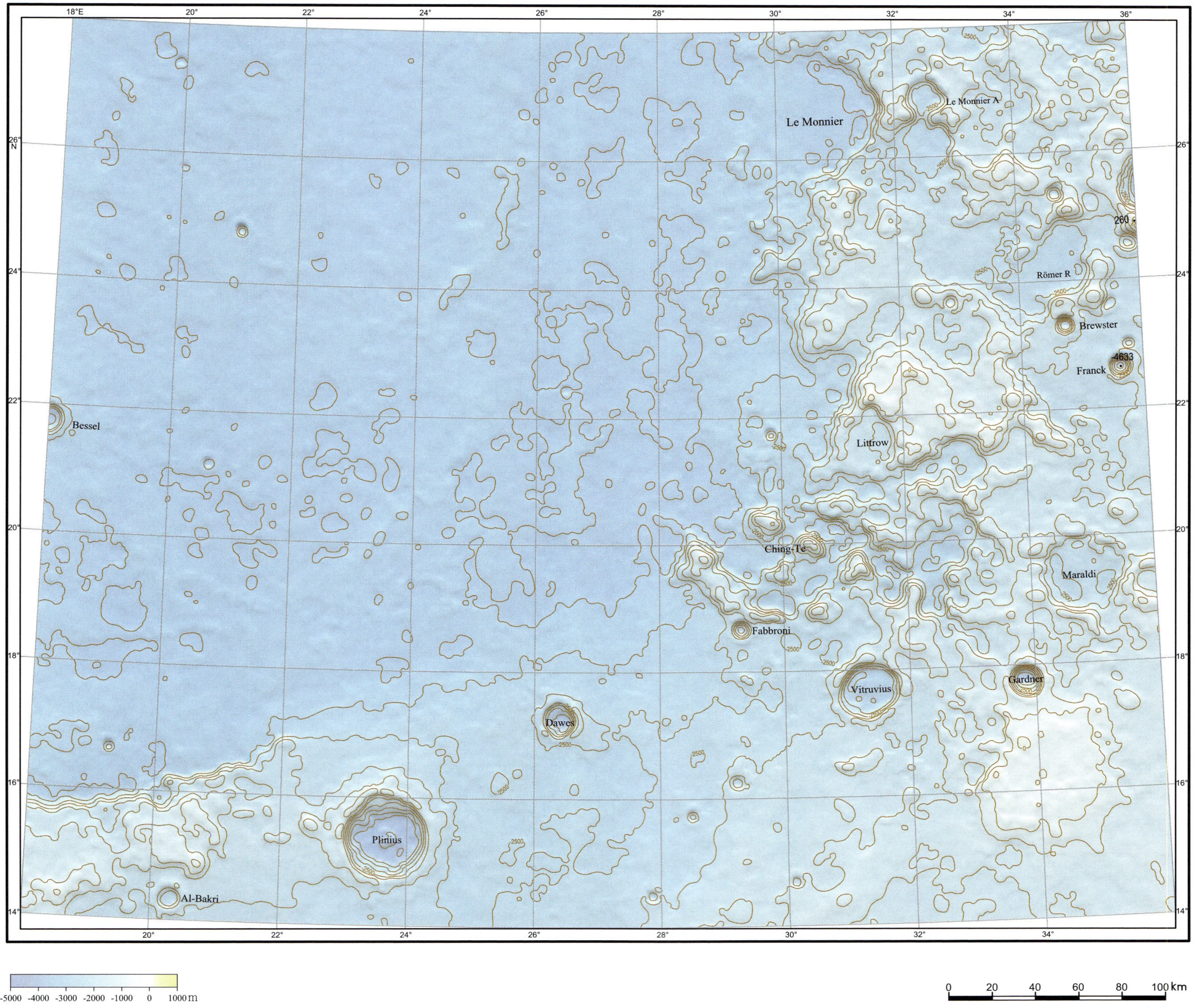

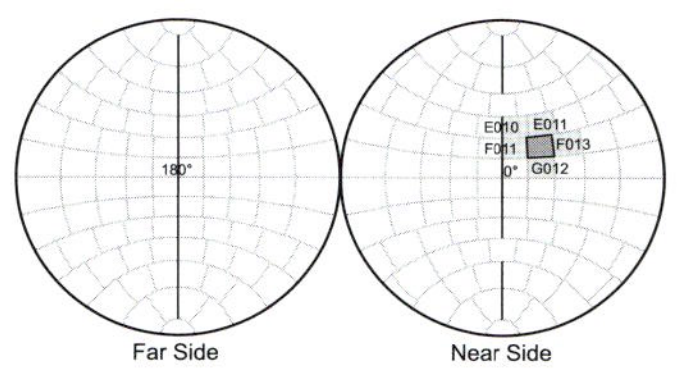

Image Code **F013**

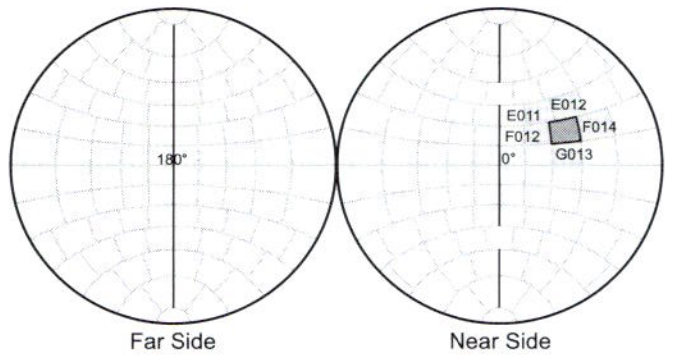

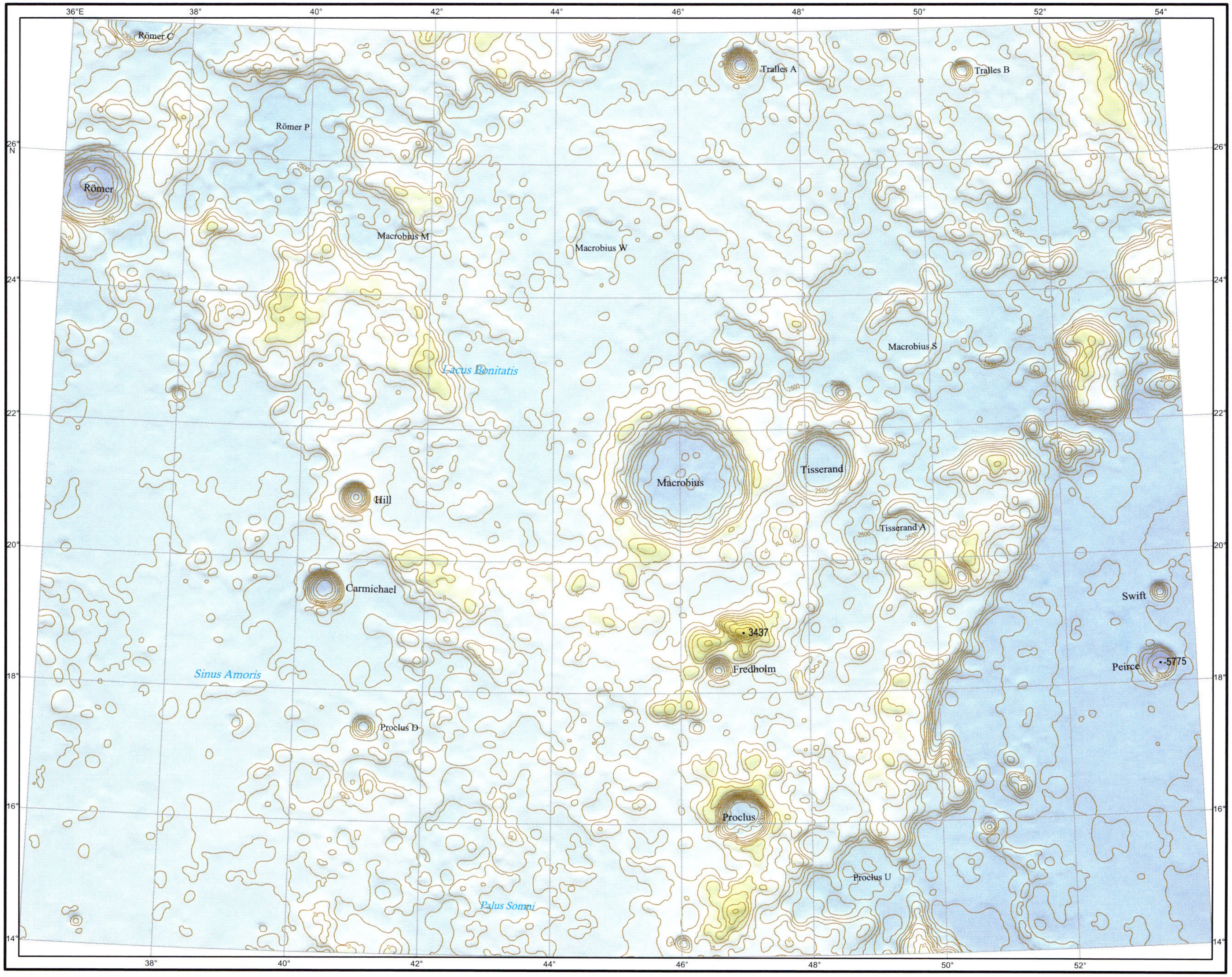

0 20 40 60 80 100 km

-6000 -5000 -4000 -3000 -2000 -1000 0 1000 2000 3000 4000 m

Image Code F014

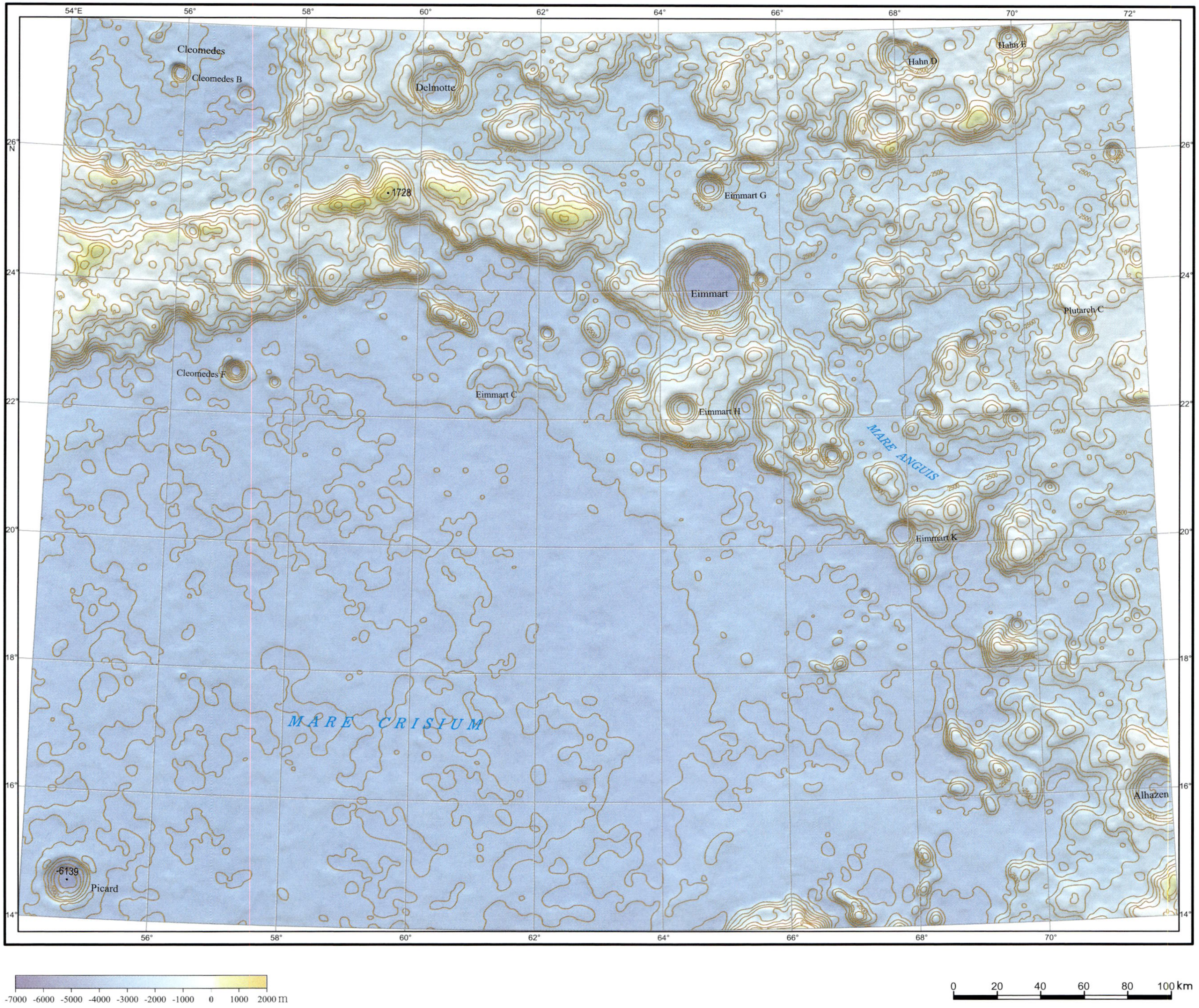

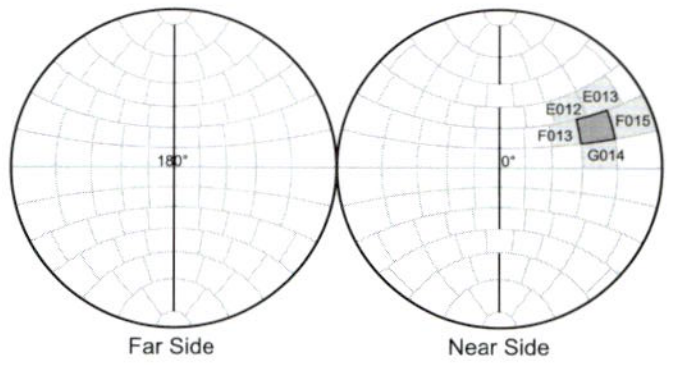

Image Code **F015**

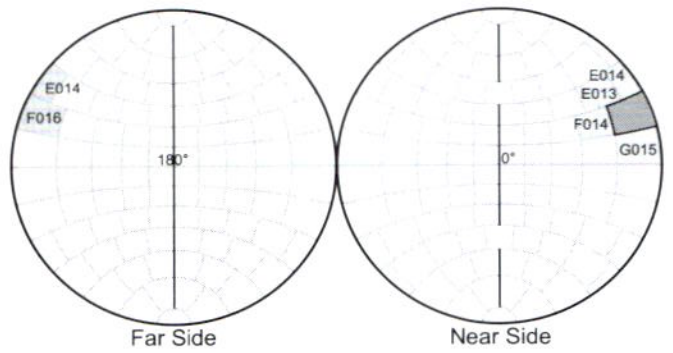

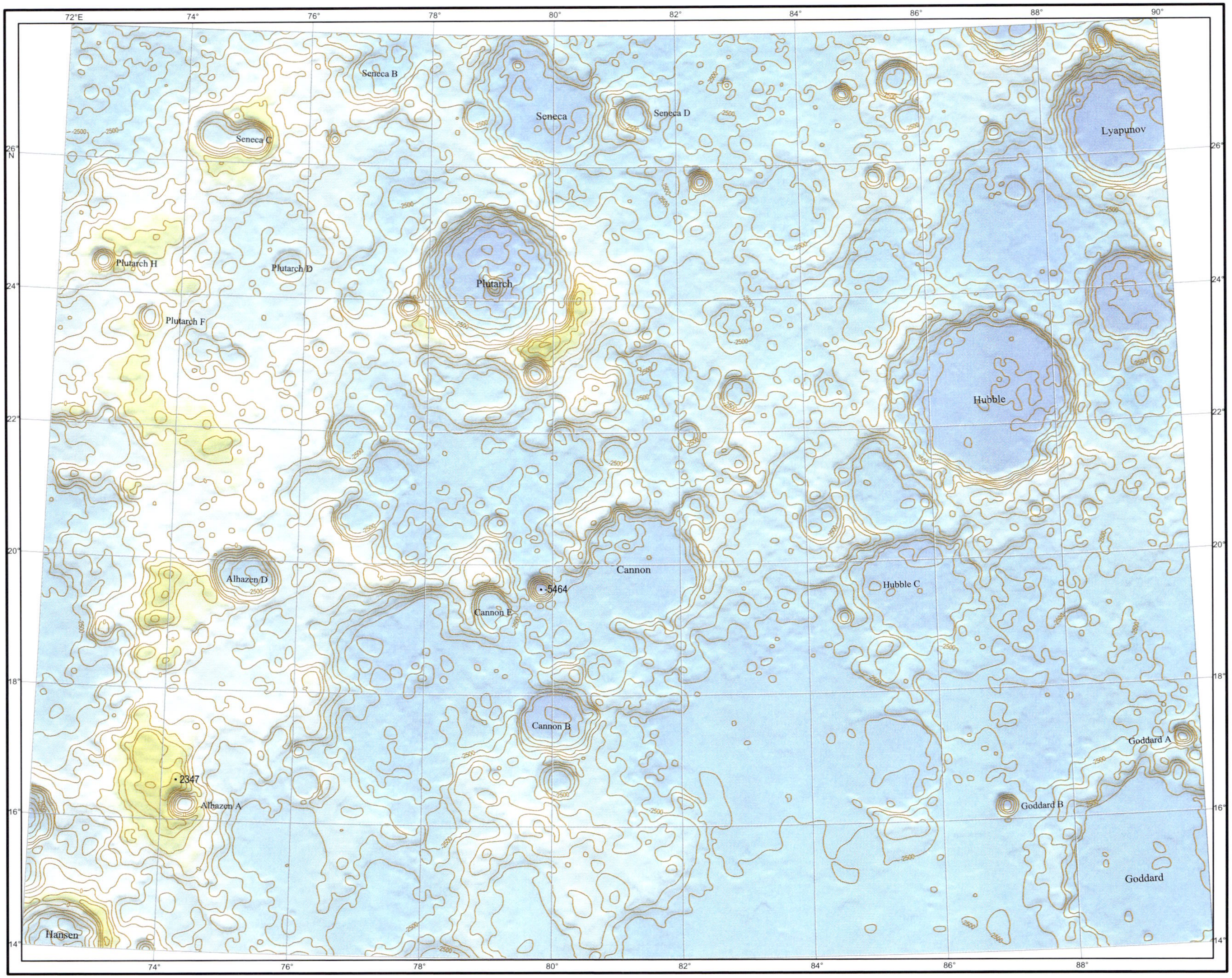

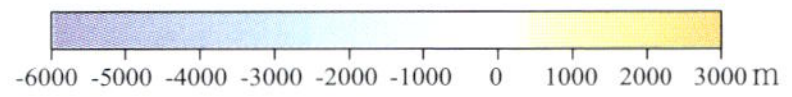

Image Code **F016**

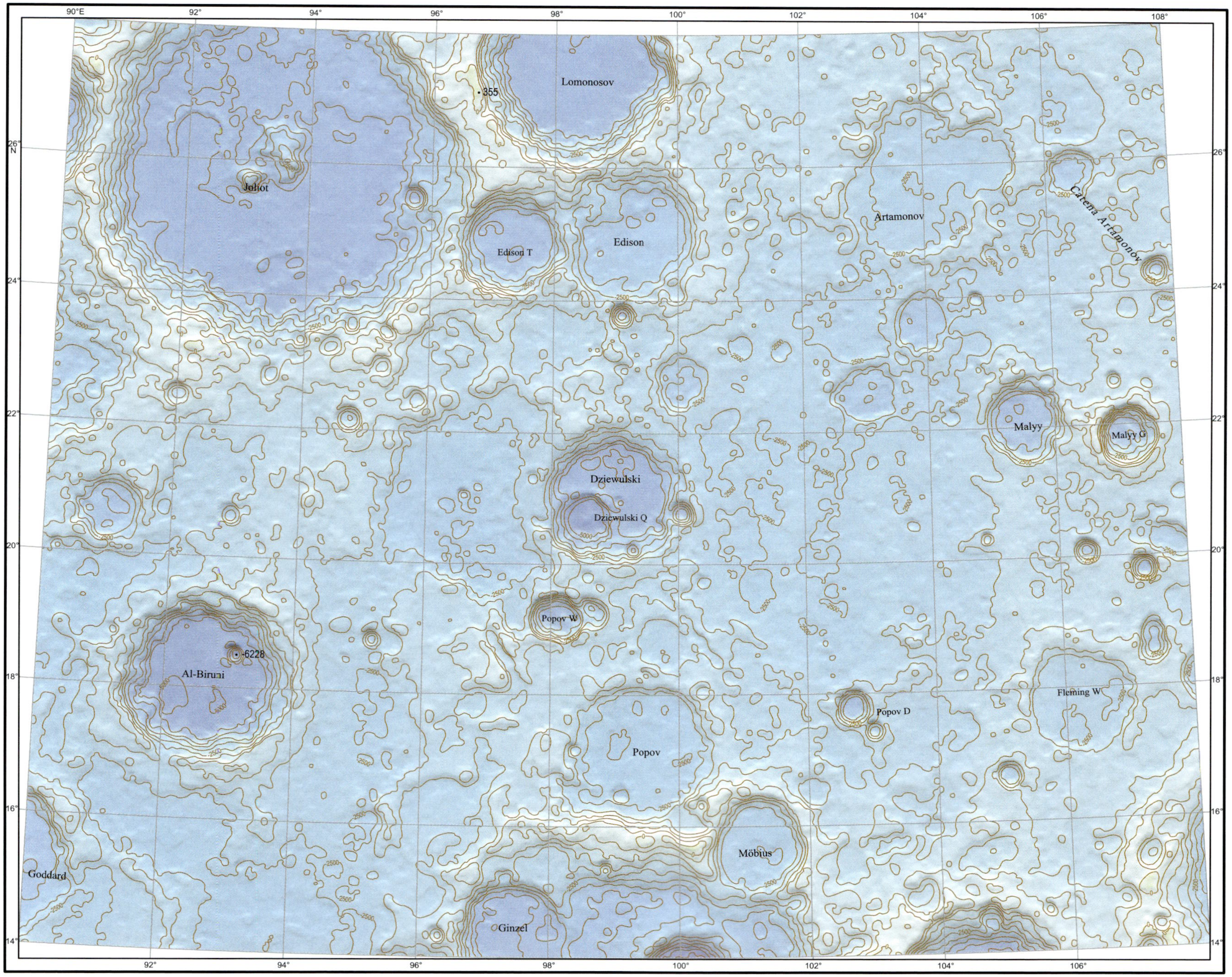

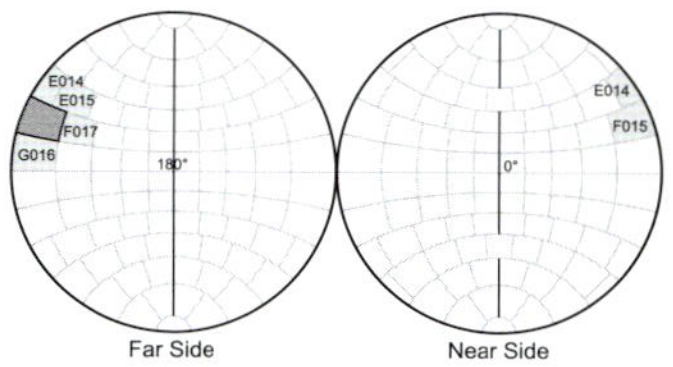

Image Code **F017**

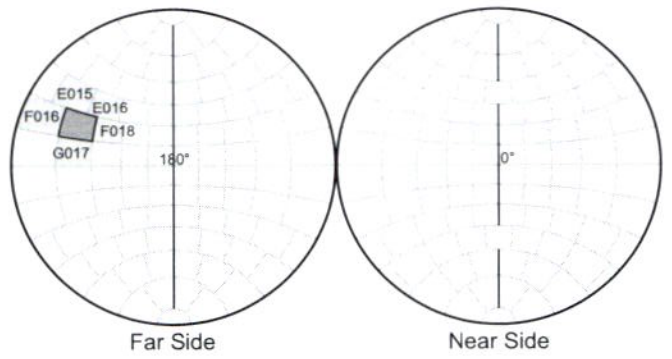

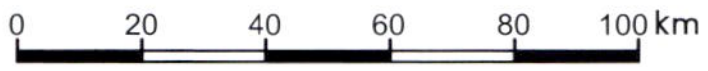

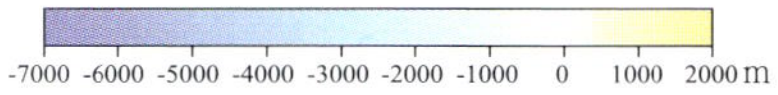

Image Code F018

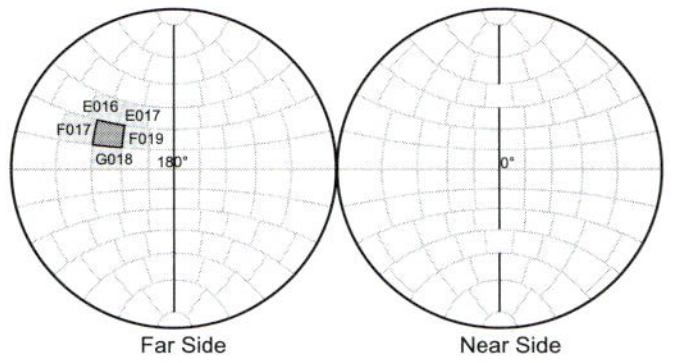

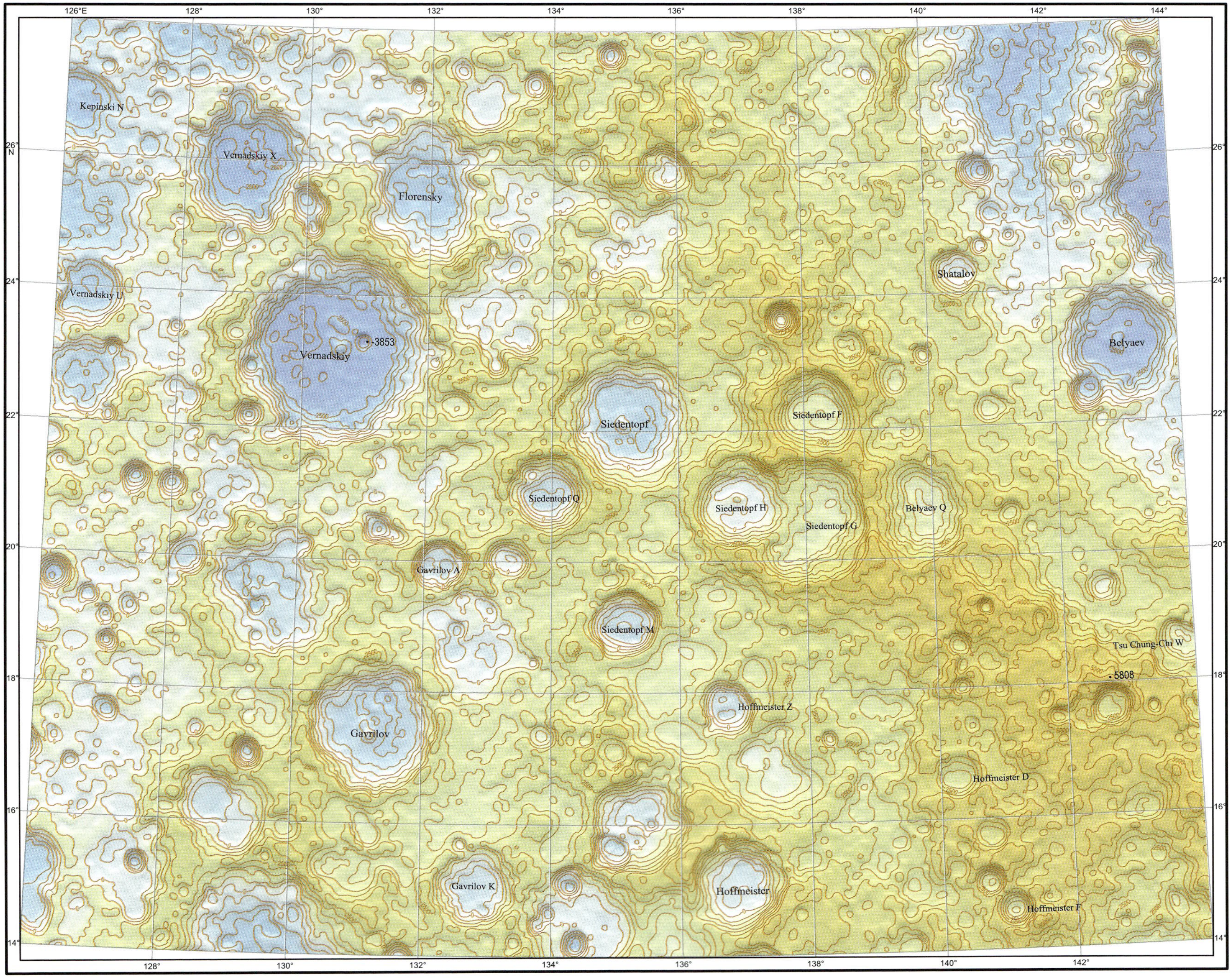

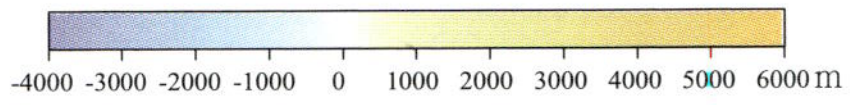

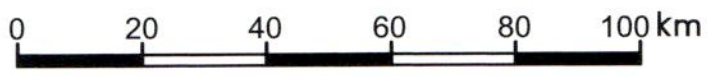

Image Code F019

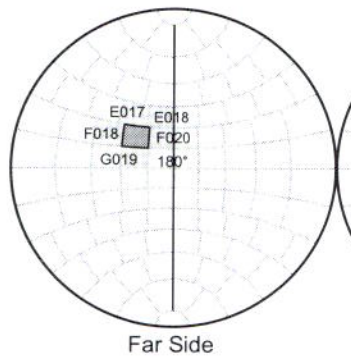

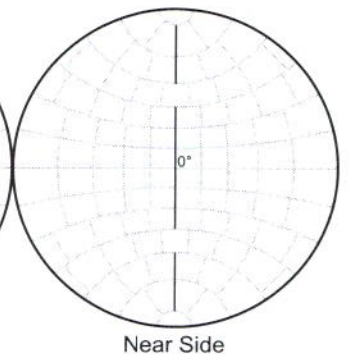

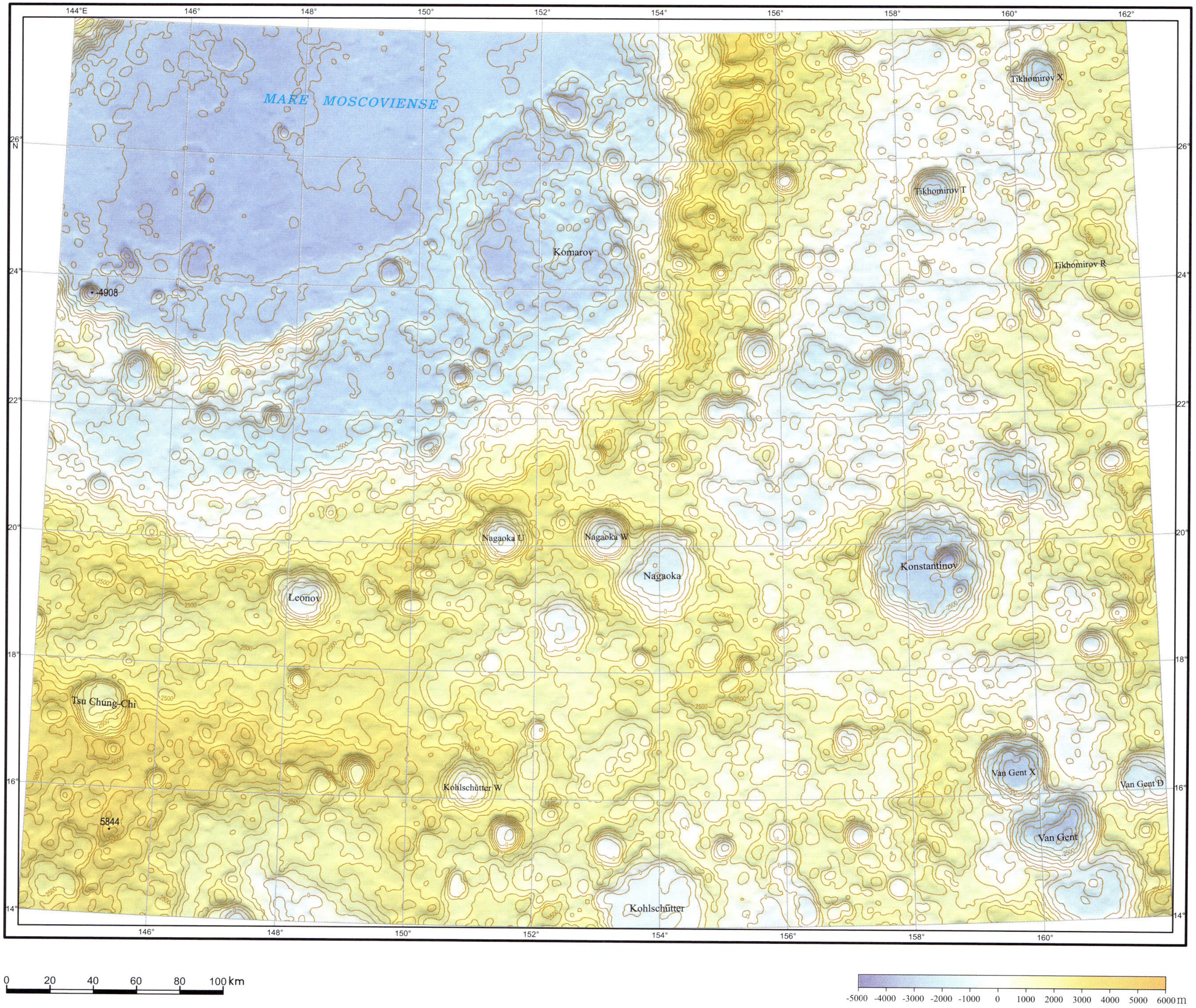

Image Code F020

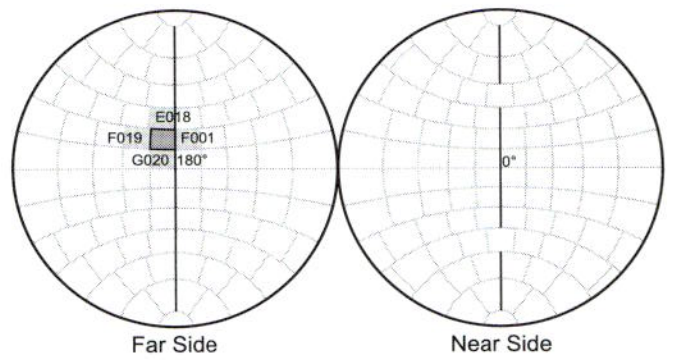

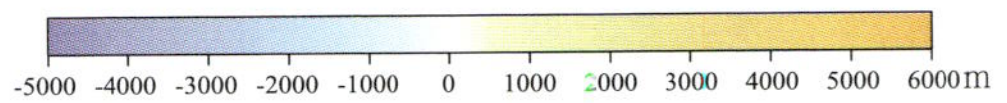

Image Code G001

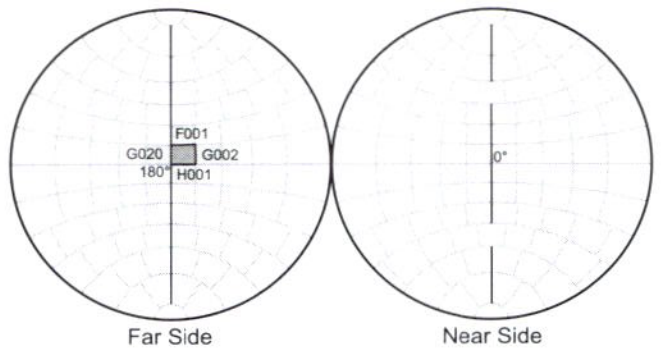

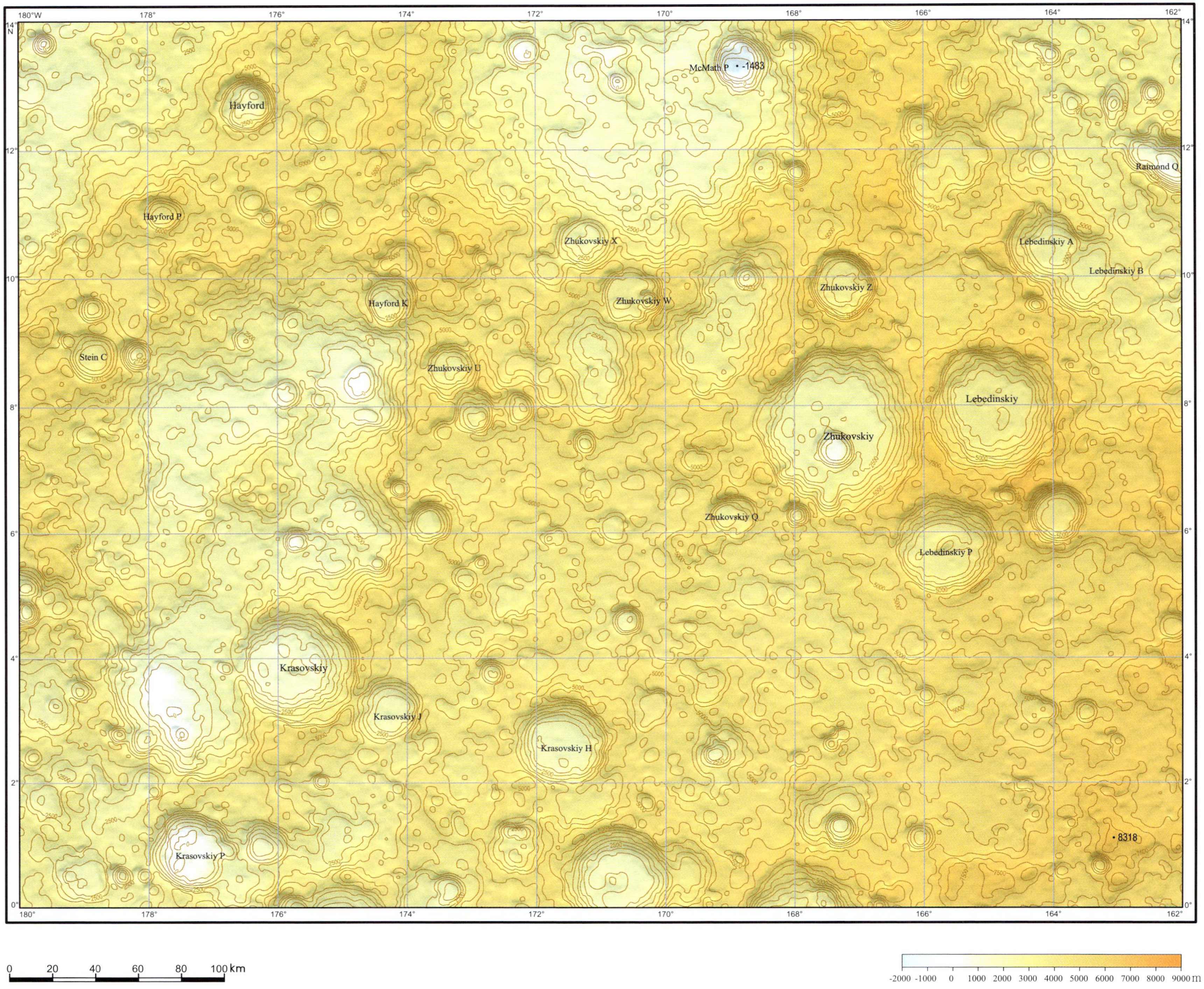

Image Code **G002**

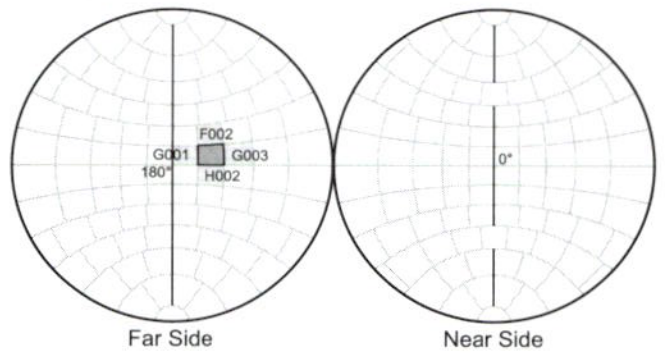

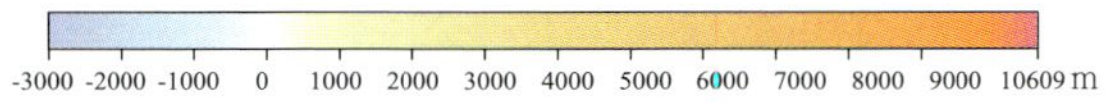

Image Code **G003**

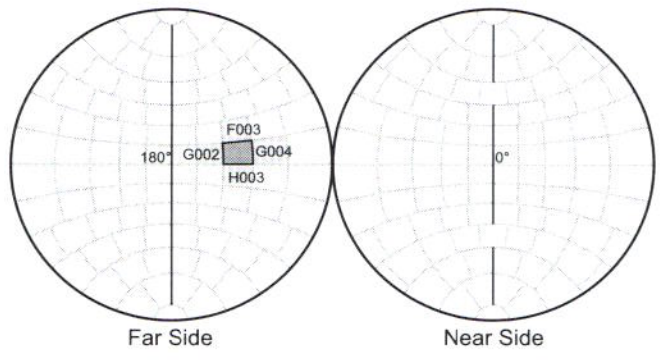

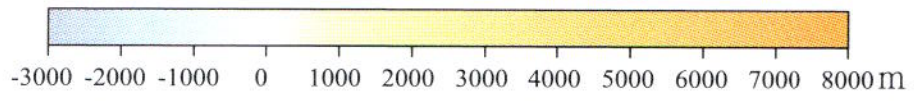

Image Code G004

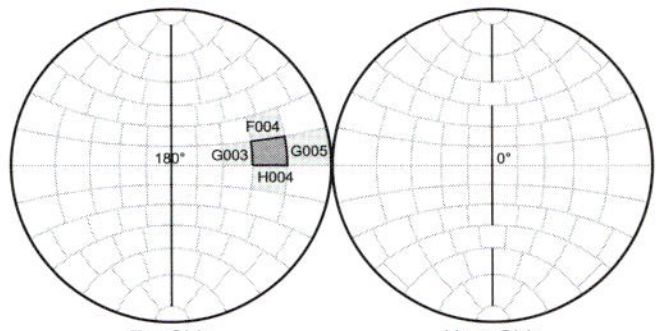

Image Code G005

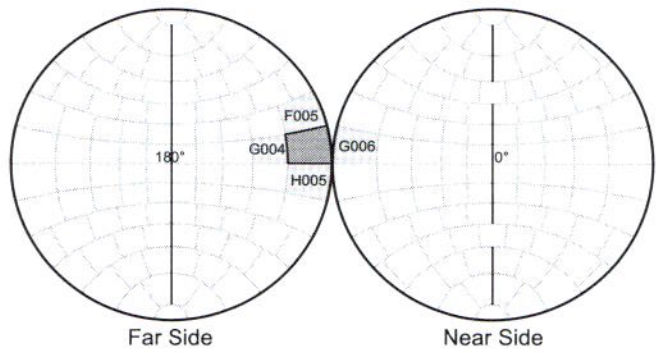

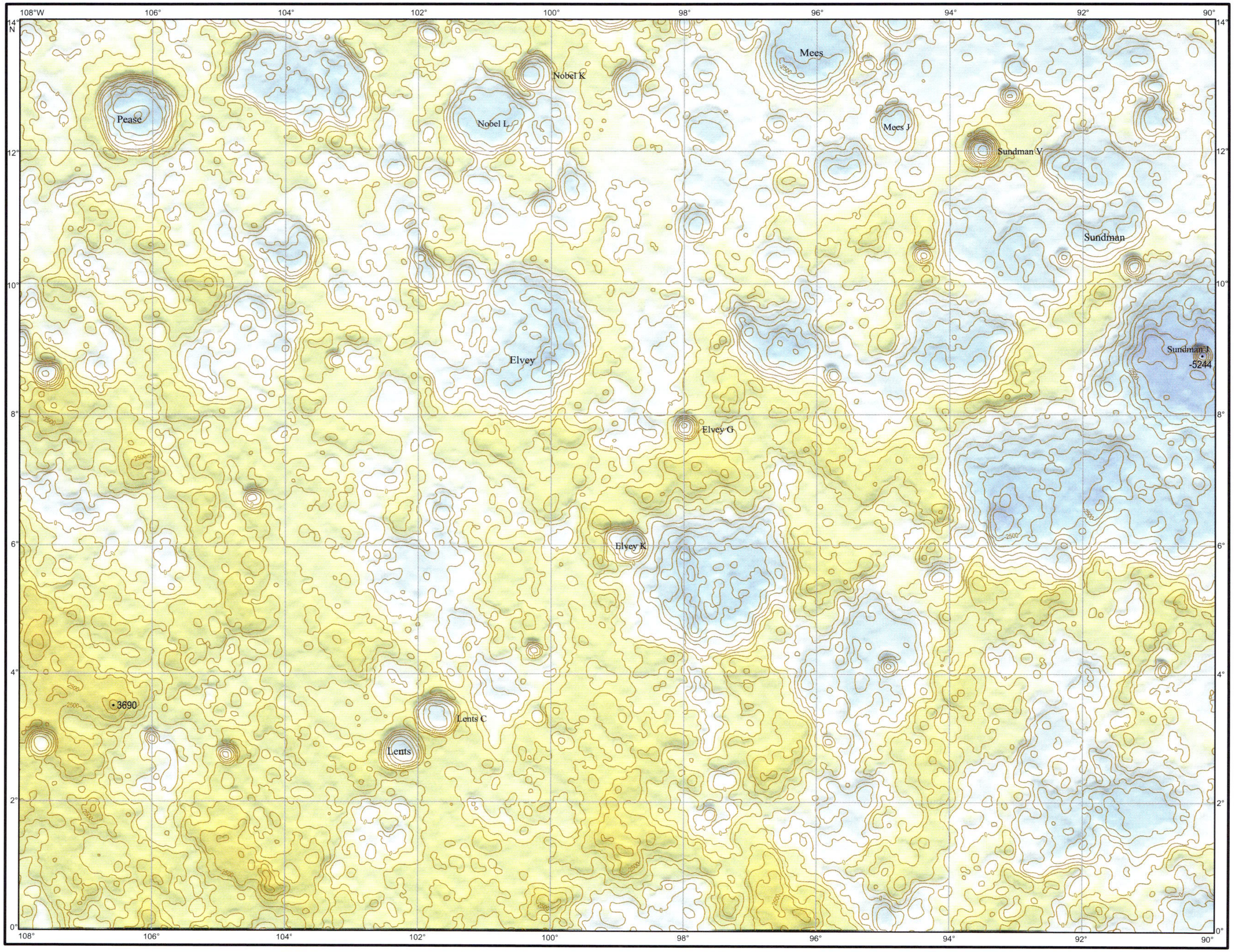

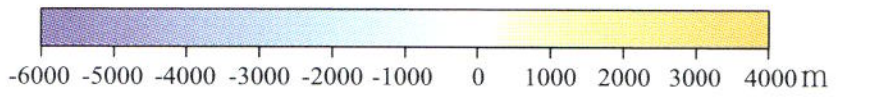

Image Code G006

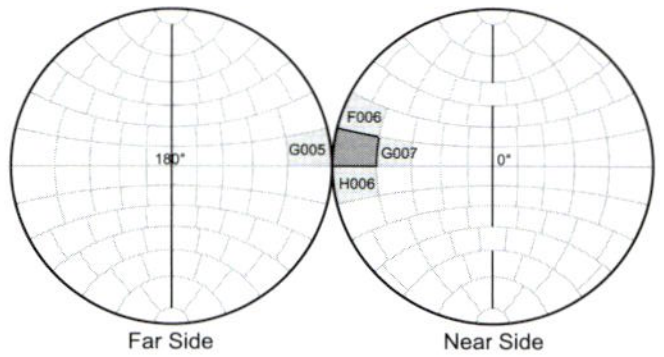

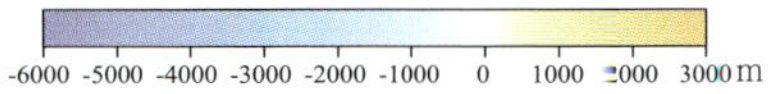

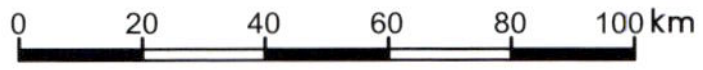

Image Code G007

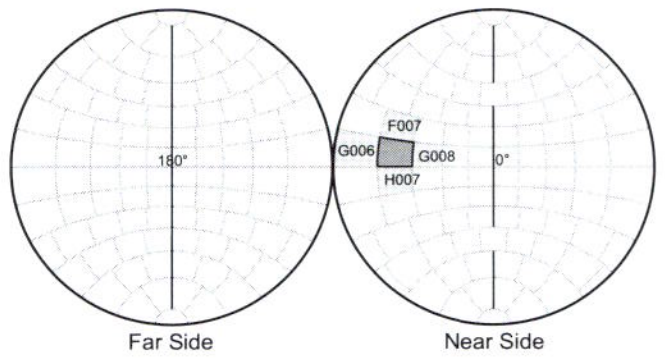

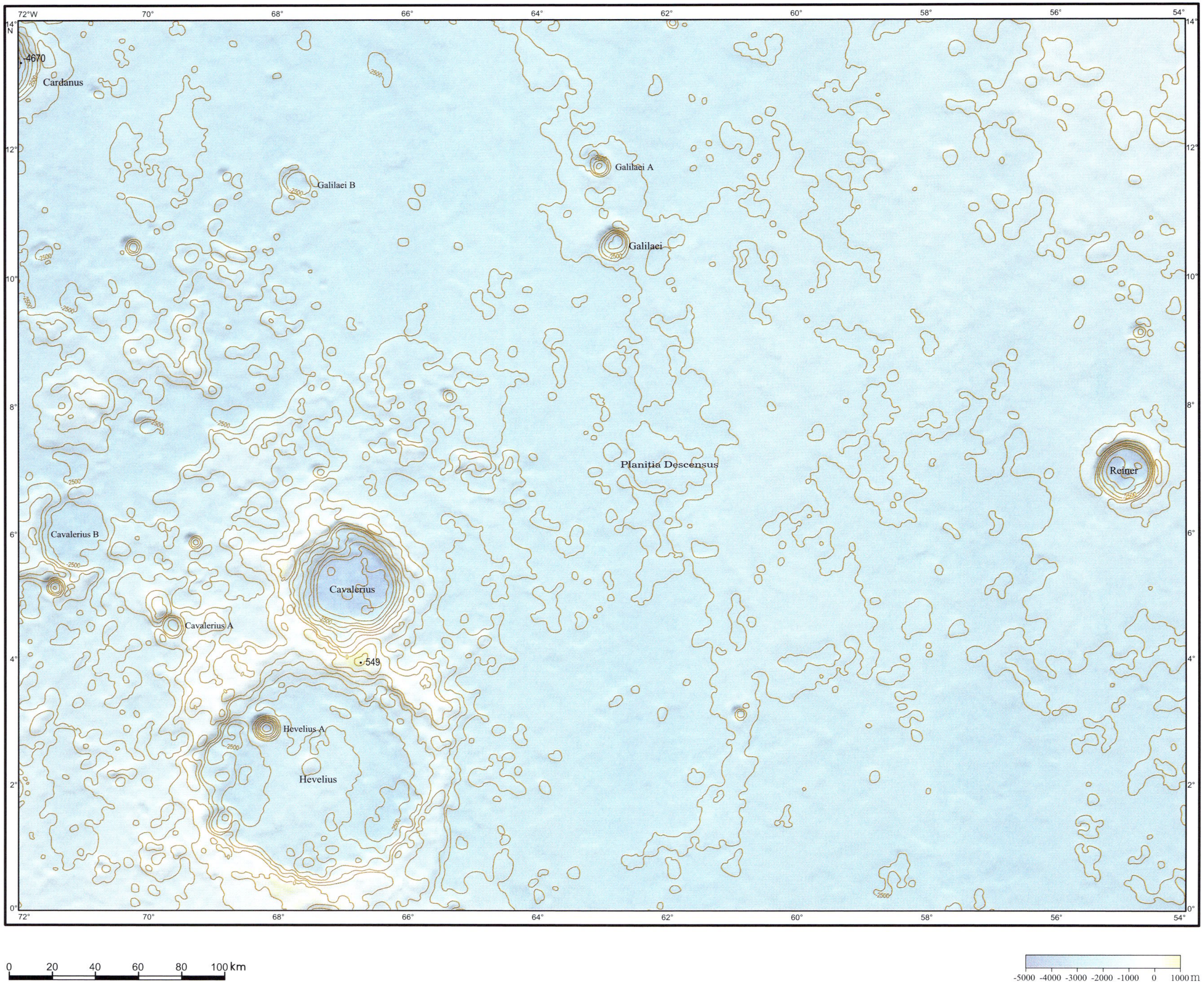

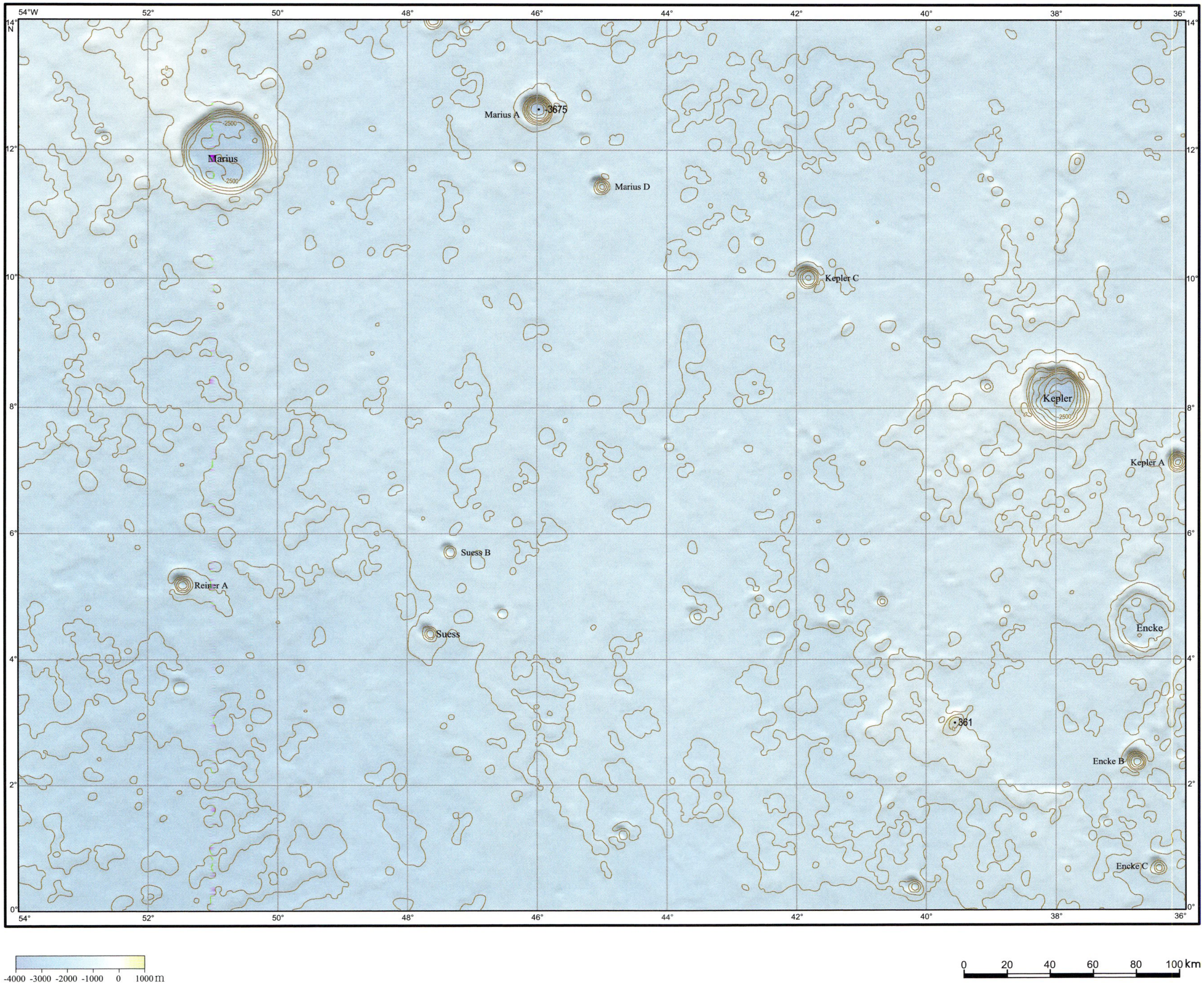

Image Code G008

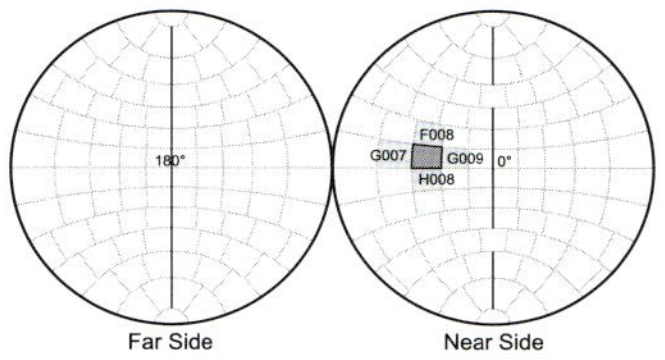

Image Code **G009**

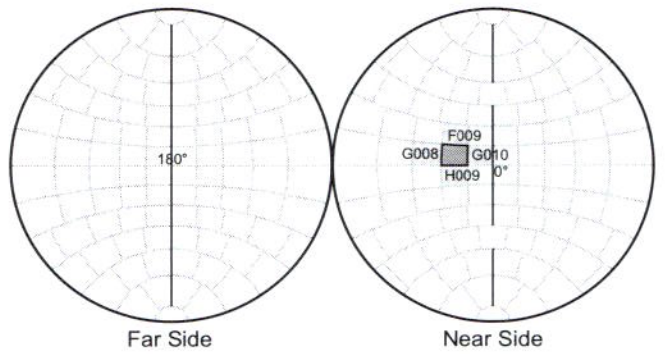

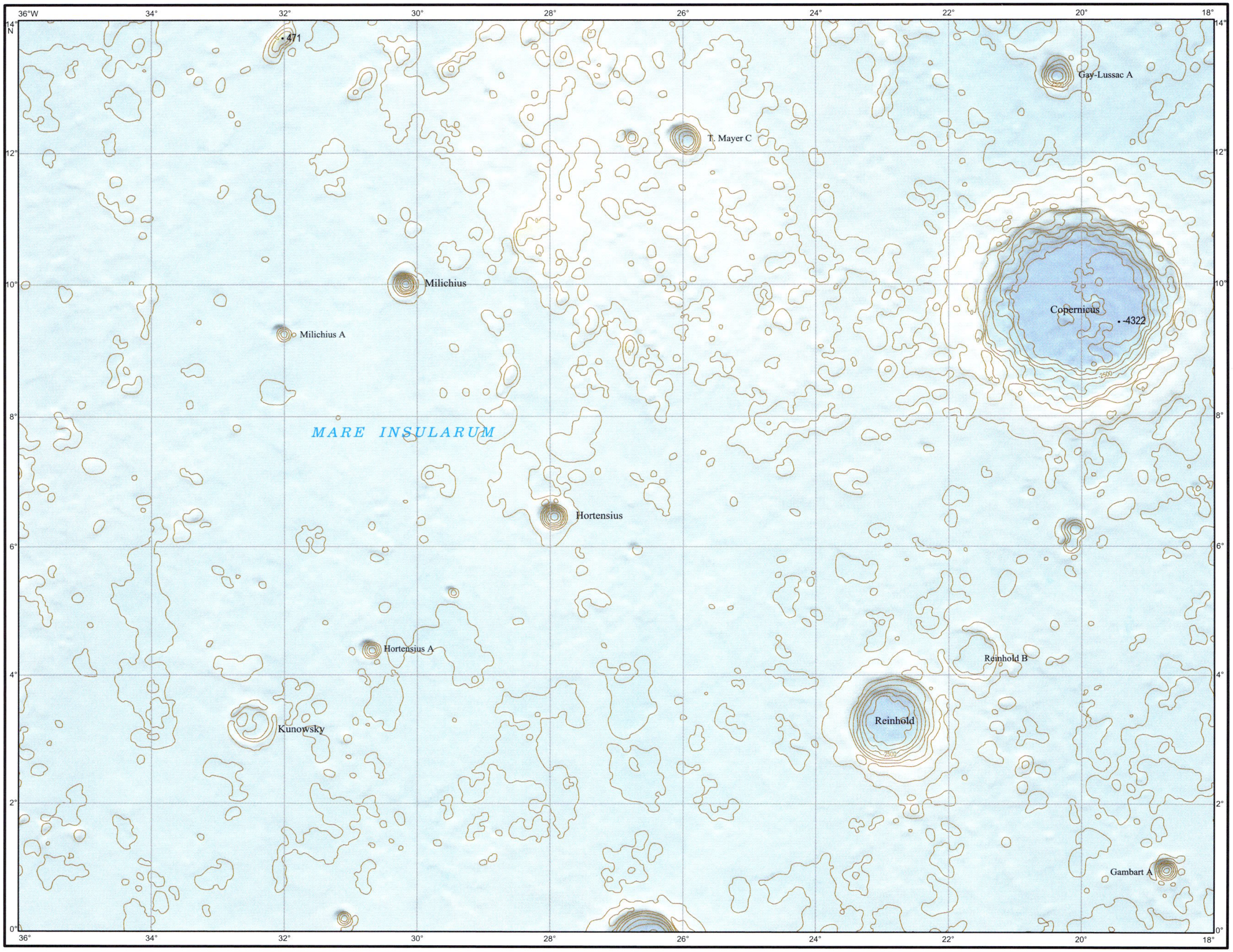

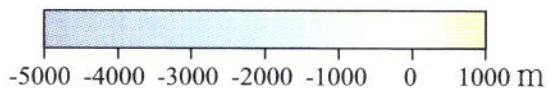

Image Code G010

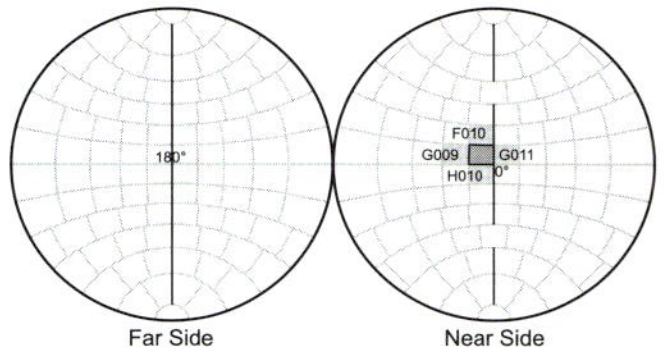

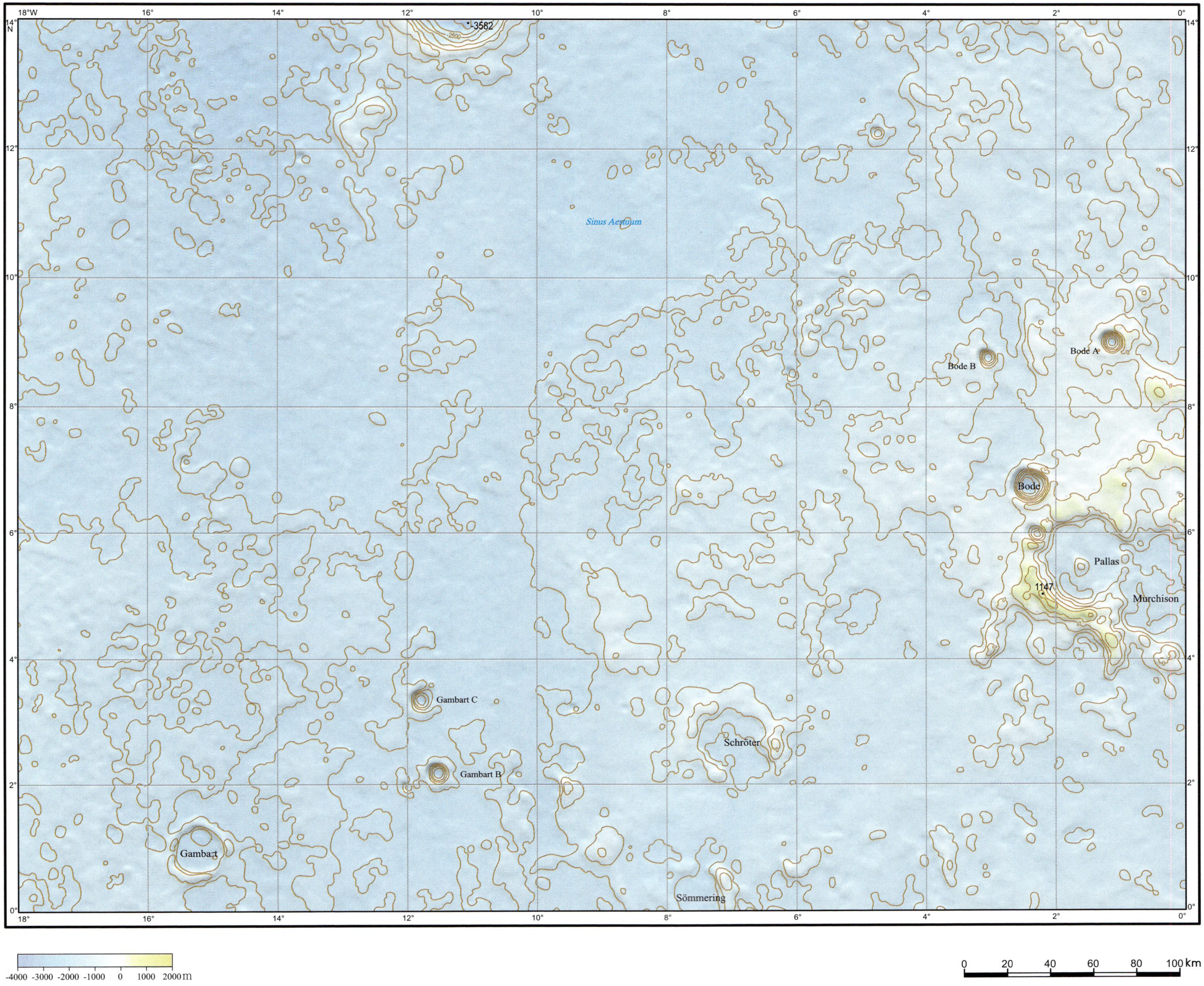

Image Code G011

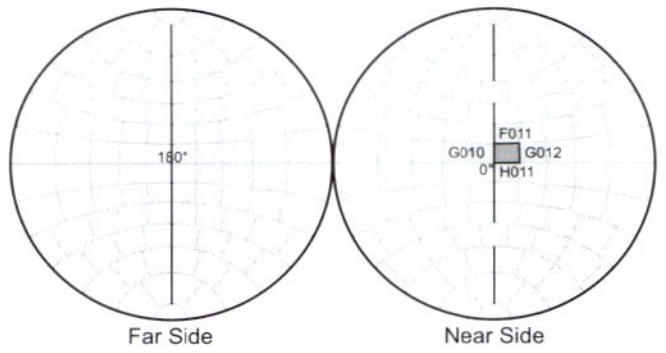

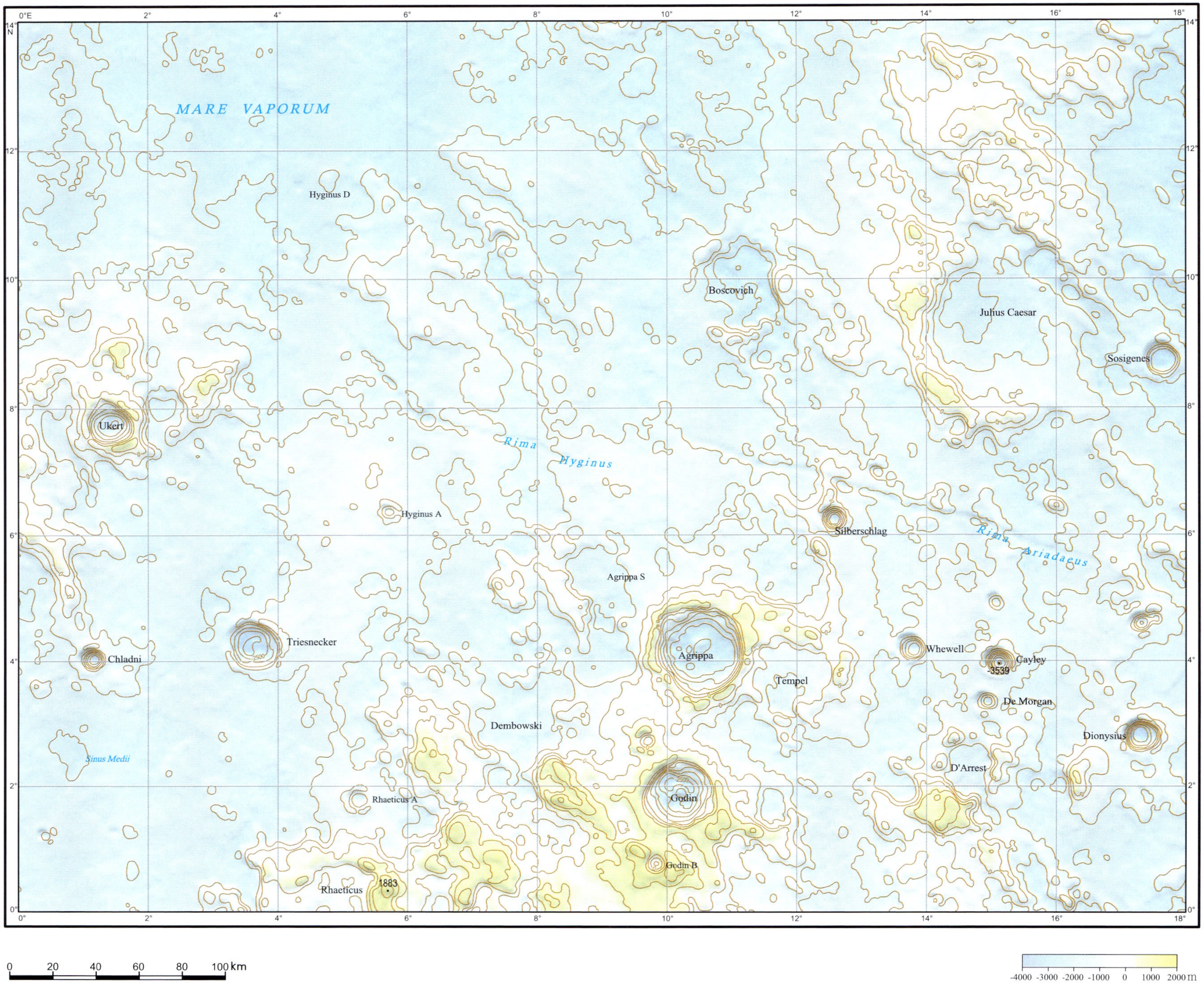

Image Code G012

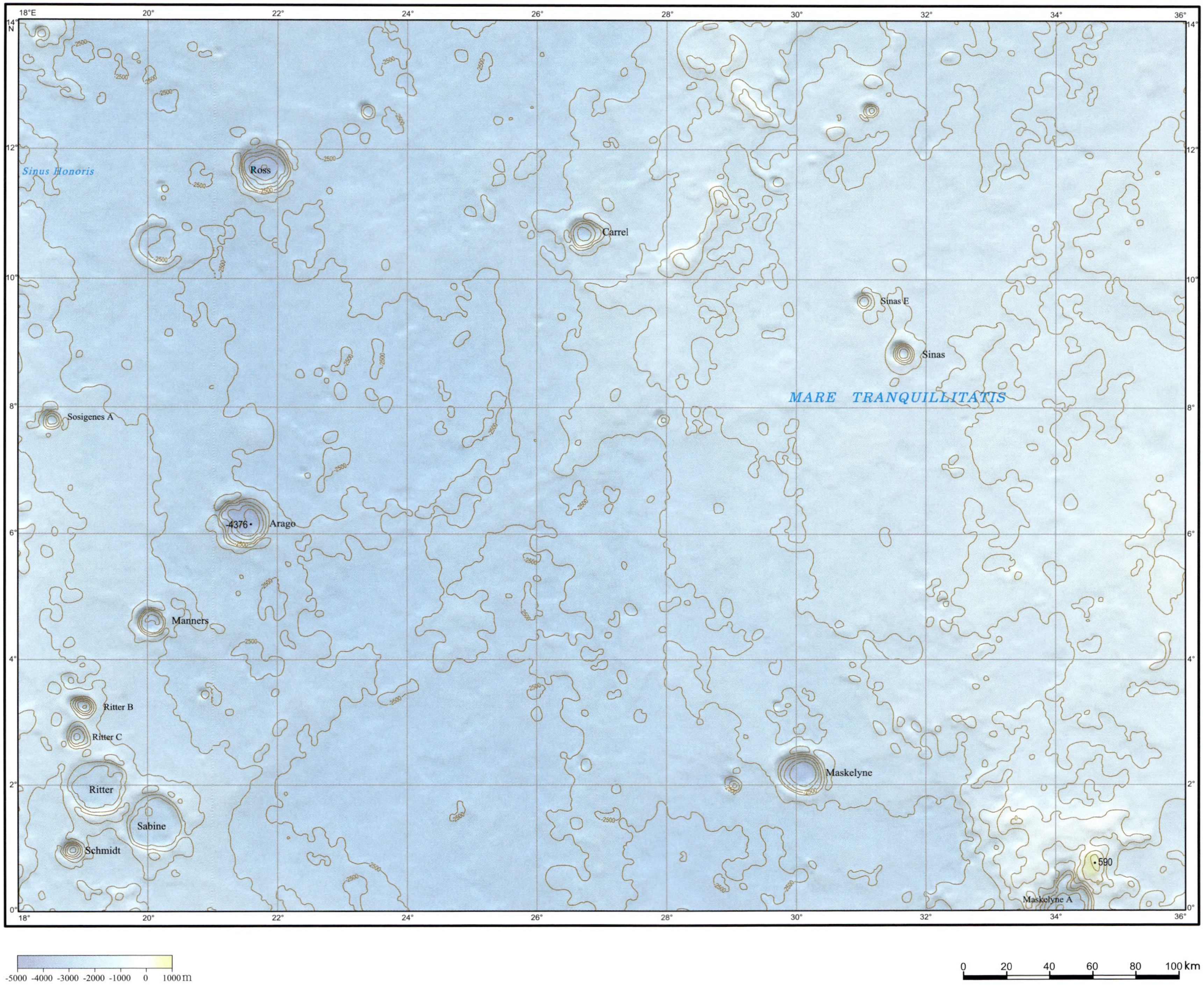

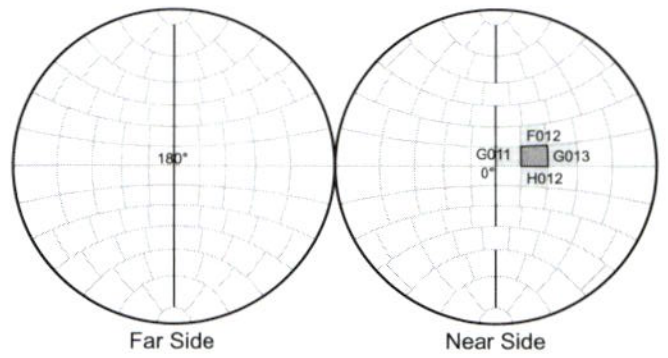

Image Code G013

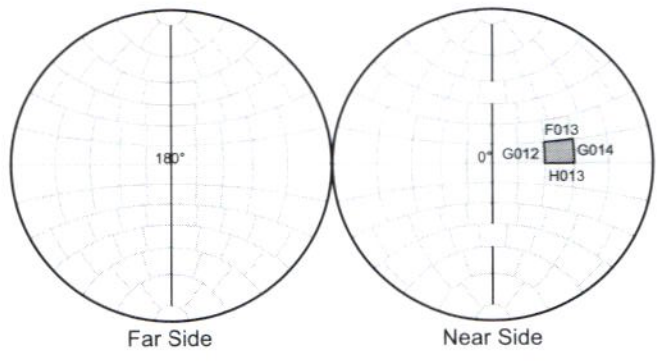

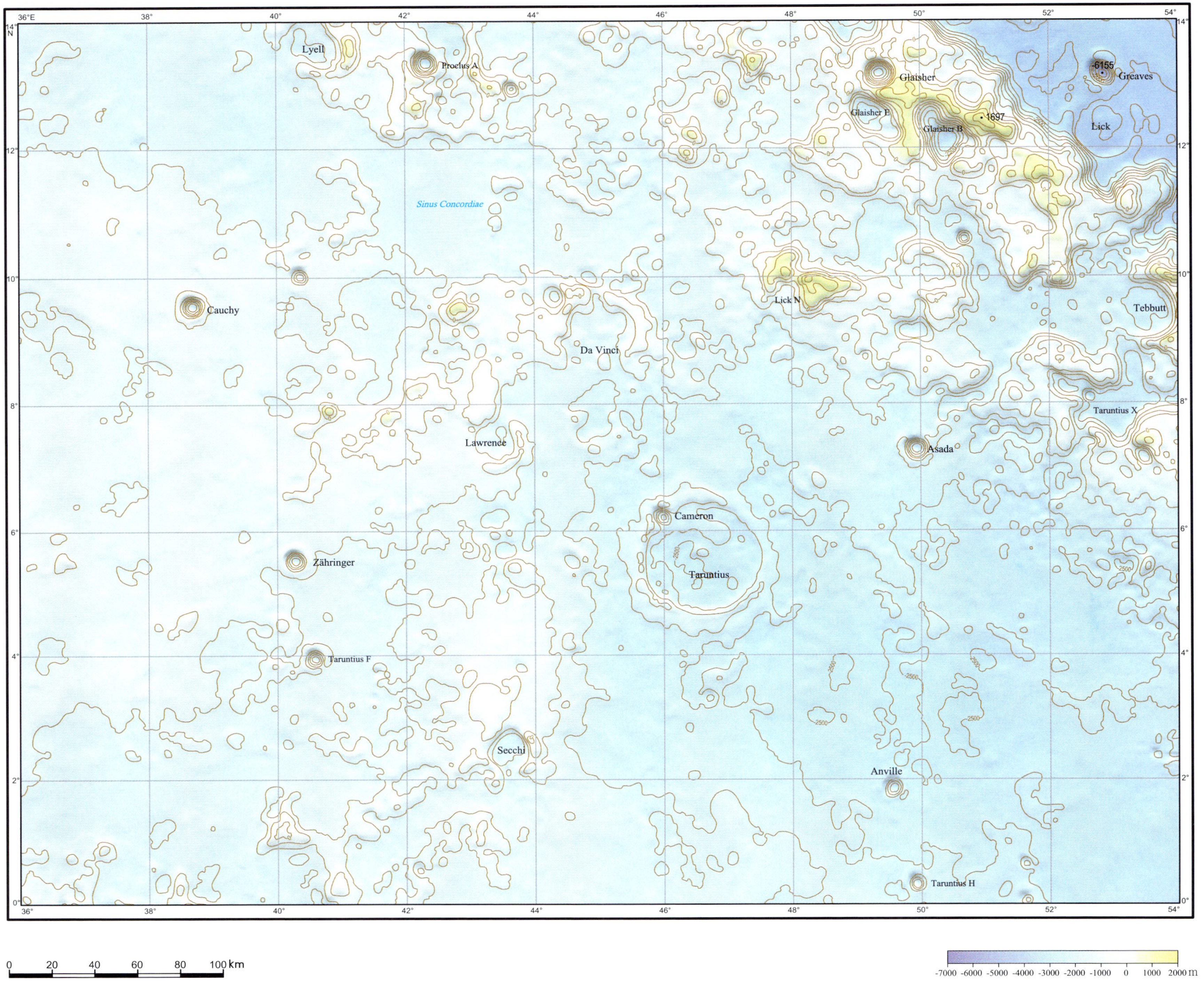

Image Code G014

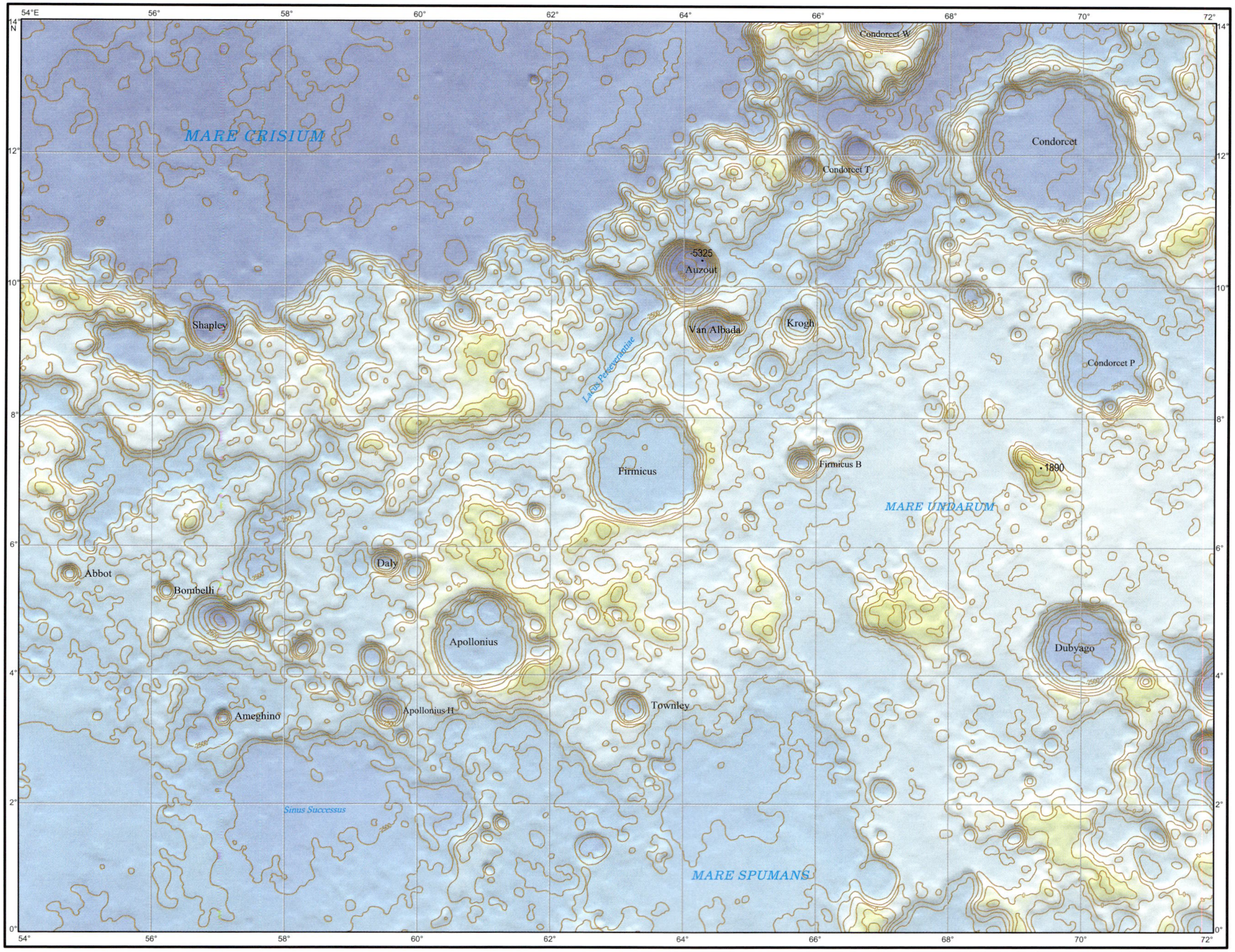

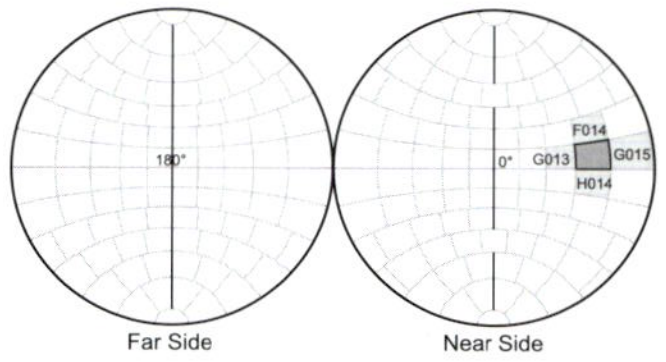

-6000 -5000 -4000 -3000 -2000 -1000 0 1000 2000 m

0 20 40 60 80 100 km

Image Code G015

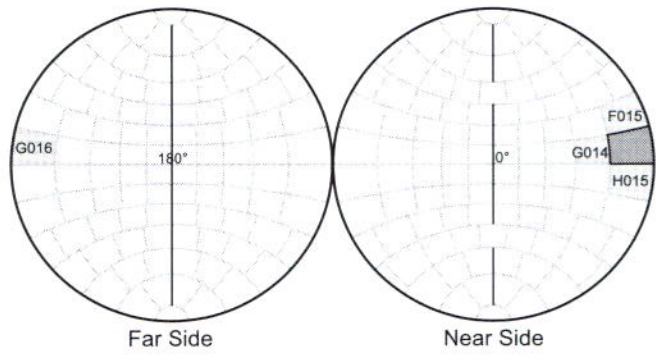

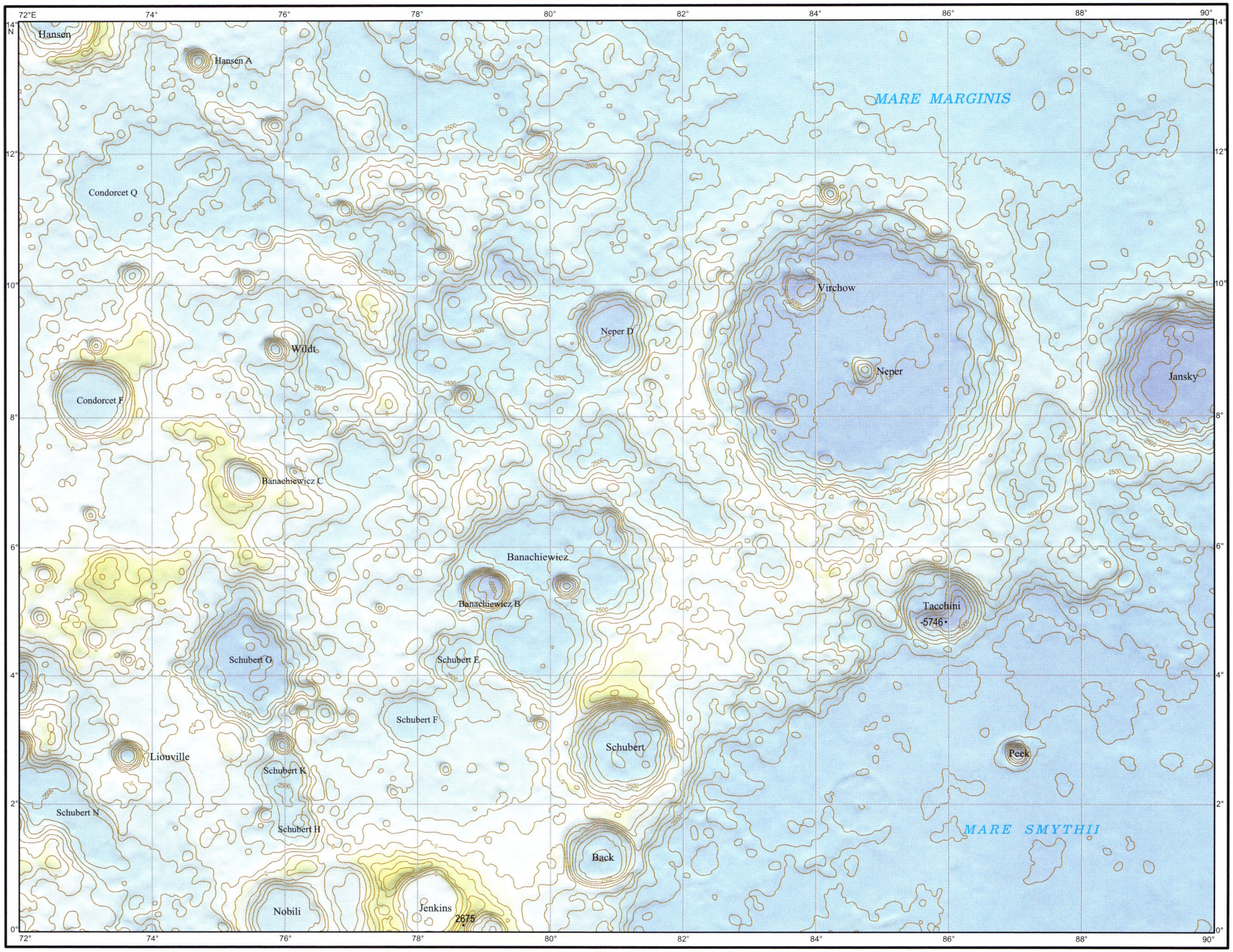

Image Code G016

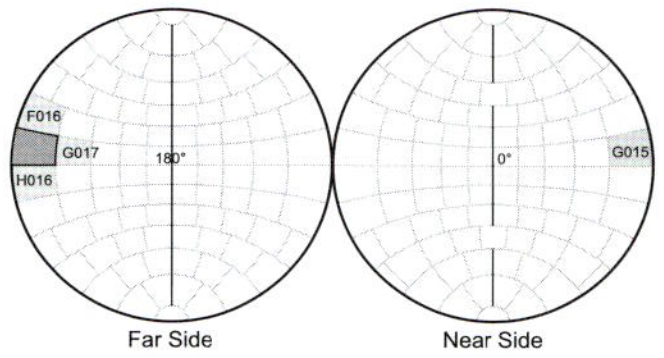

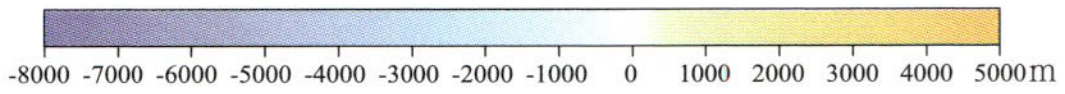

Image Code **G017**

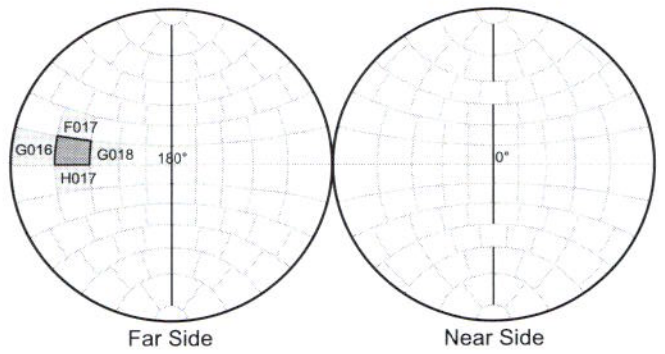

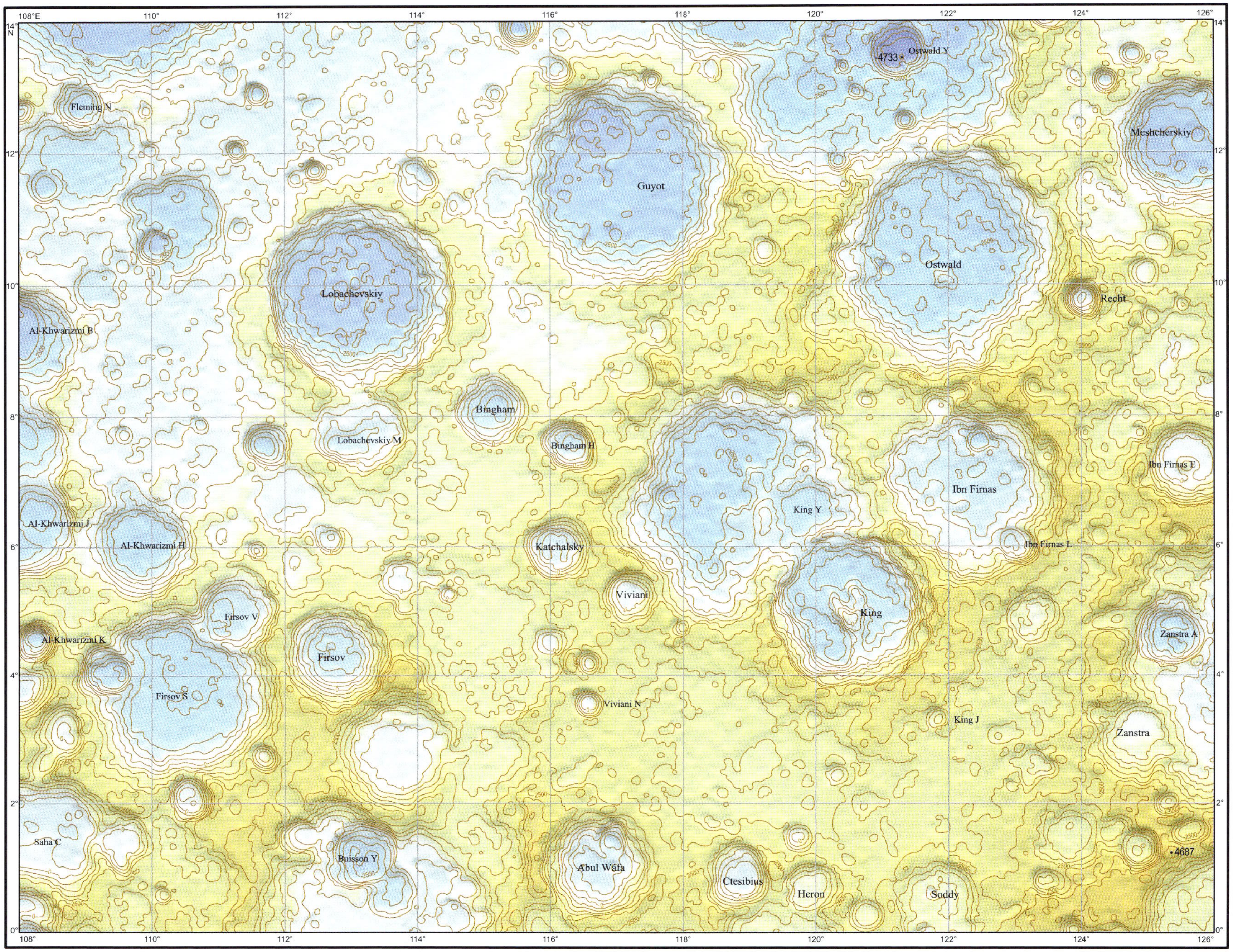

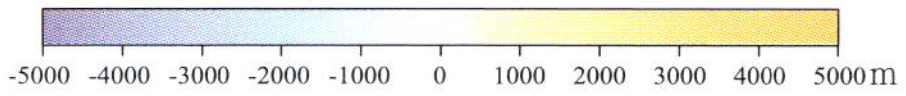

Image Code **G018**

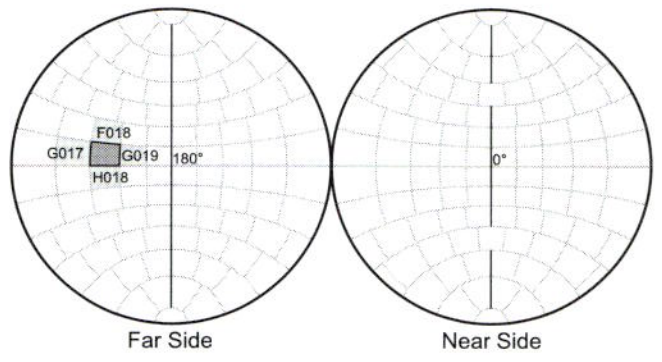

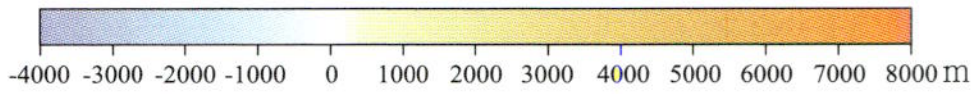

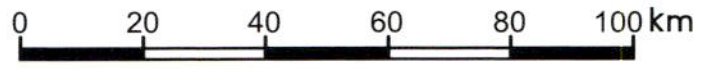

Image Code G019

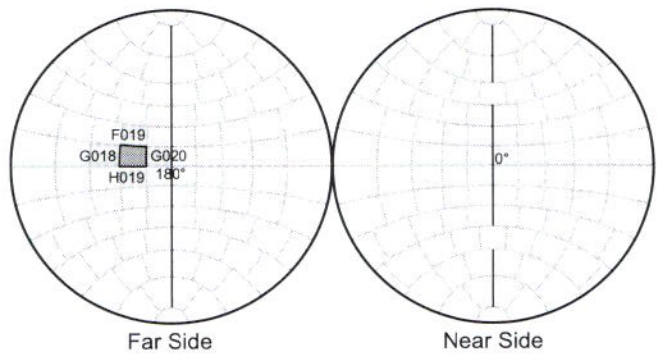

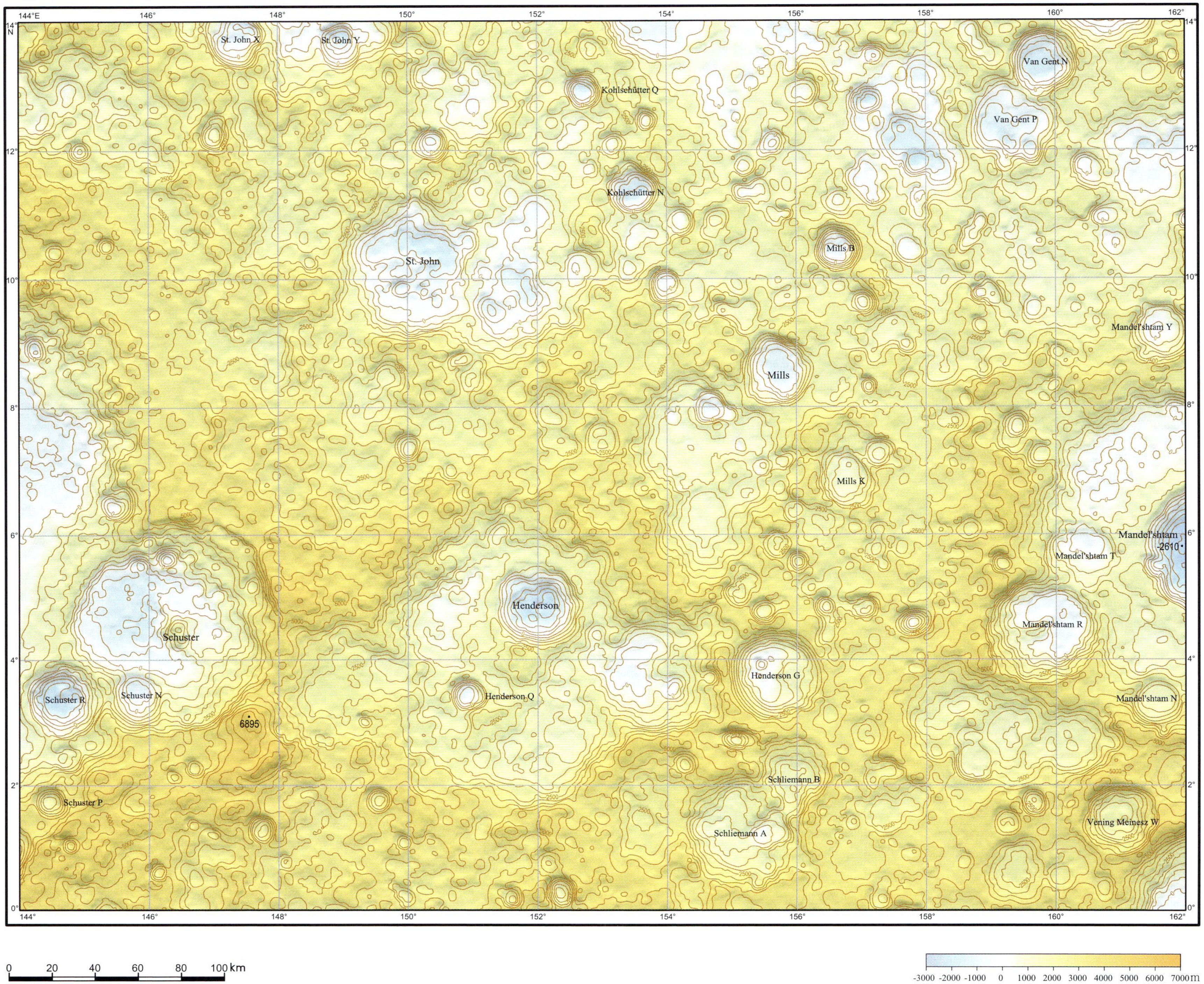

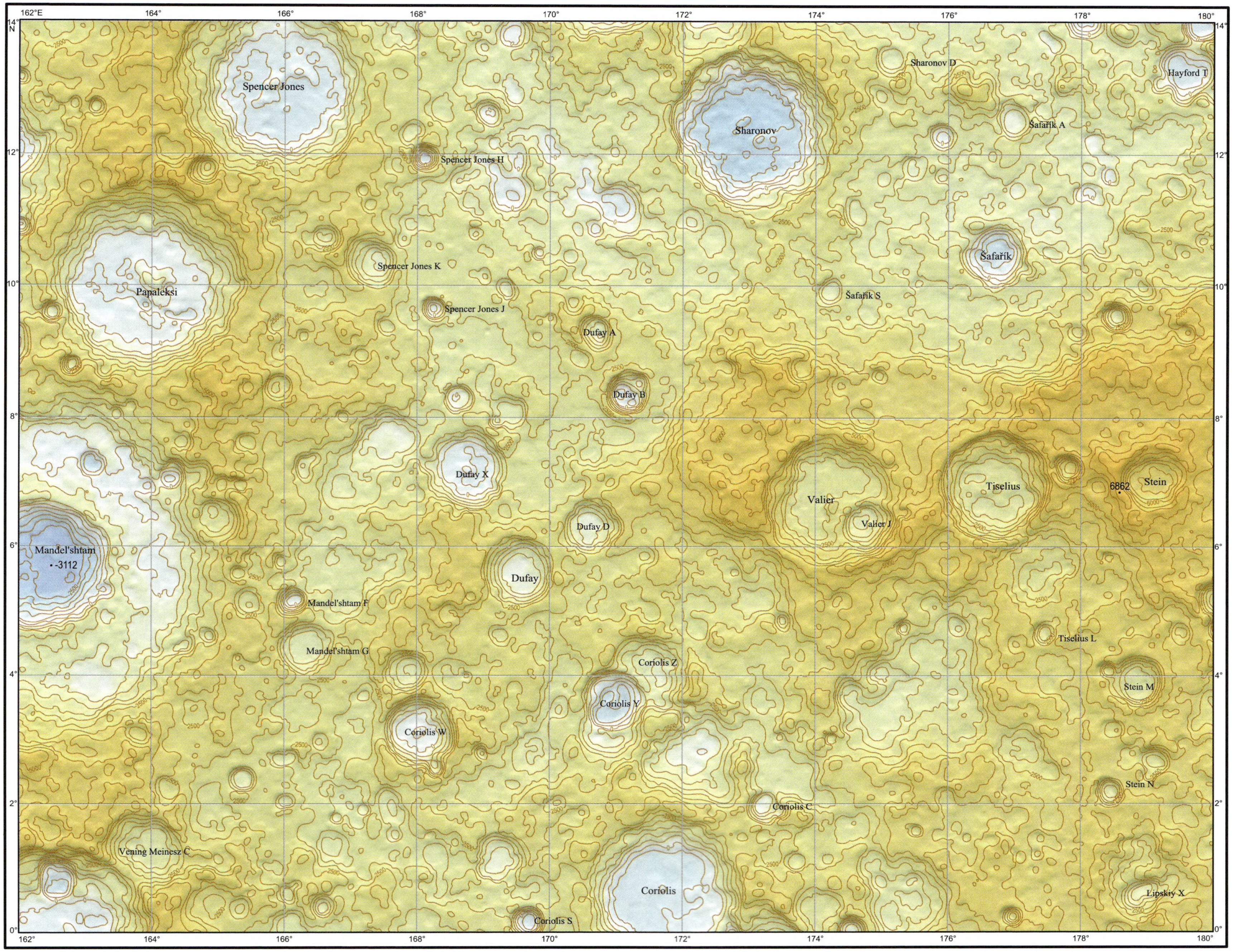

Image Code **G020**

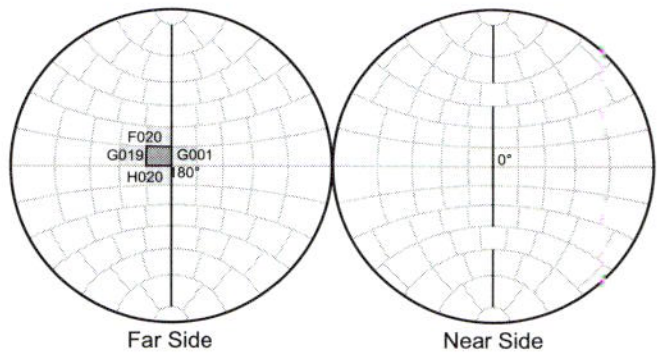

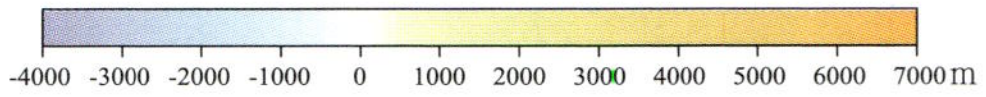

Image Code **H001**

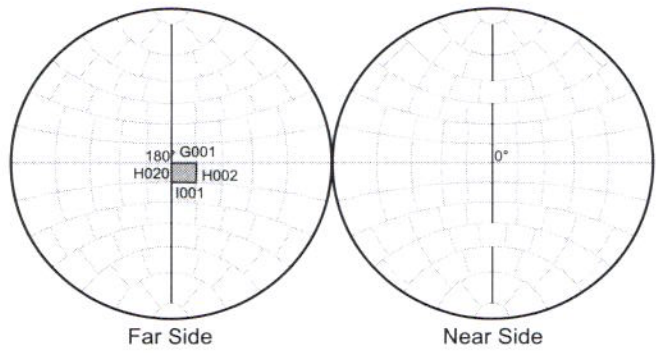

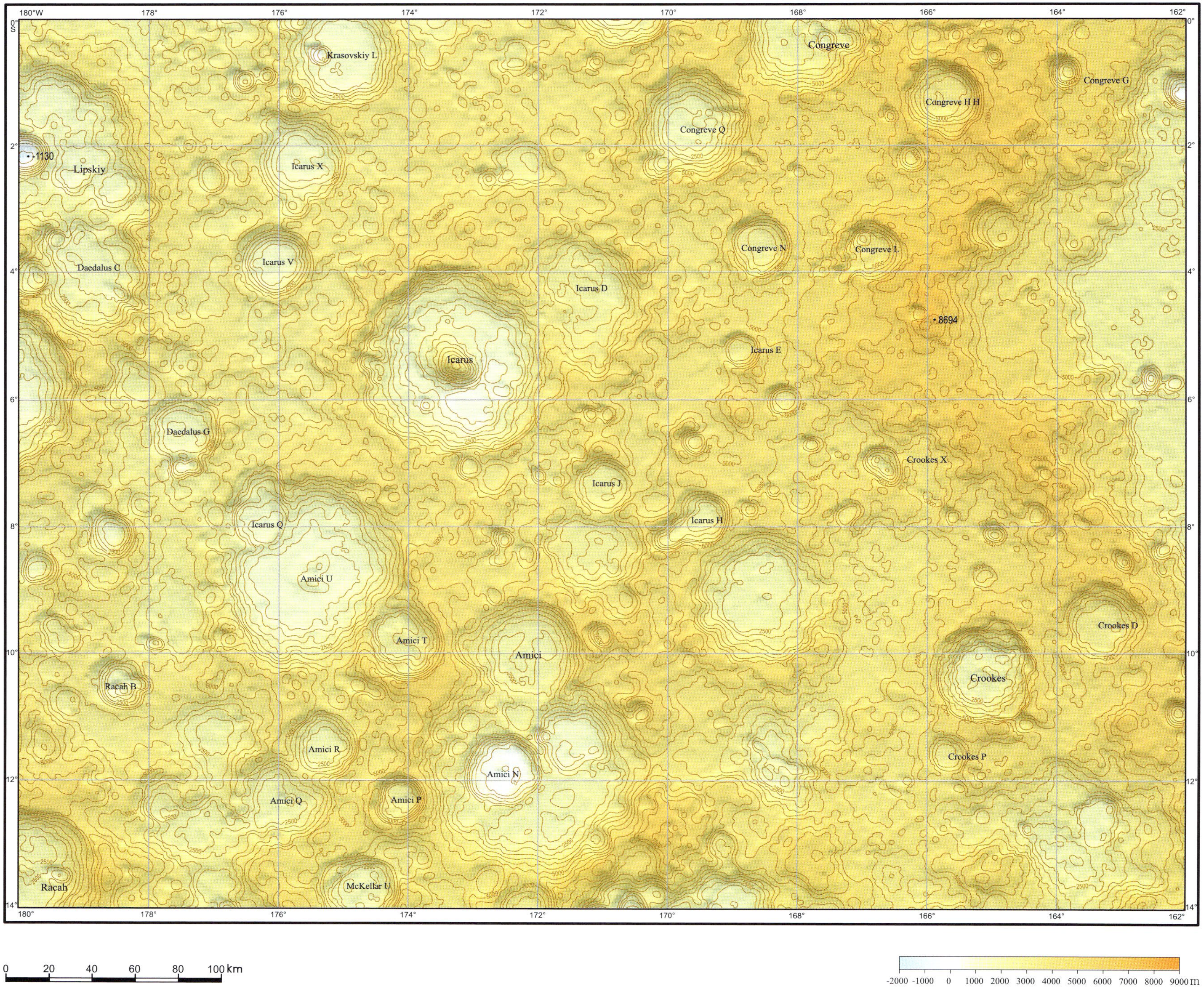

Image Code H002

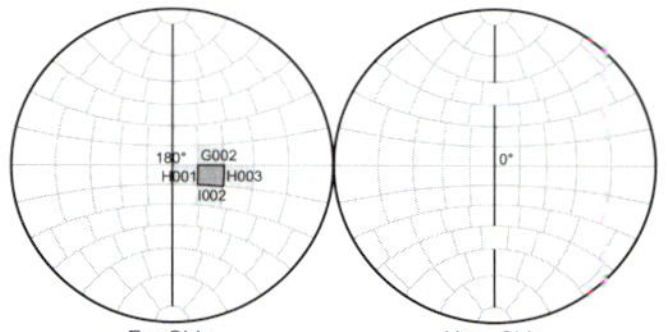

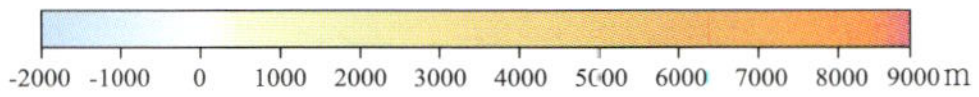

Image Code H003

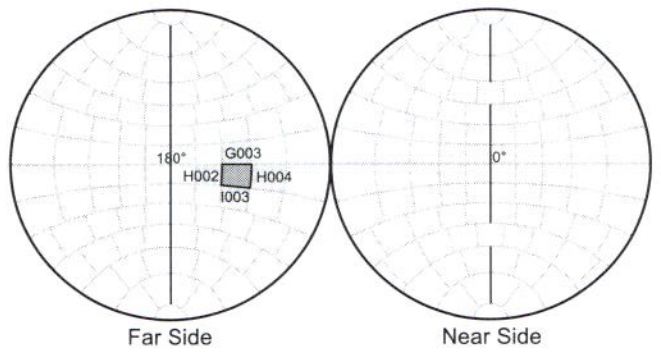

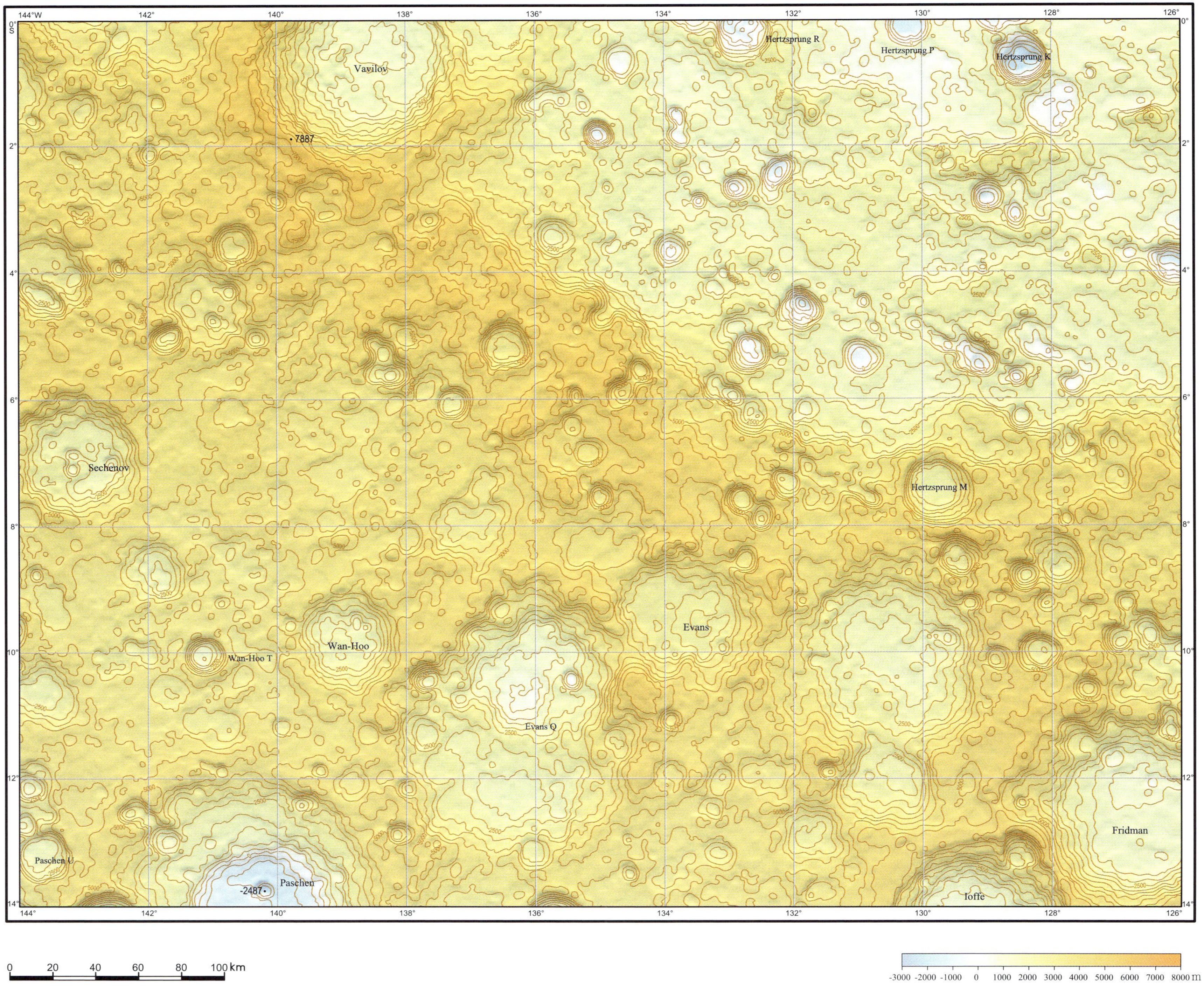

Image Code H004

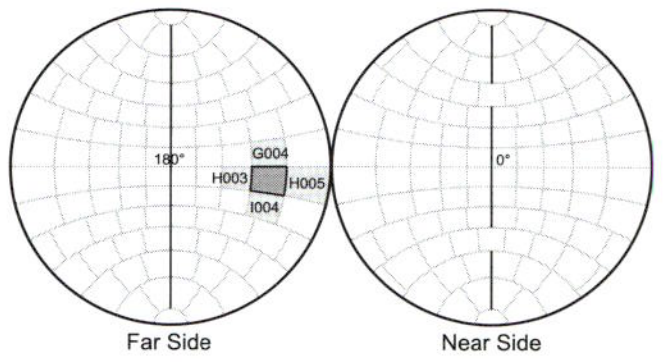

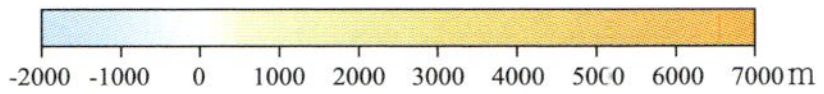

Image Code H013

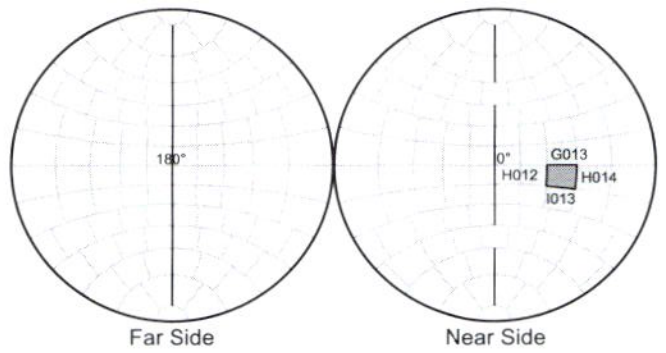

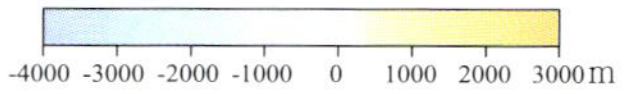

Image Code **H014**

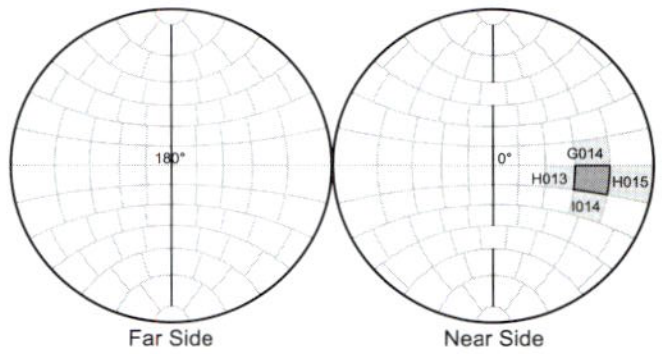

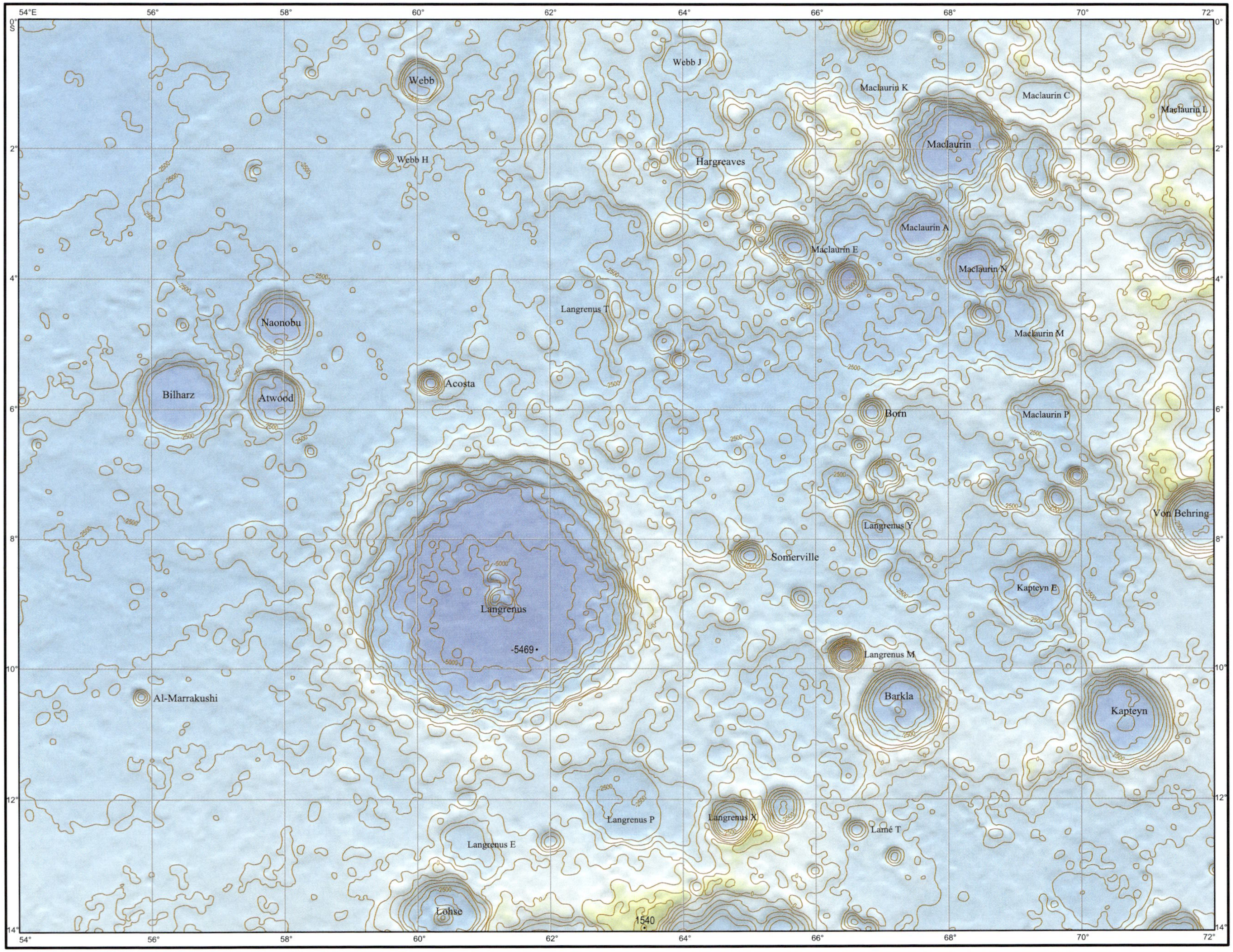

-6000 -5000 -4000 -3000 -2000 -1000 0 1000 2000m

0 20 40 60 80 100 km

Image Code **H015**

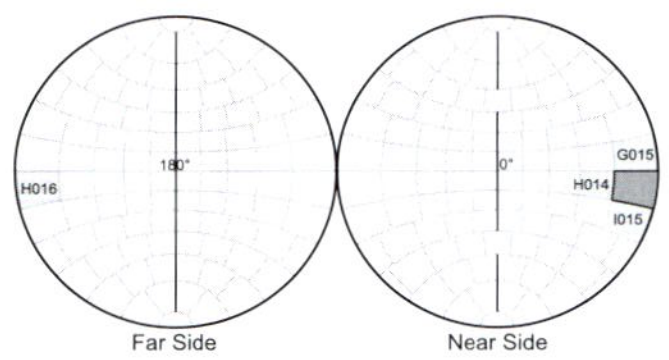

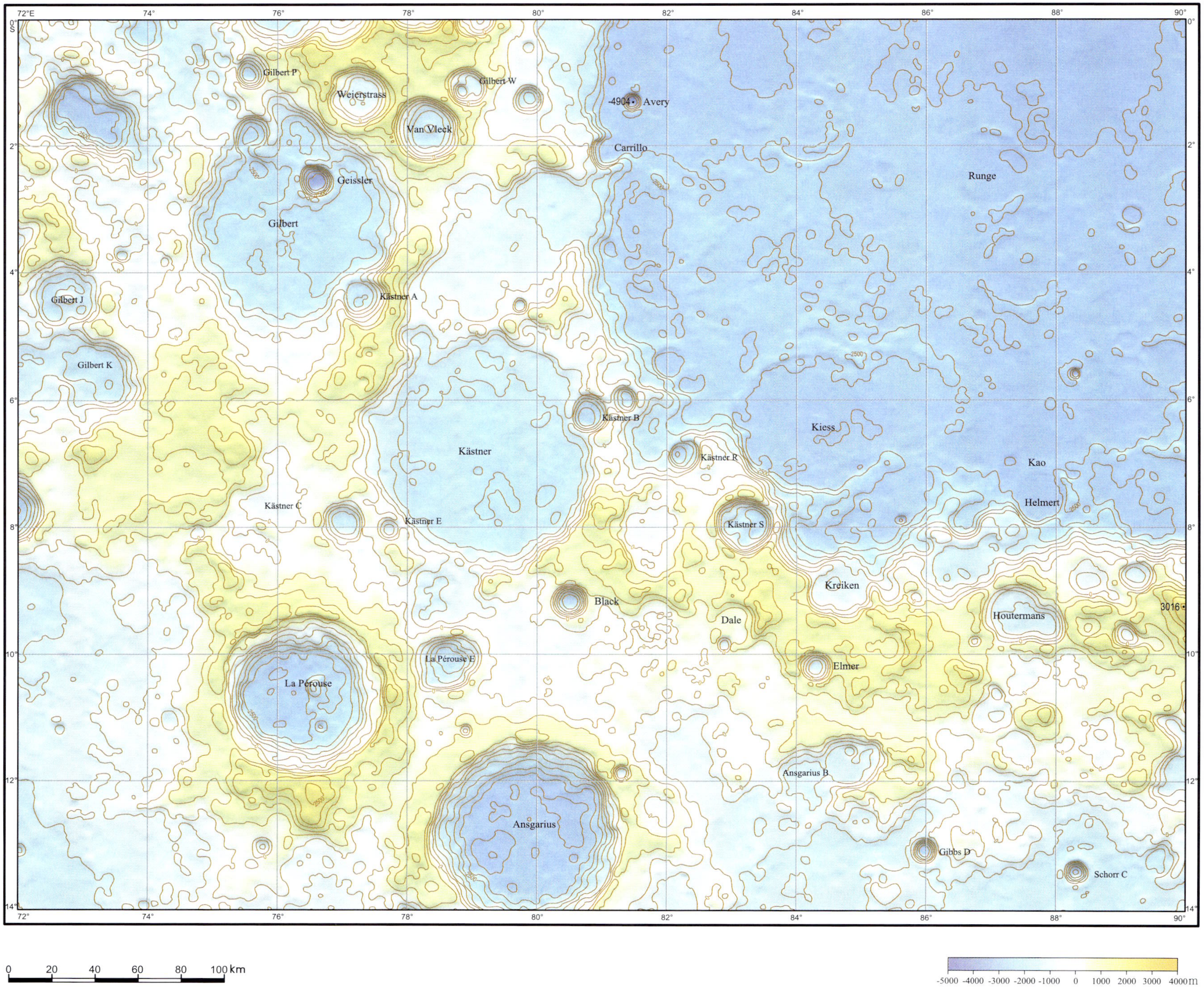

Image Code **H016**

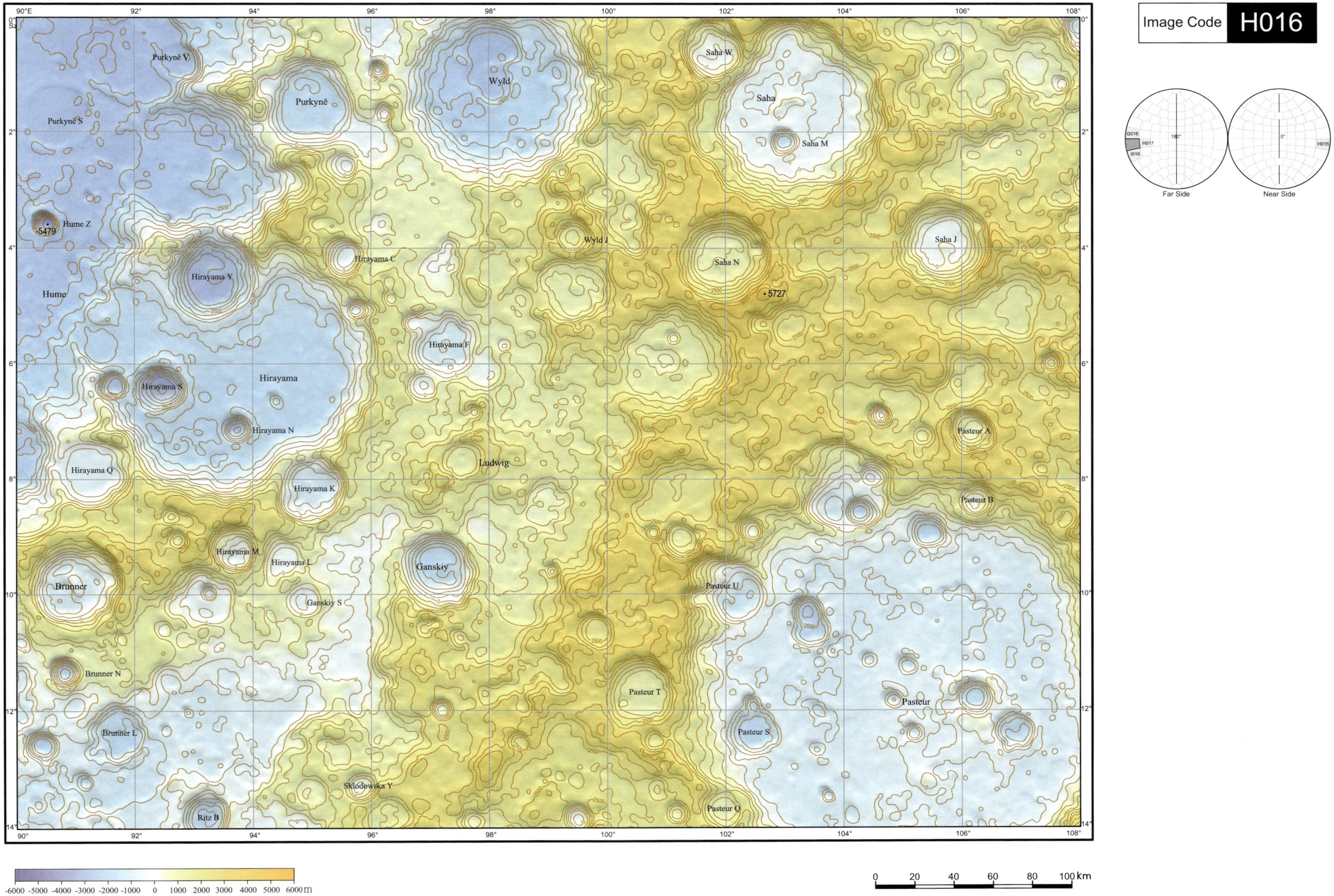

Image Code H017

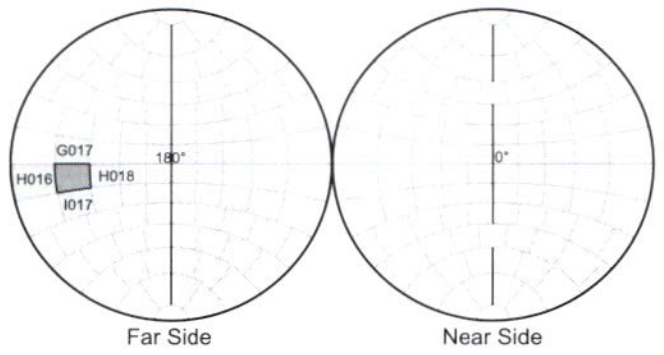

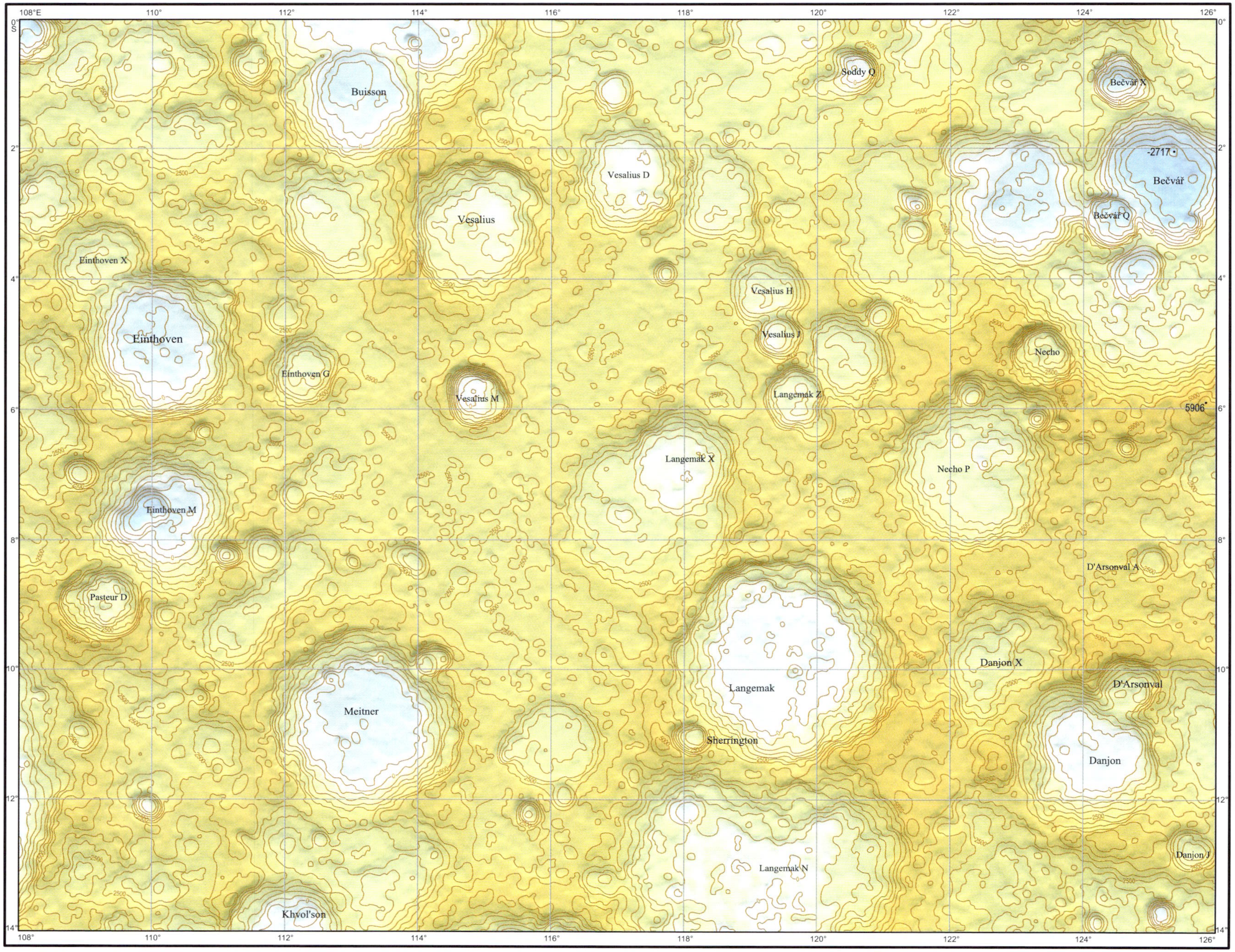

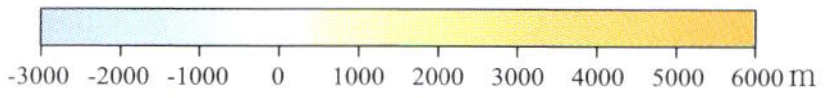

Image Code **H018**

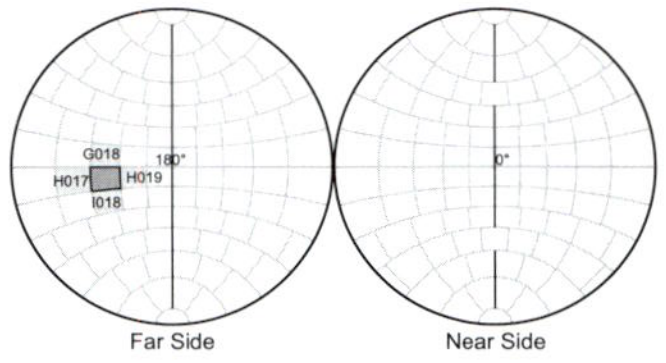

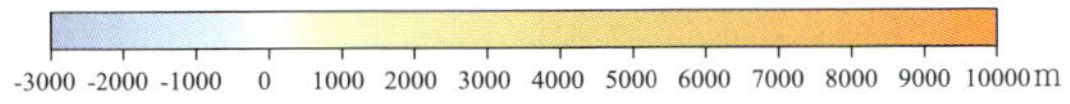

Image Code H019

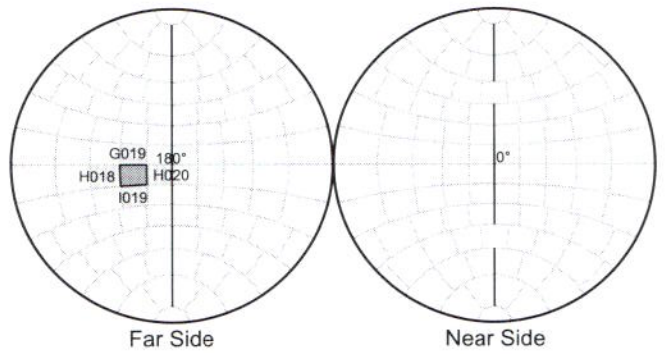

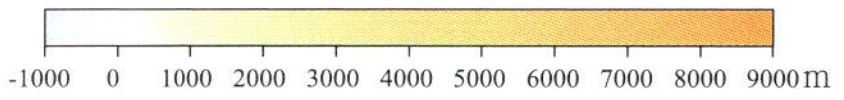

Image Code H020

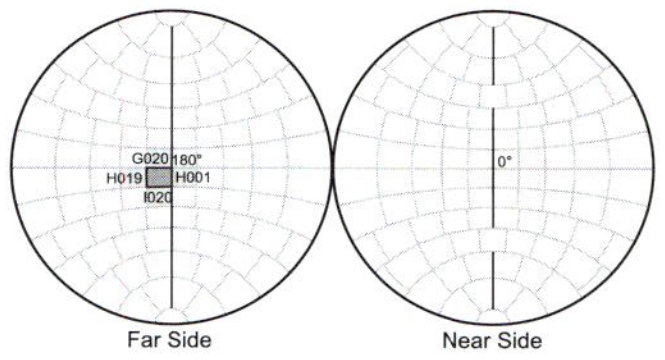

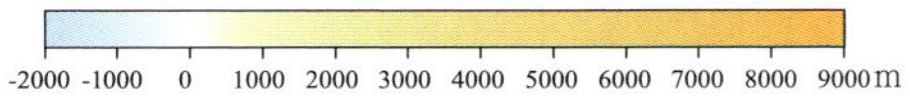

Image Code I001

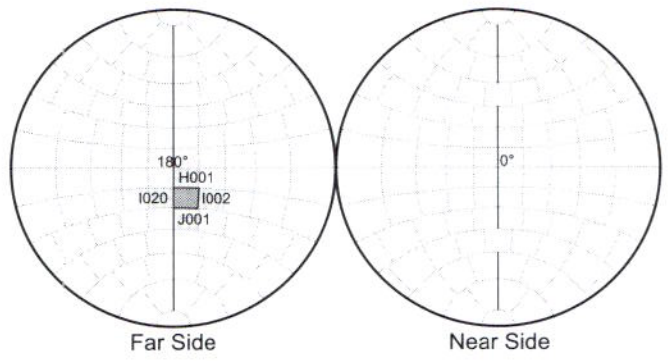

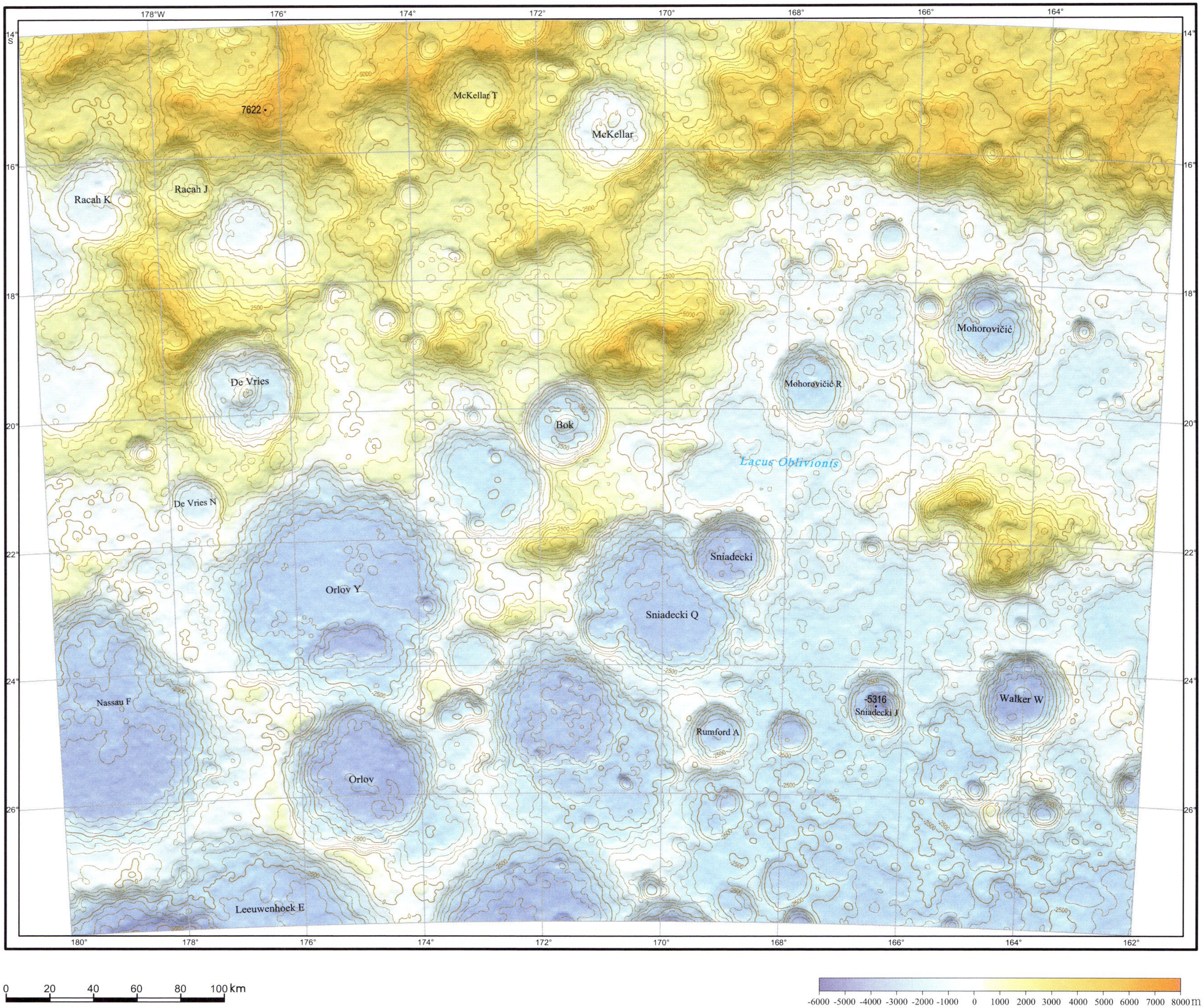

Image Code I002

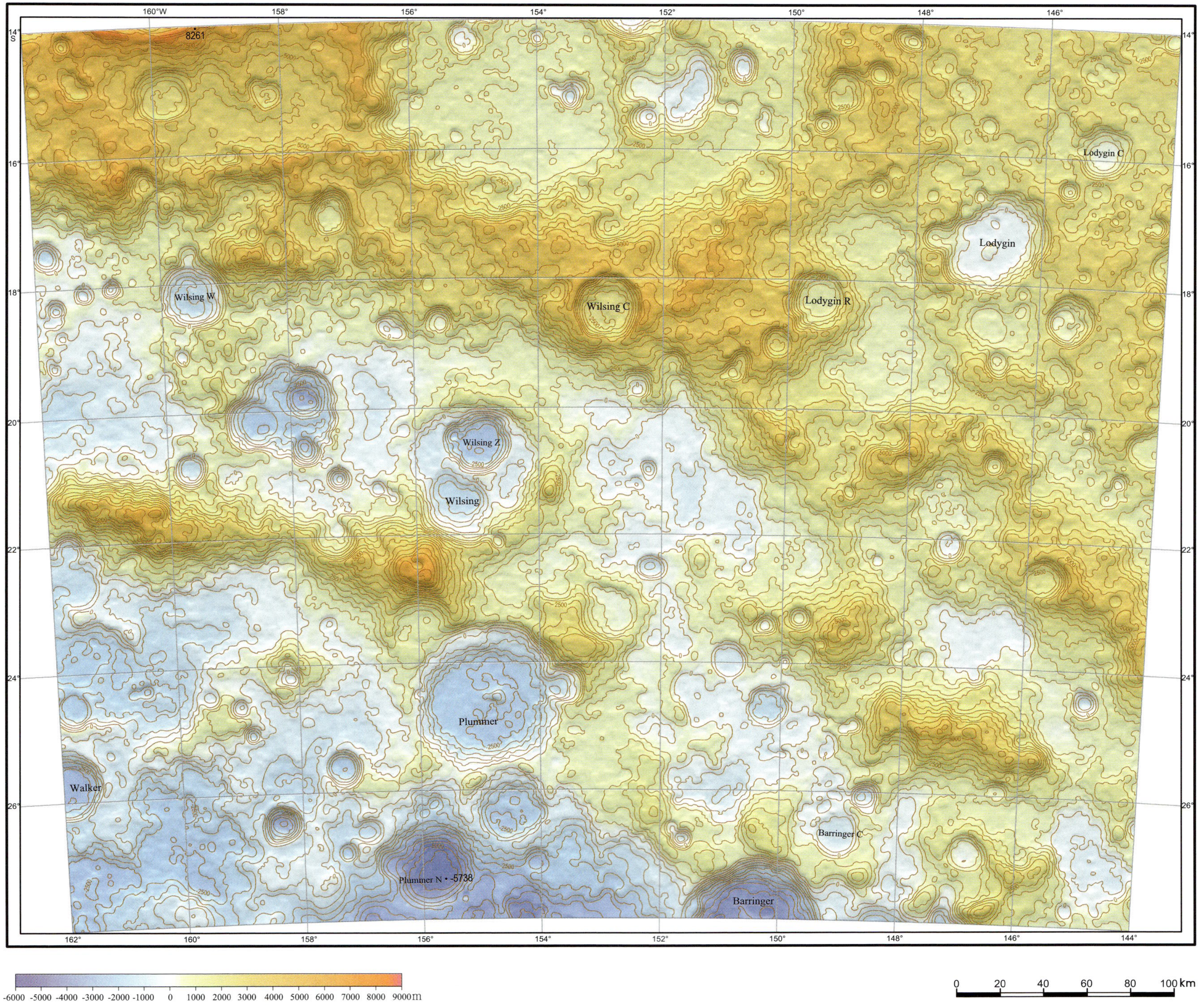

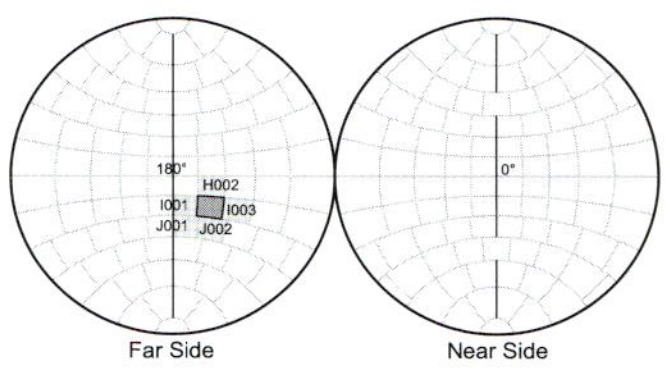

Image Code I003

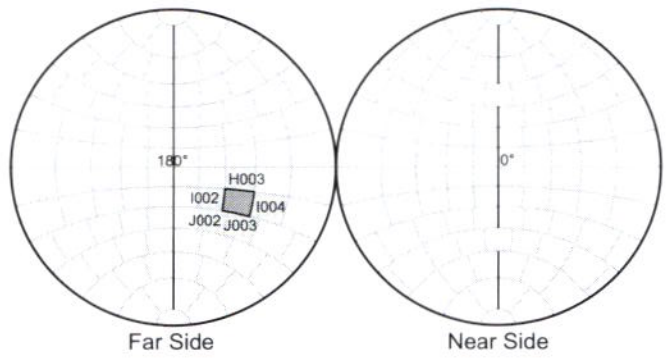

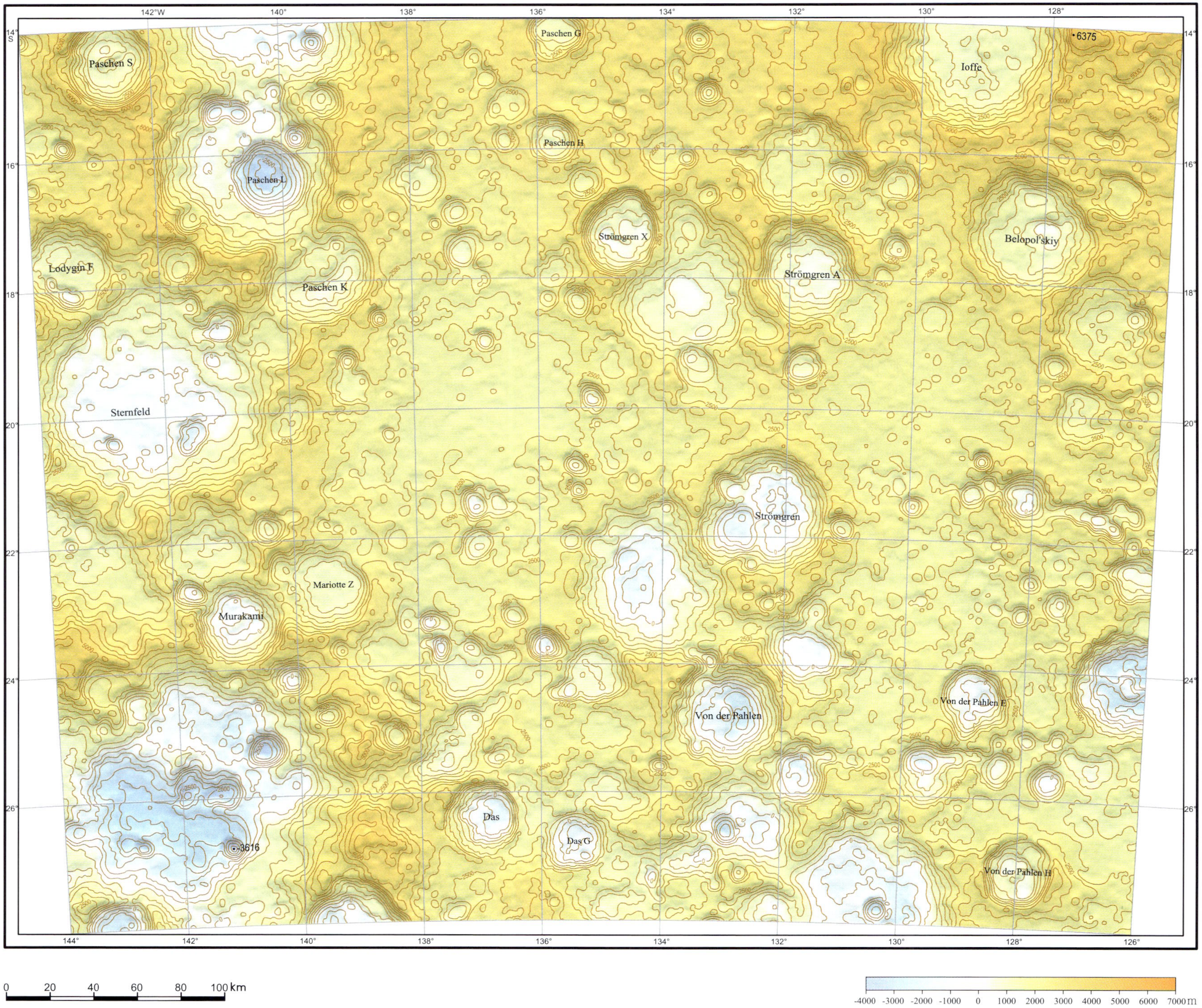

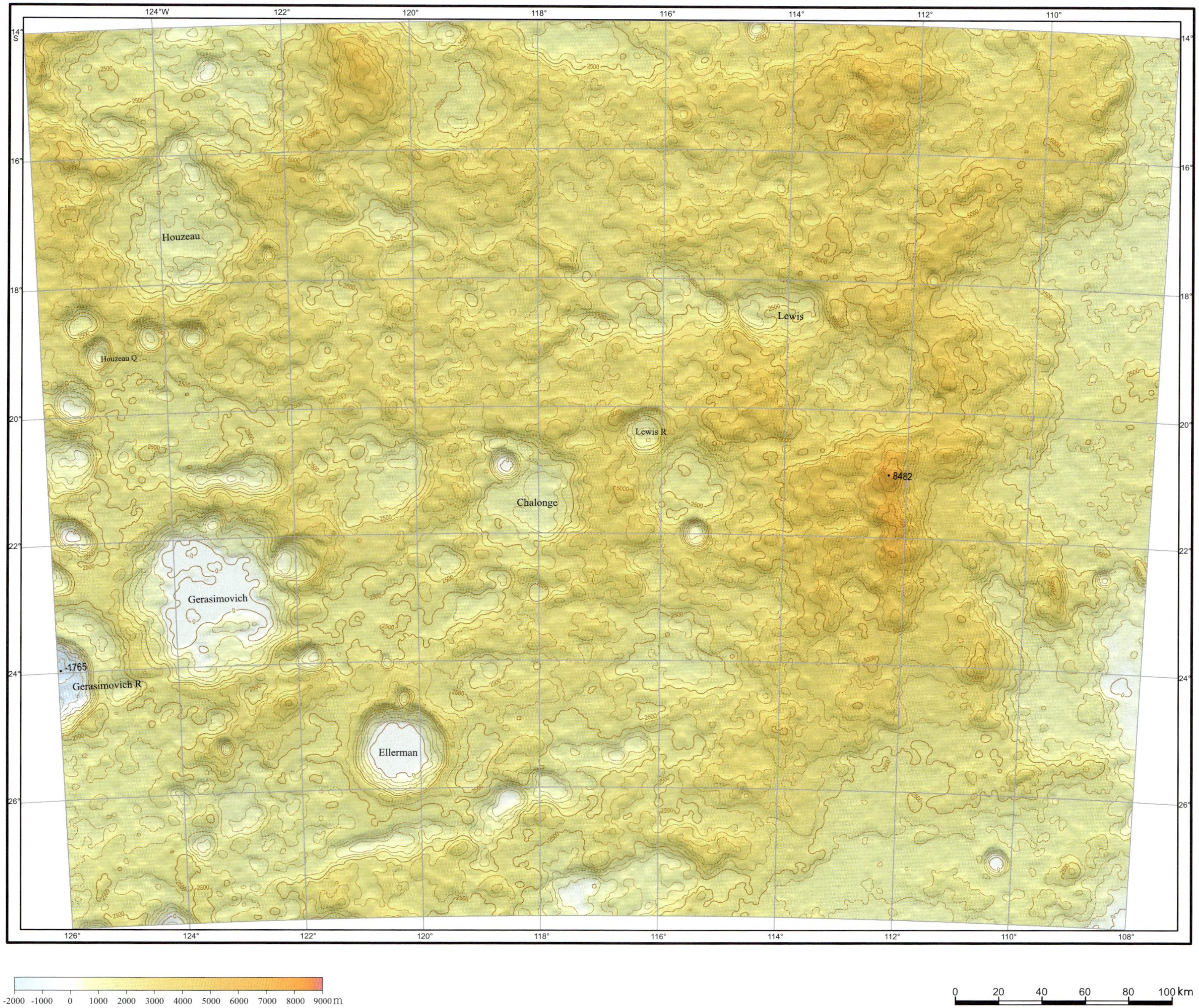

Image Code I004

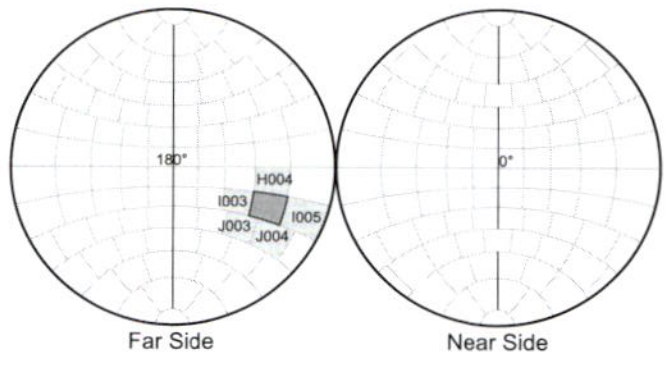

Image Code **I005**

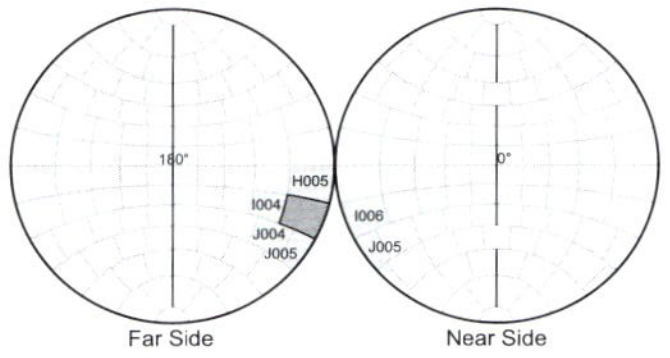

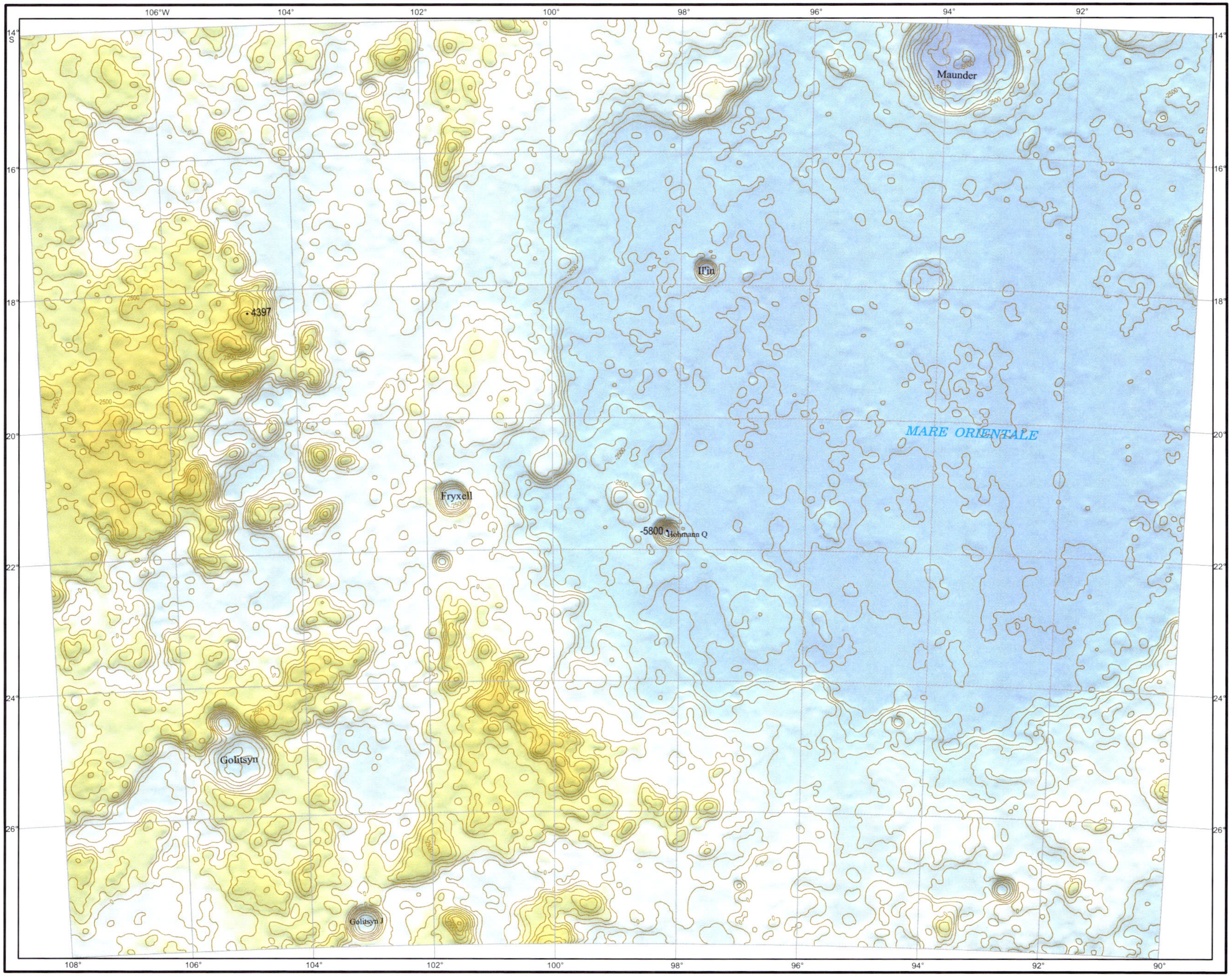

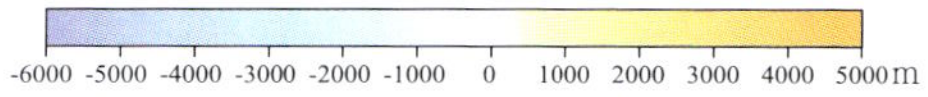

Image Code I006

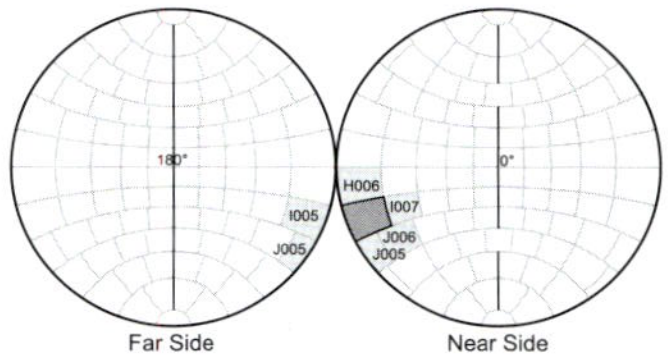

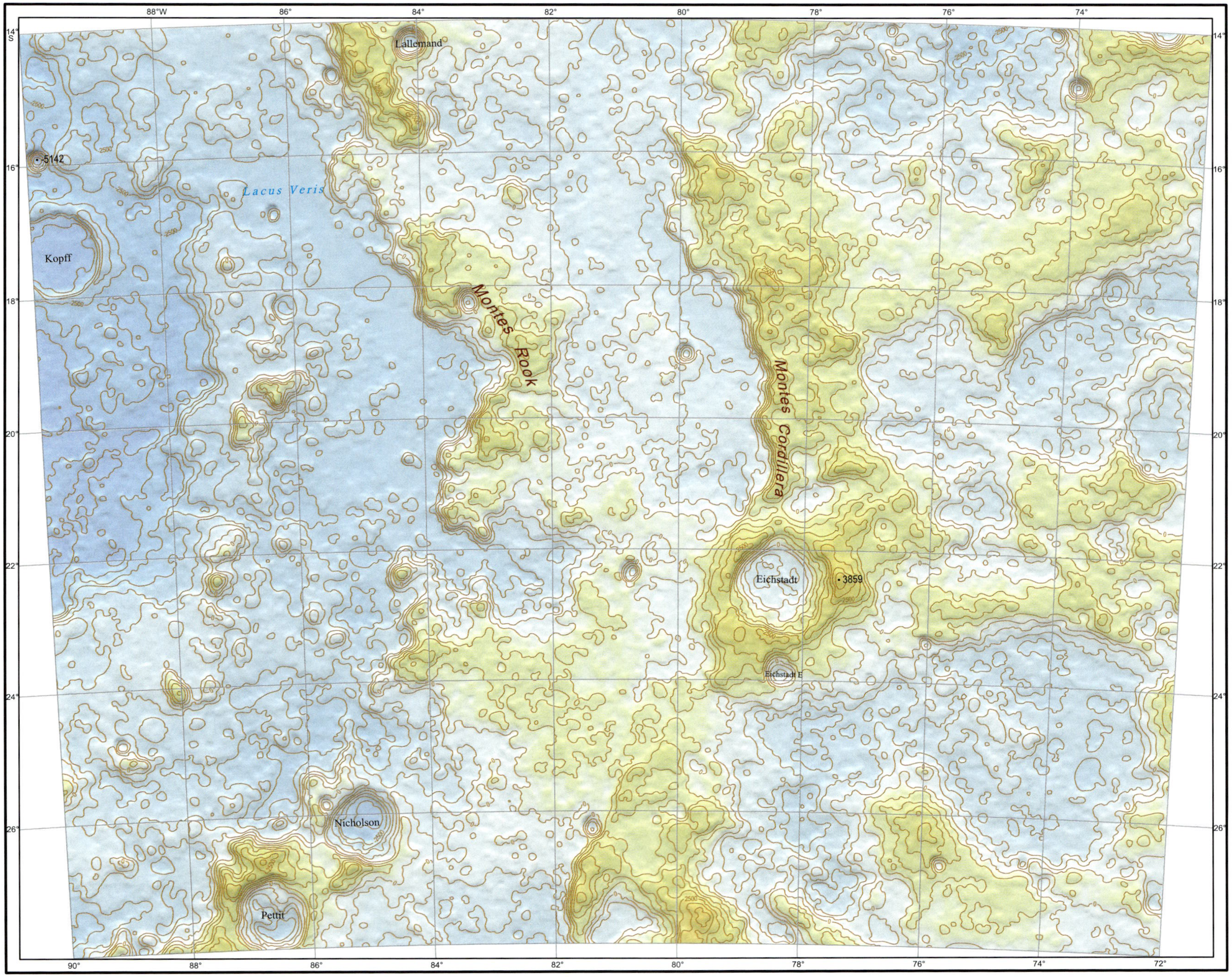

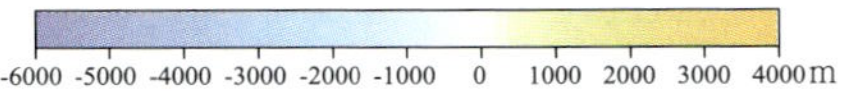

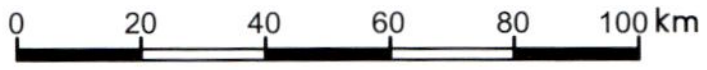

Image Code **I007**

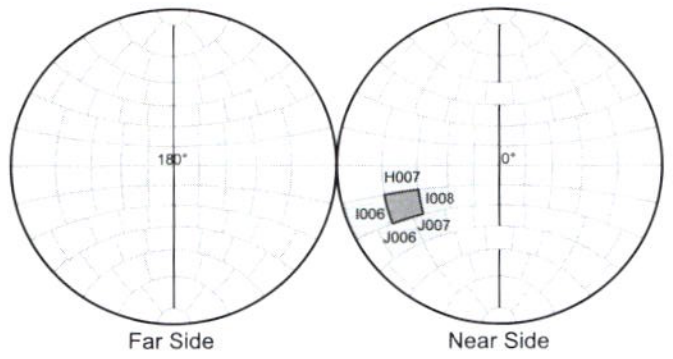

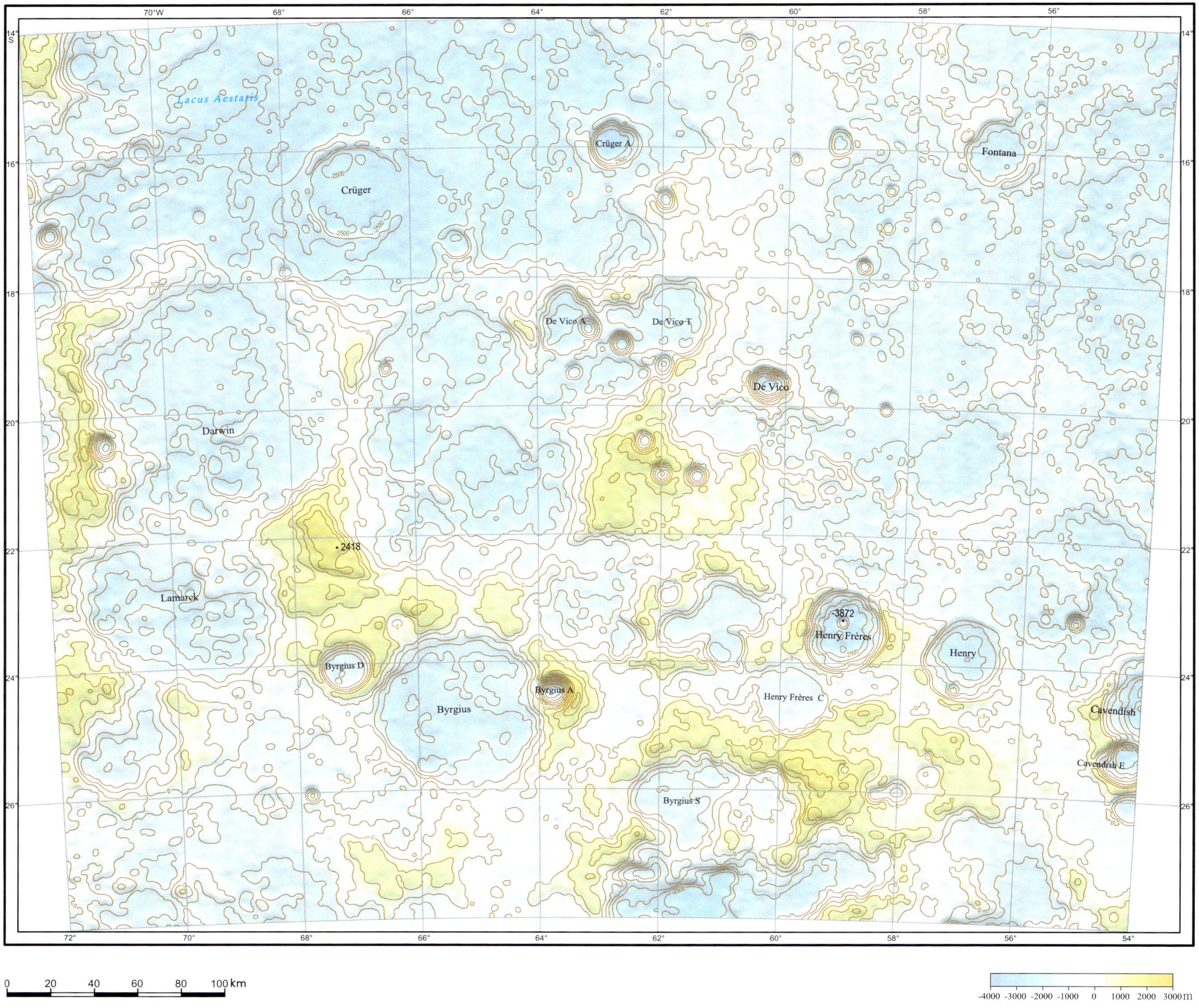

Image Code I008

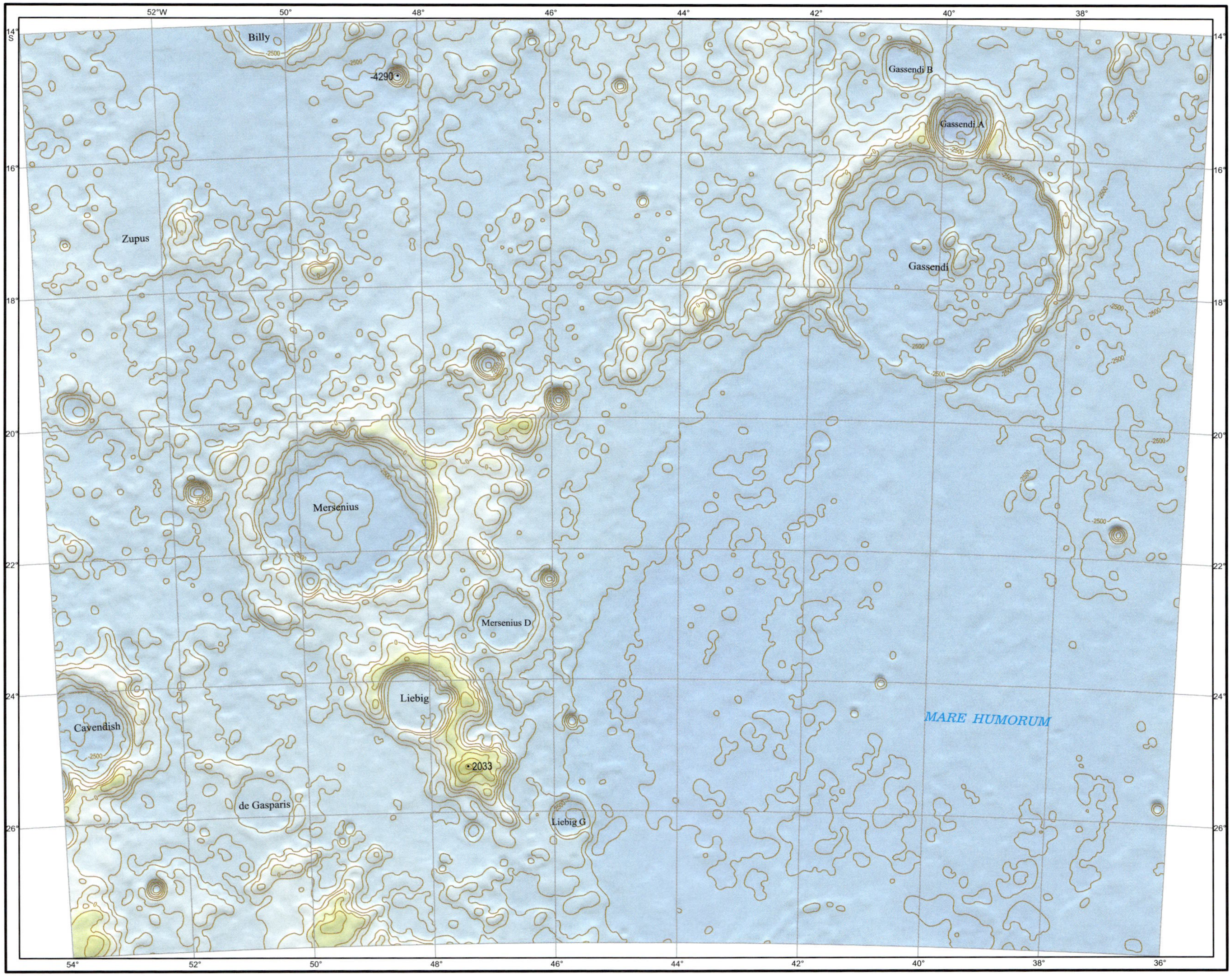

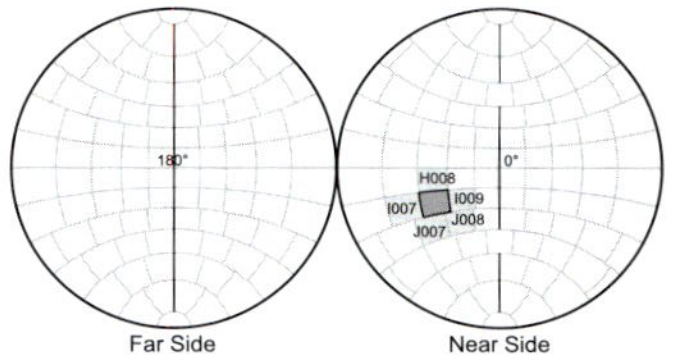

-5000 -4000 -3000 -2000 -1000 0 1000 2000 3000m

0 20 40 60 80 100 km

Image Code I009

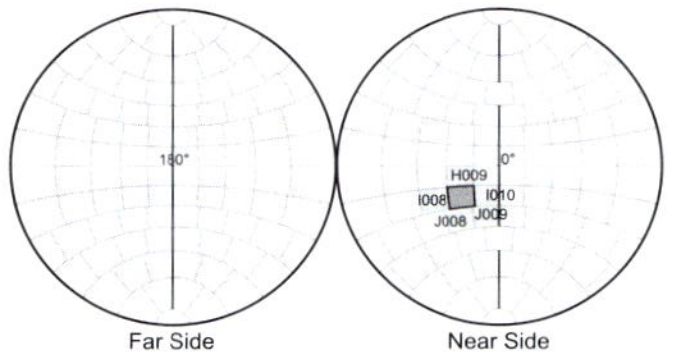

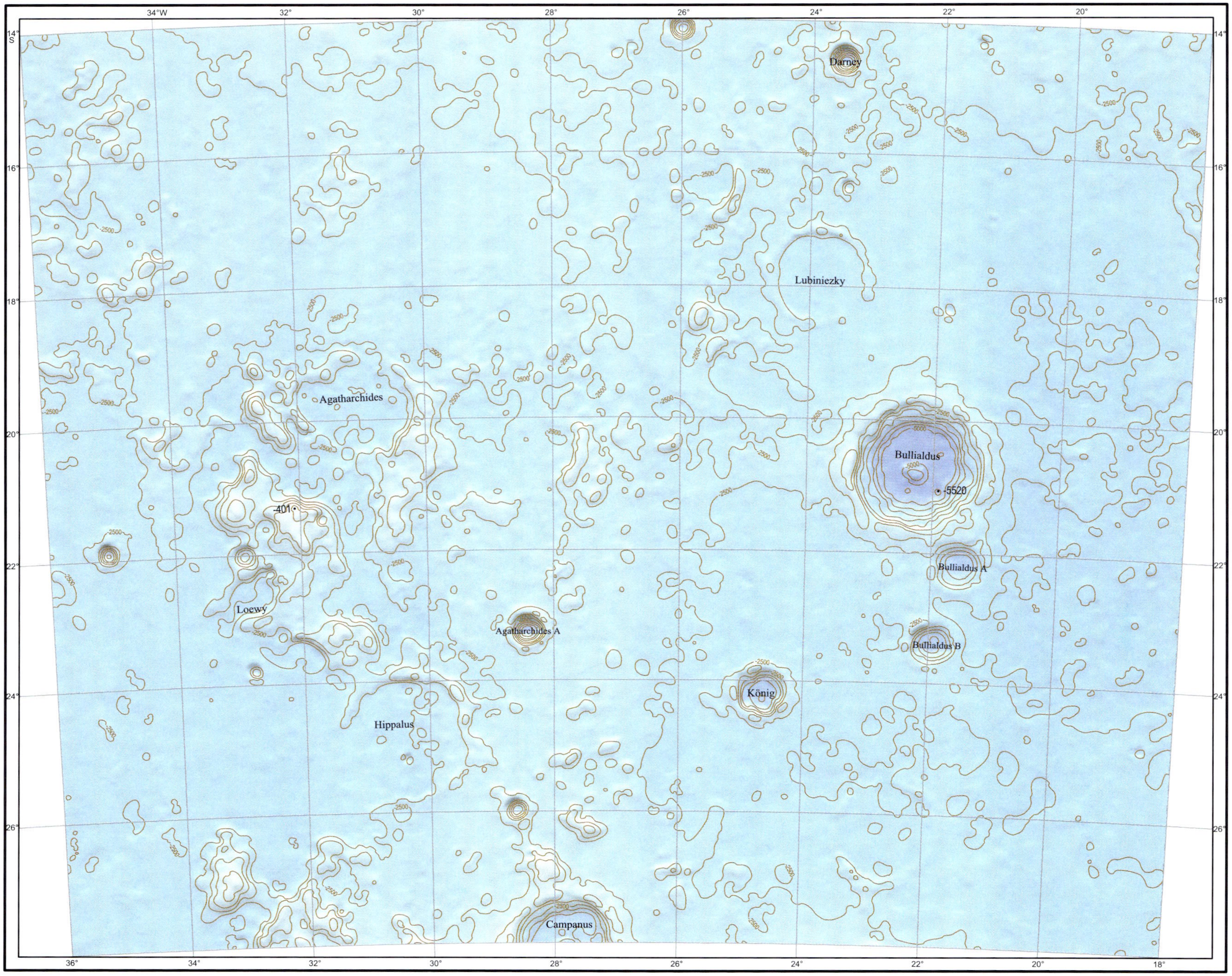

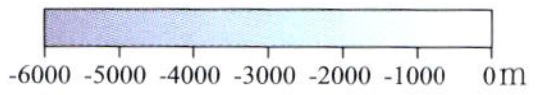

Image Code I010

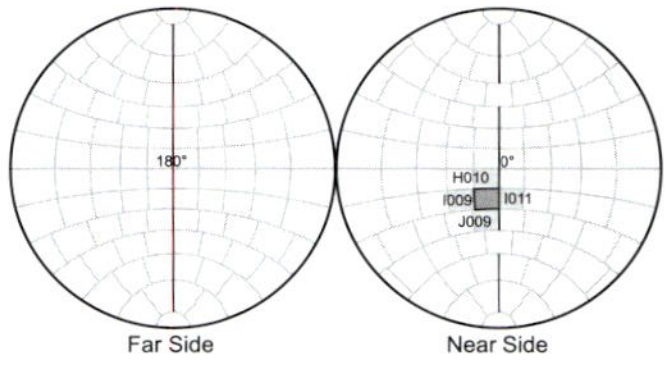

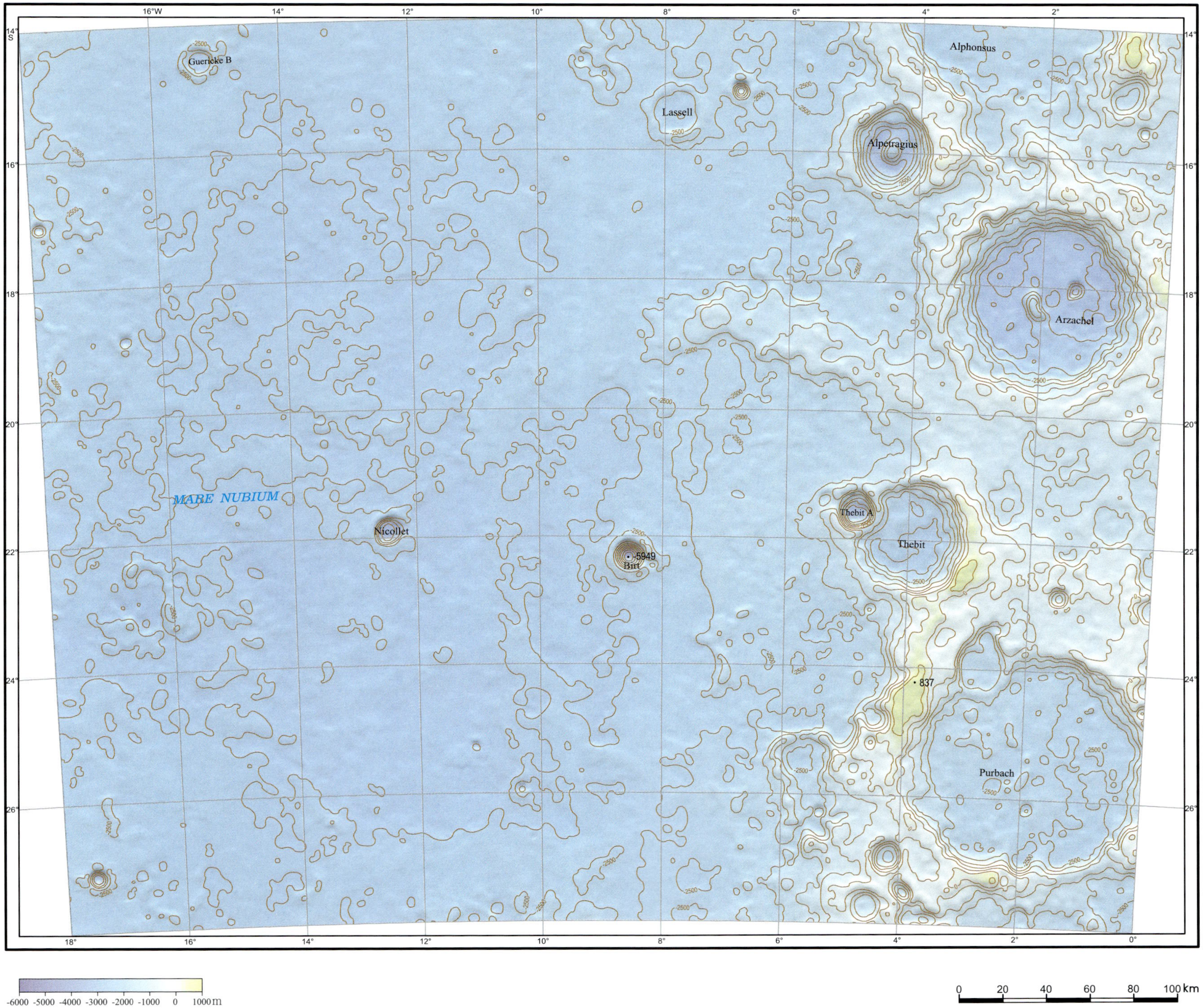

Image Code **I011**

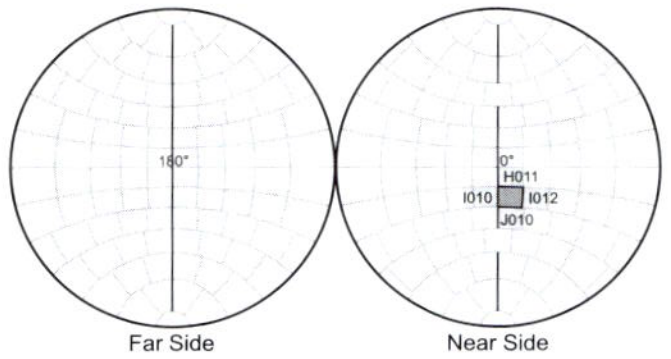

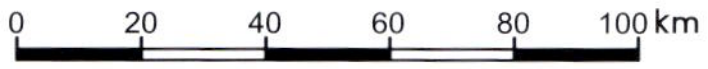

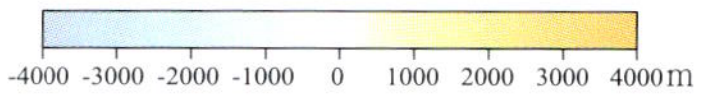

Image Code I012

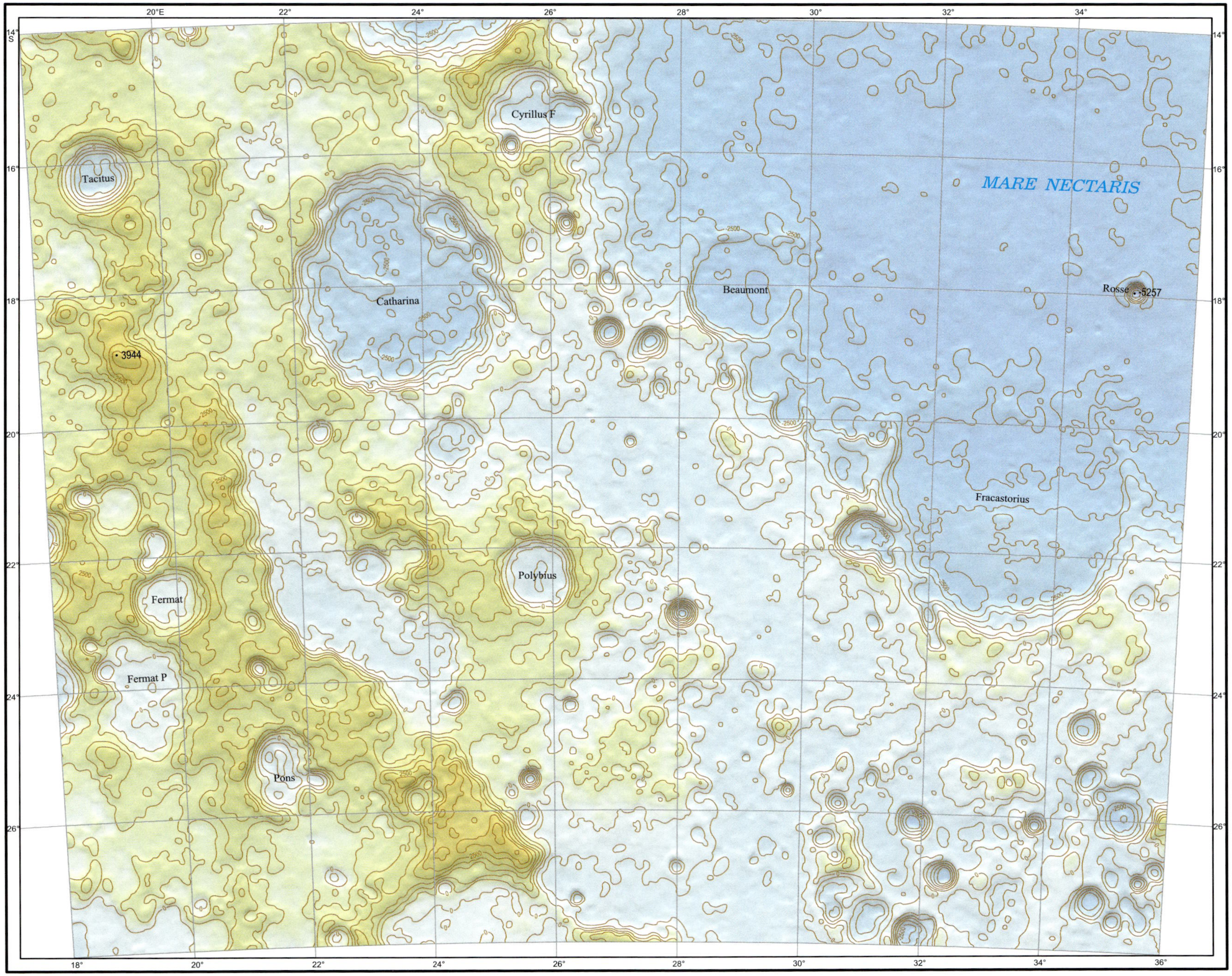

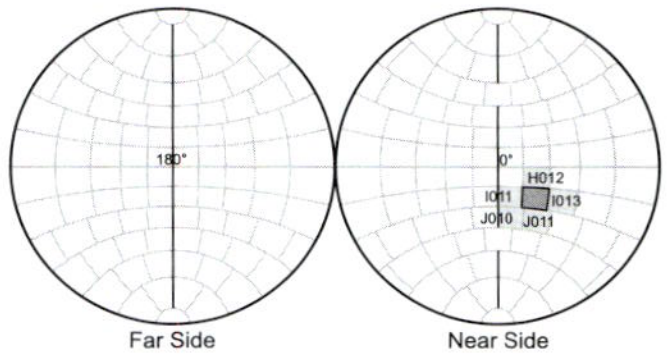

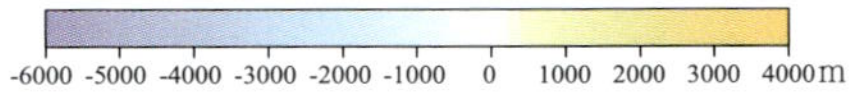

Image Code **I013**

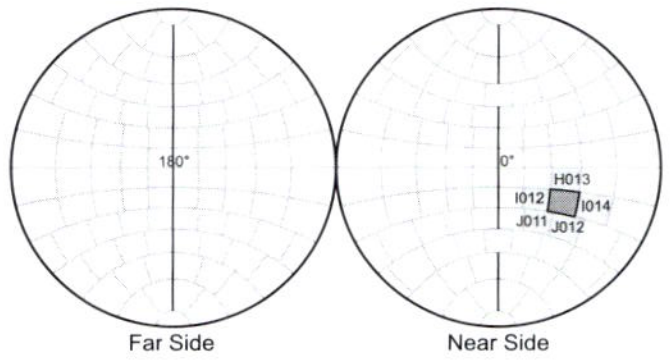

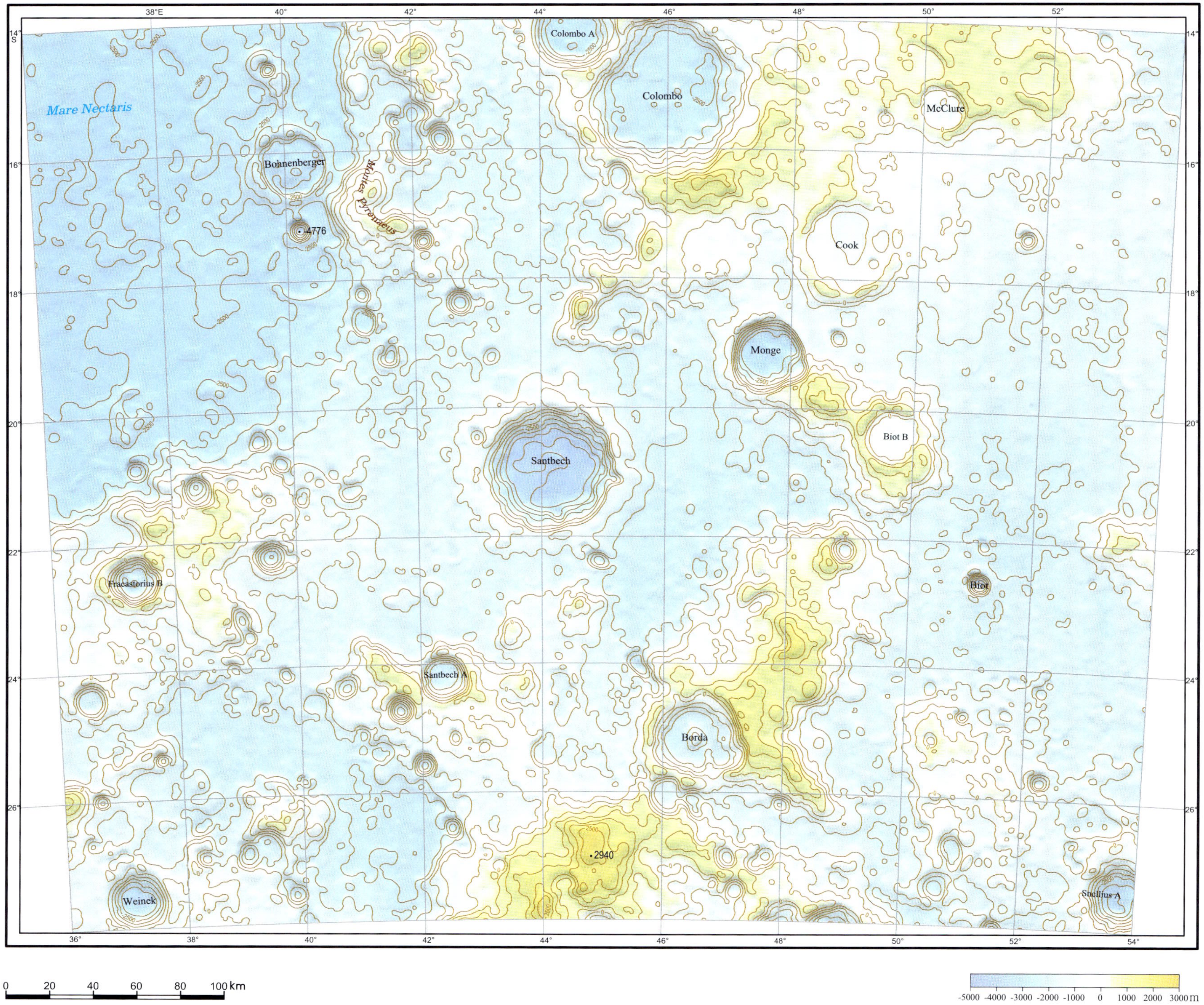

Image Code I014

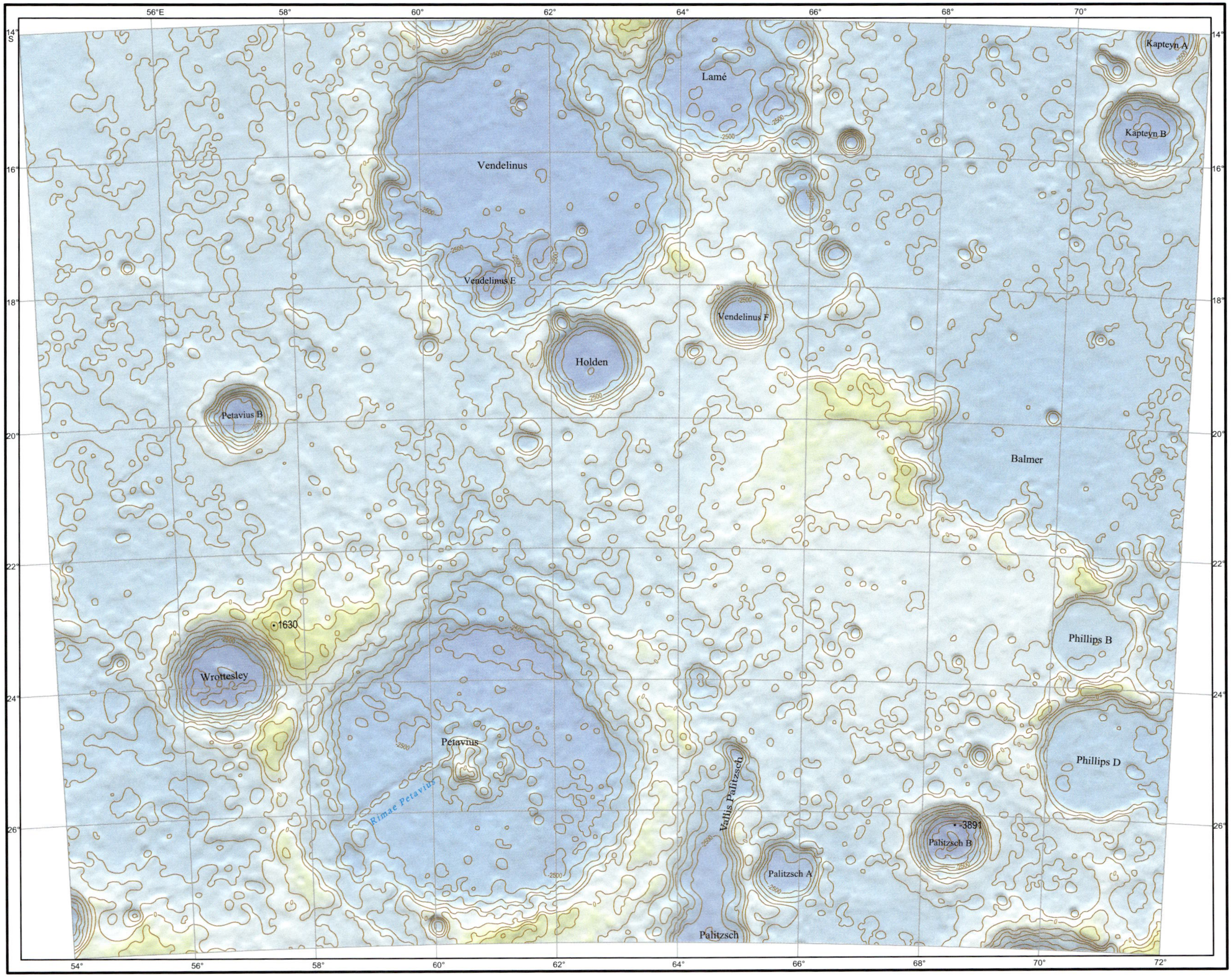

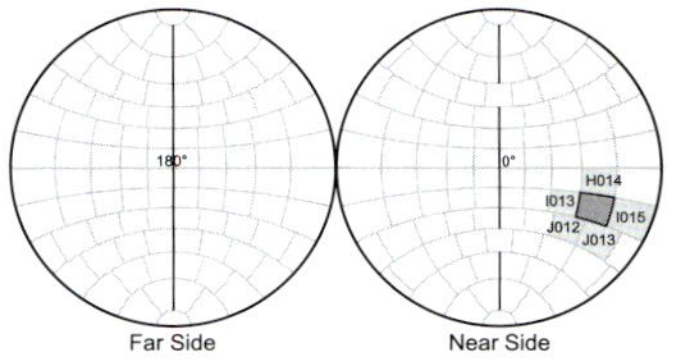

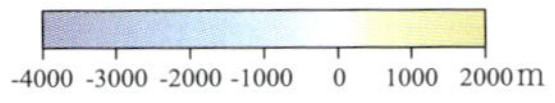

Image Code **I015**

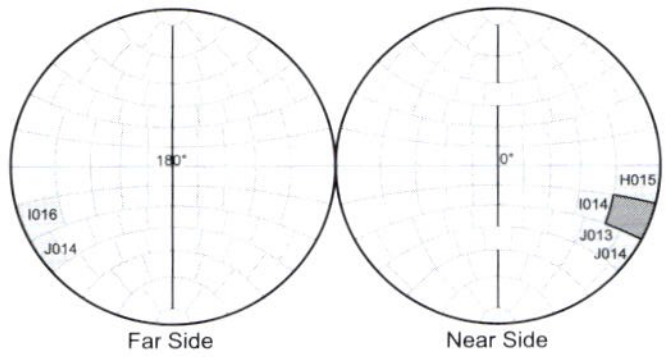

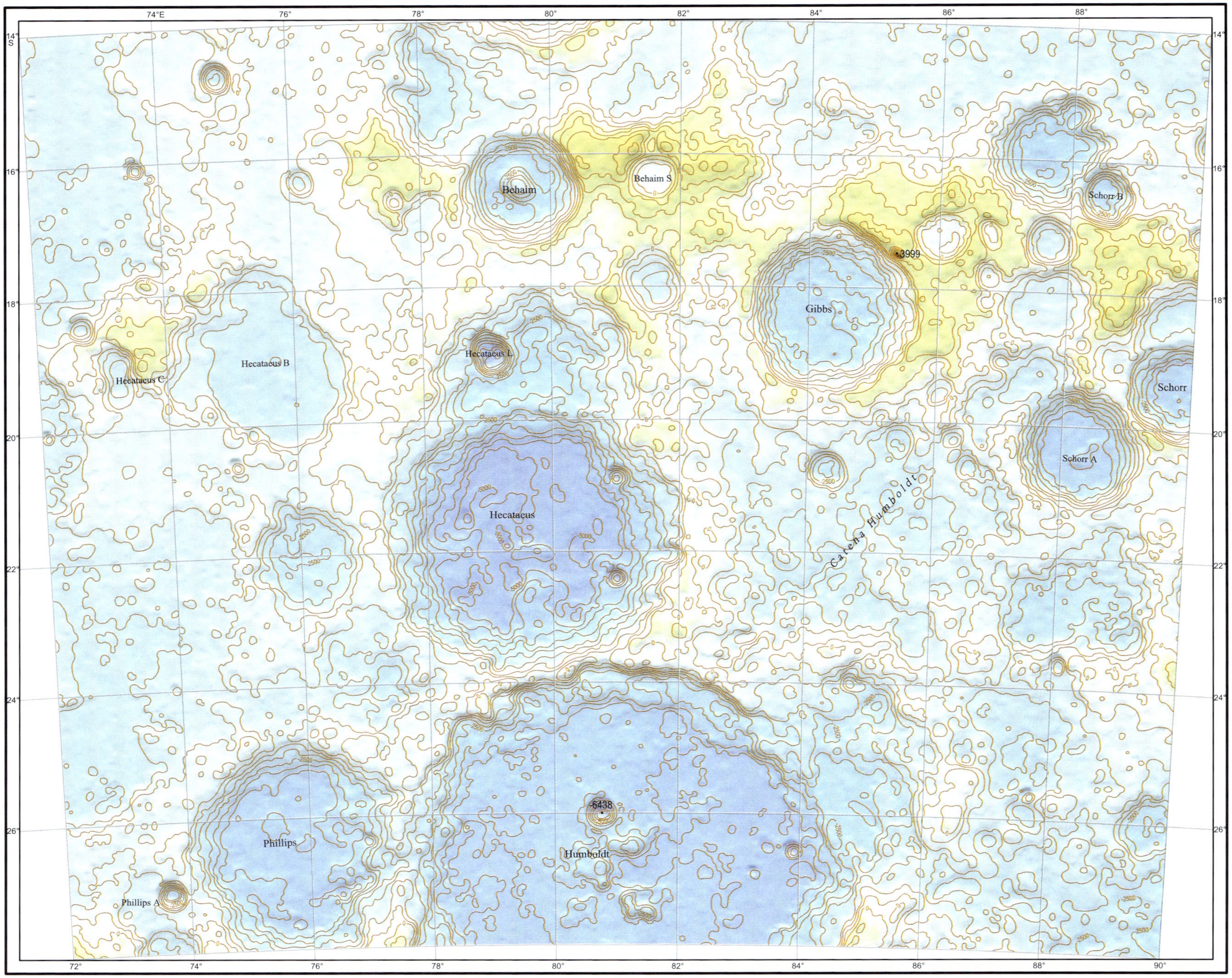

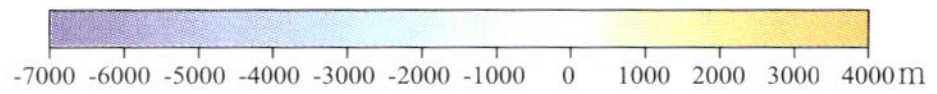

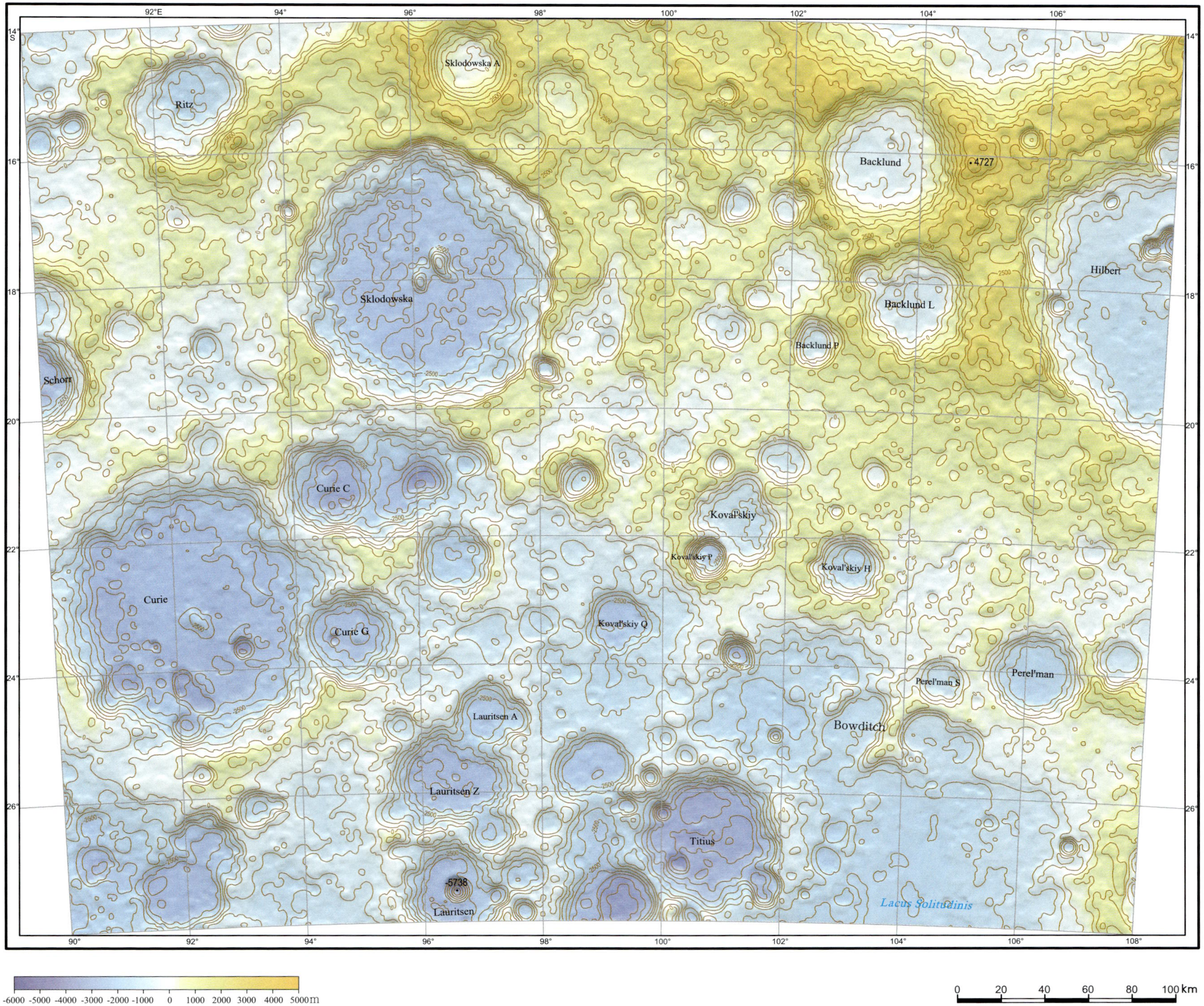

Image Code **I016**

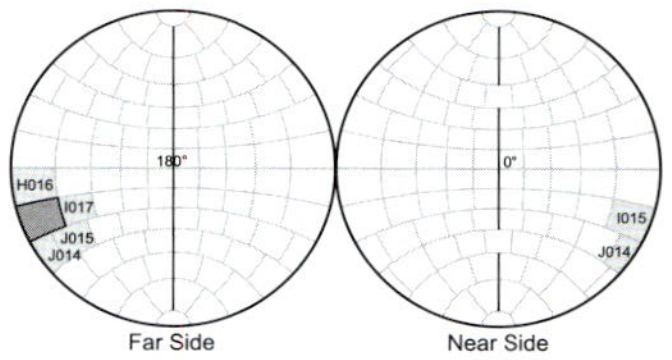

Image Code I017

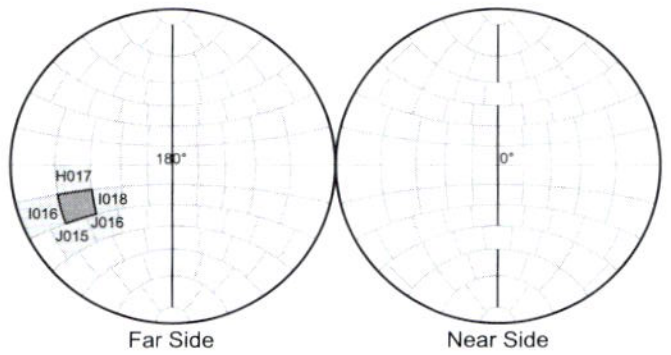

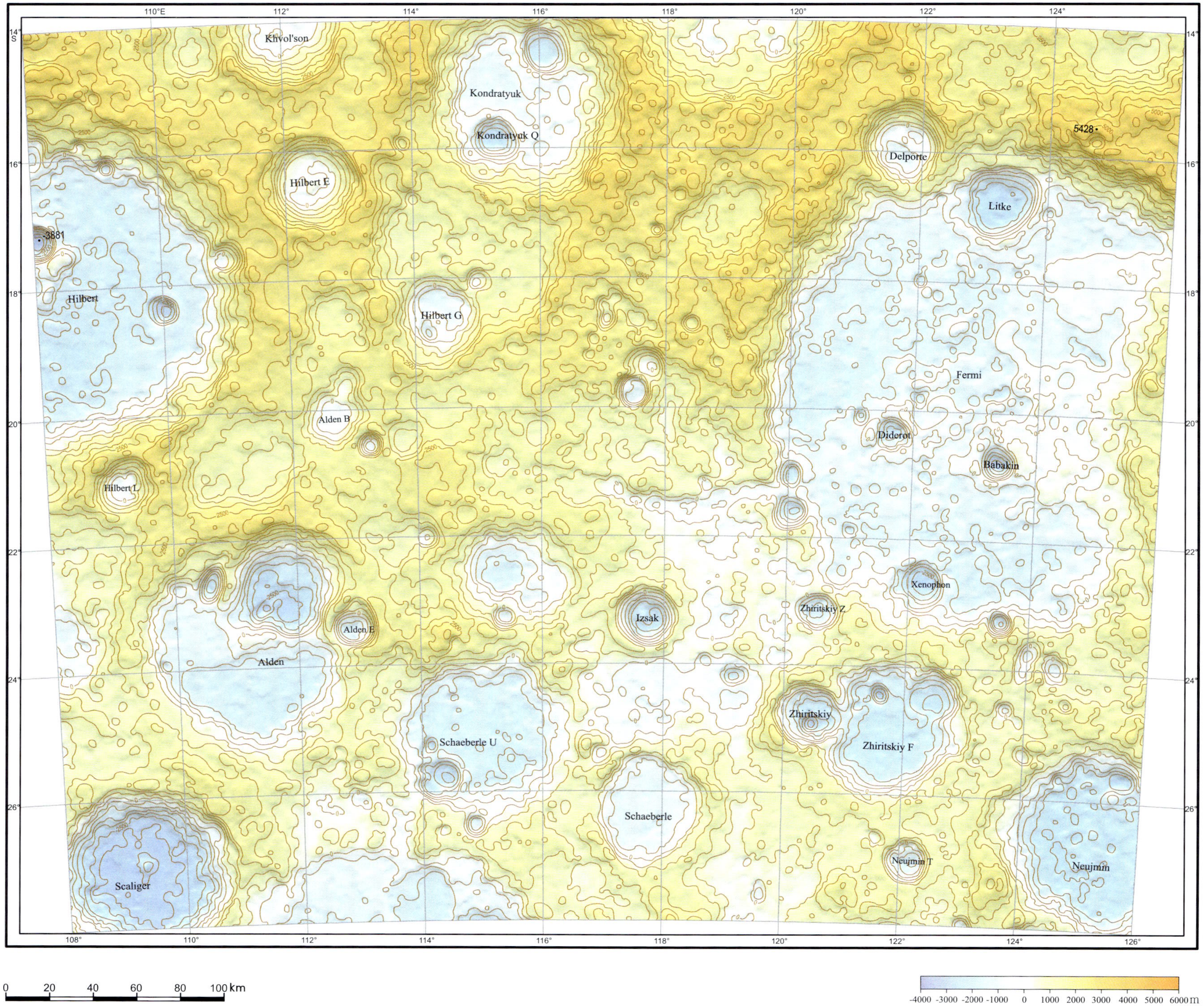

Image Code I018

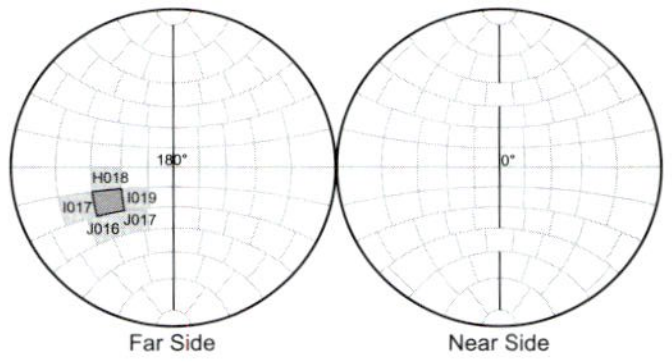

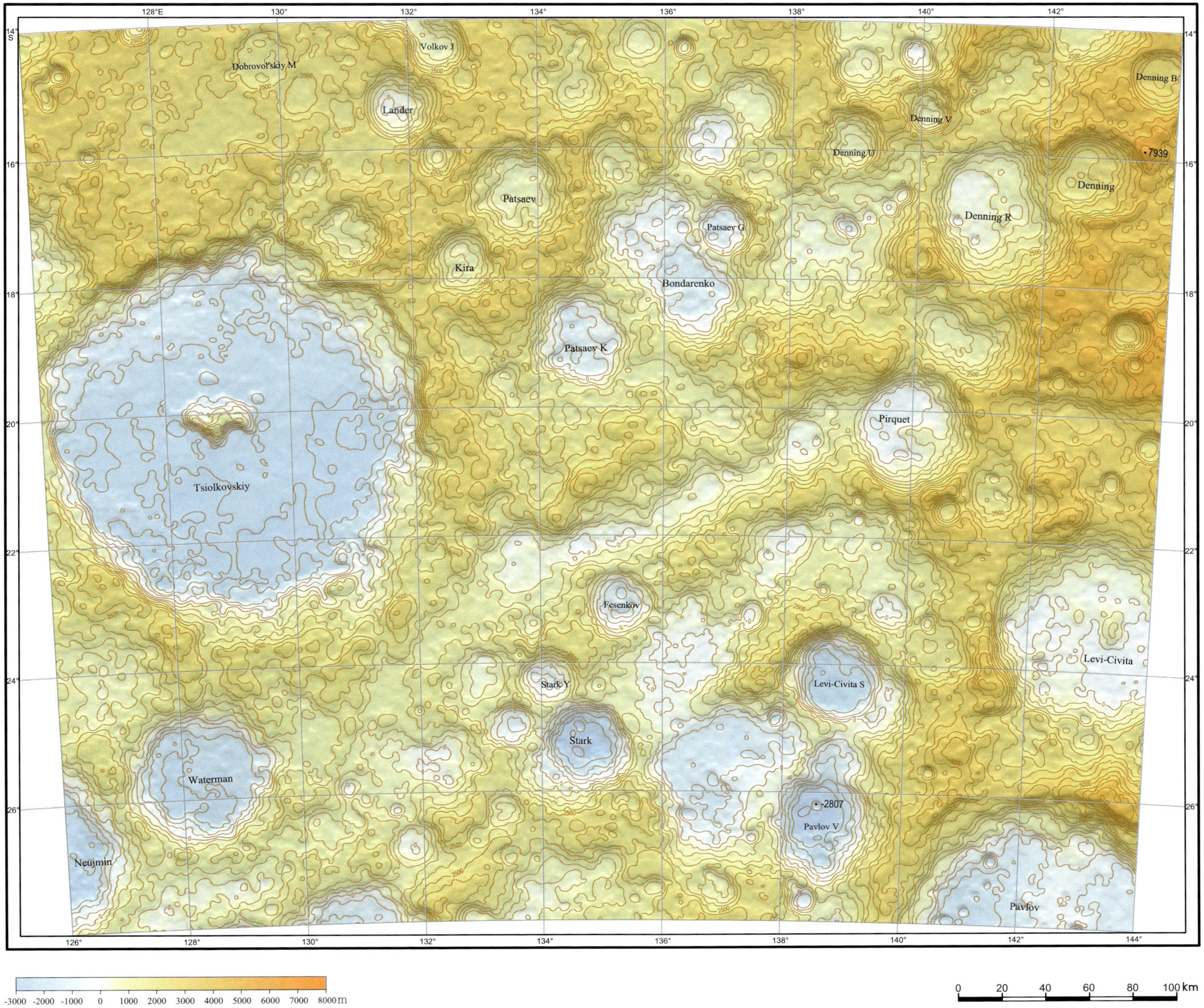

Image Code I019

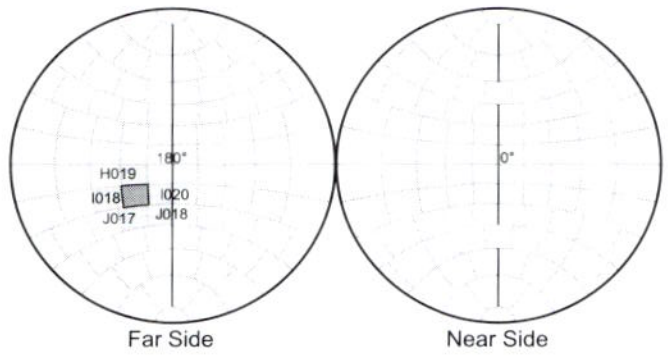

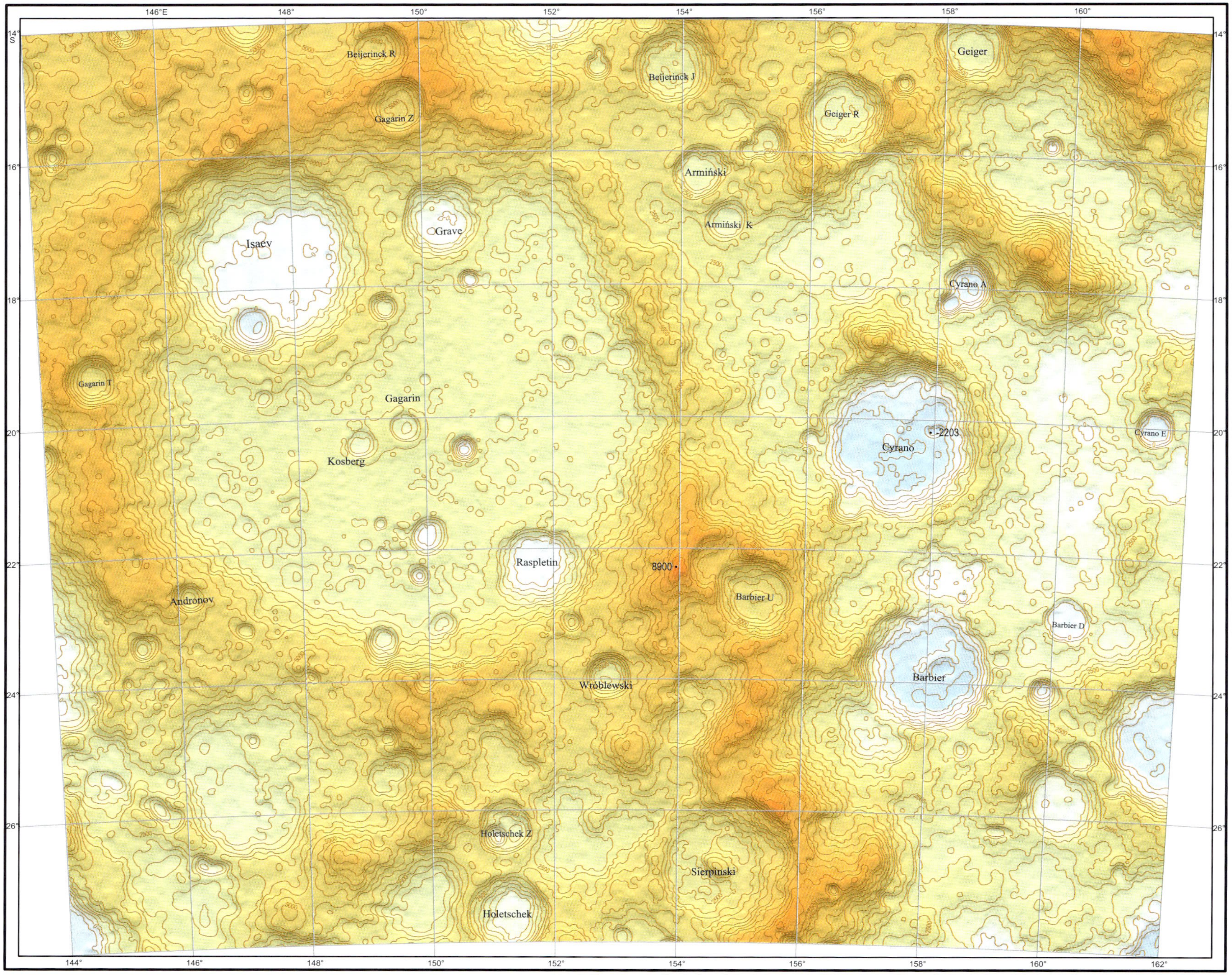

Image Code I020

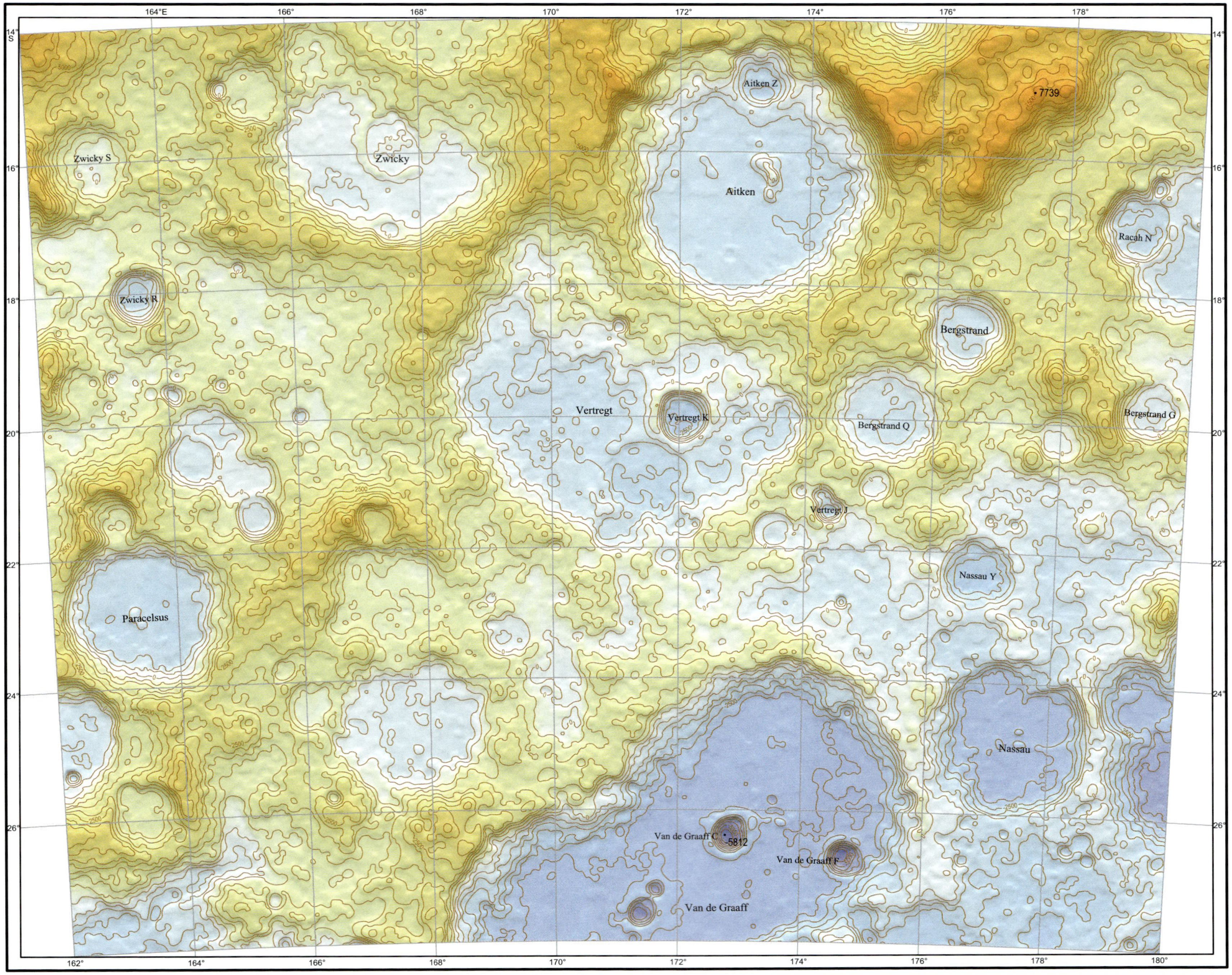

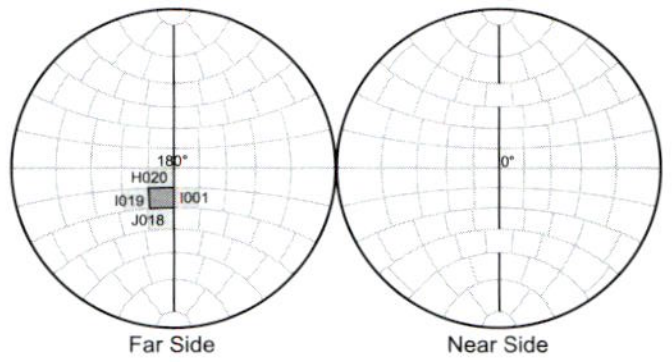

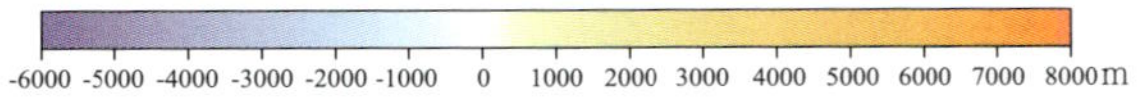

Image Code J001

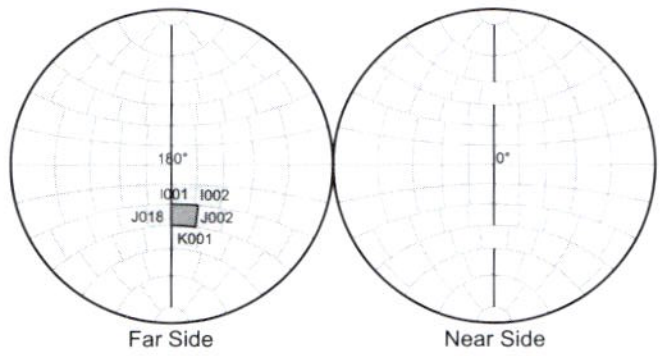

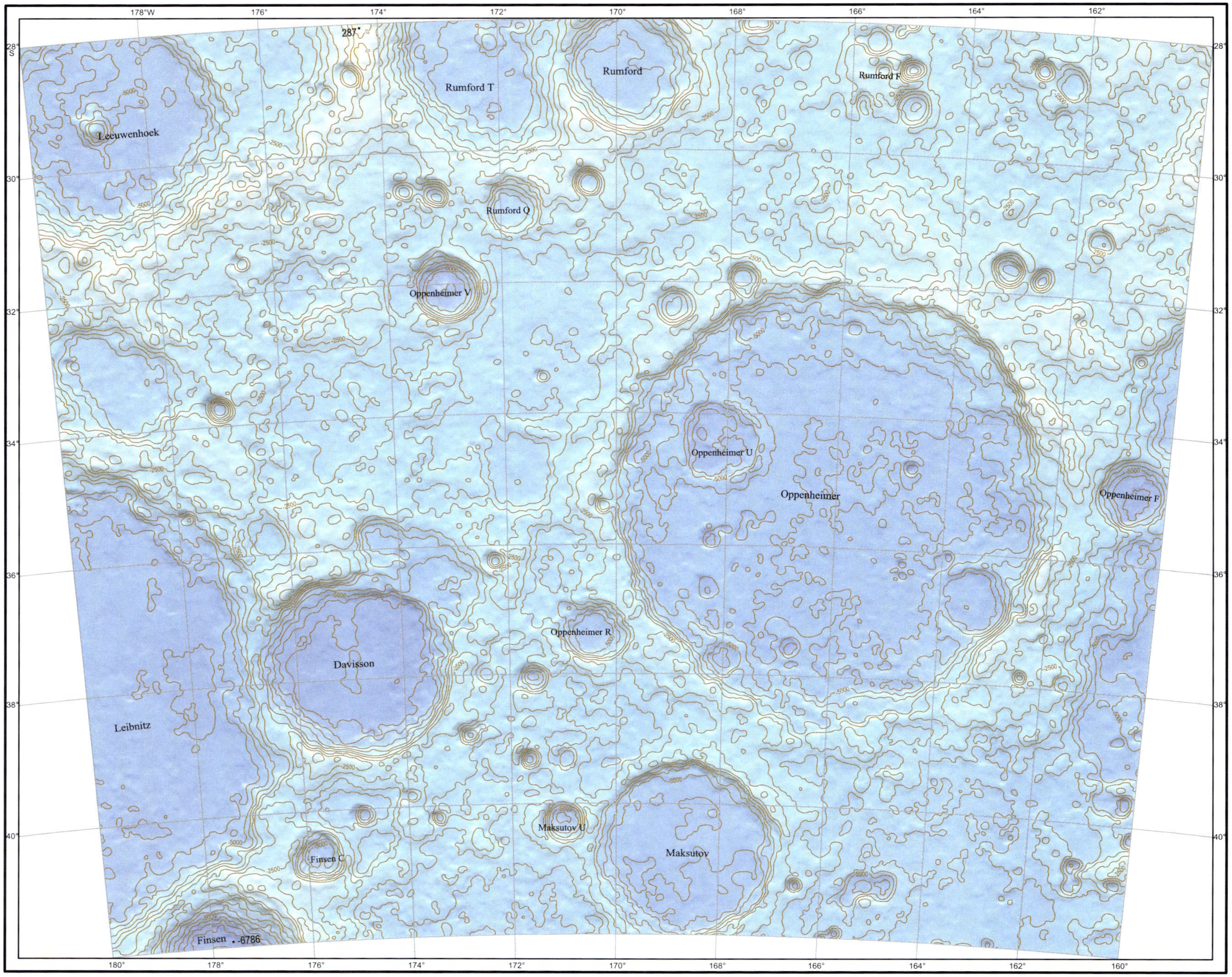

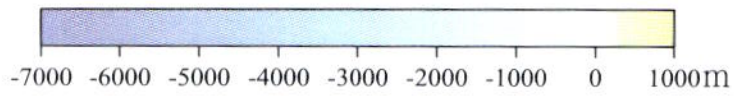

Image Code **J002**

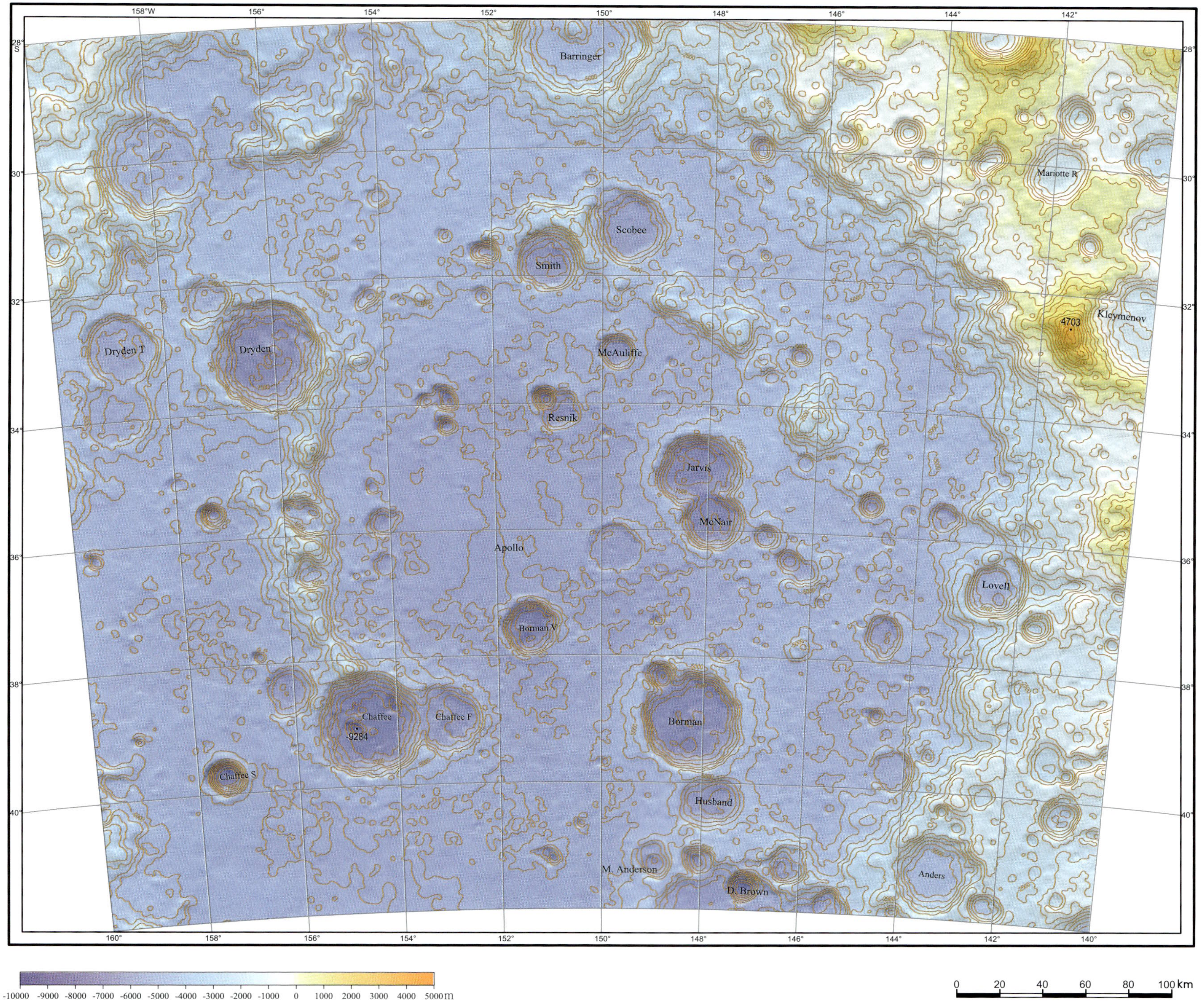

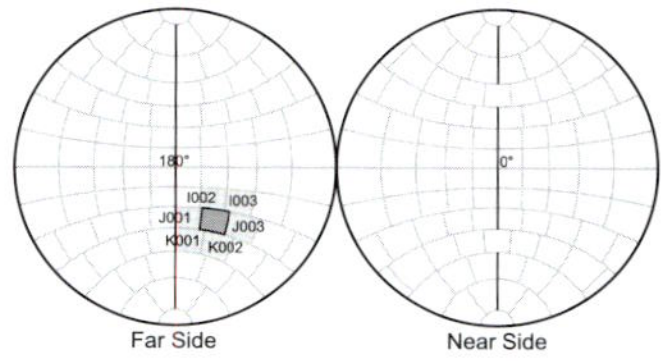

Image Code J003

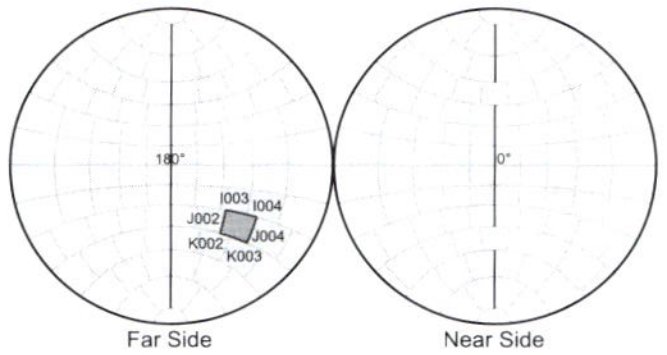

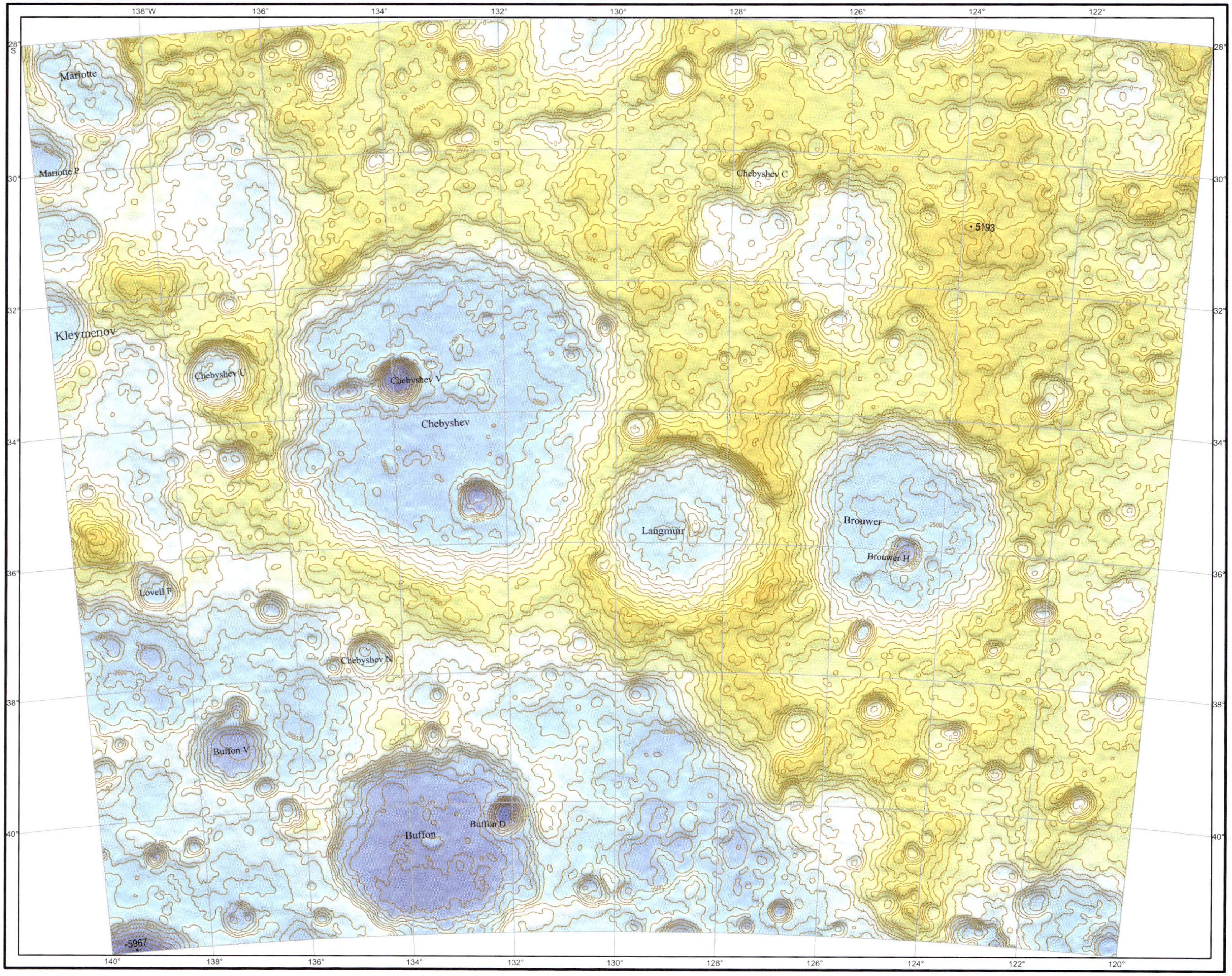

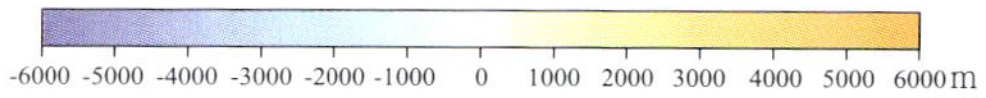

Image Code J004

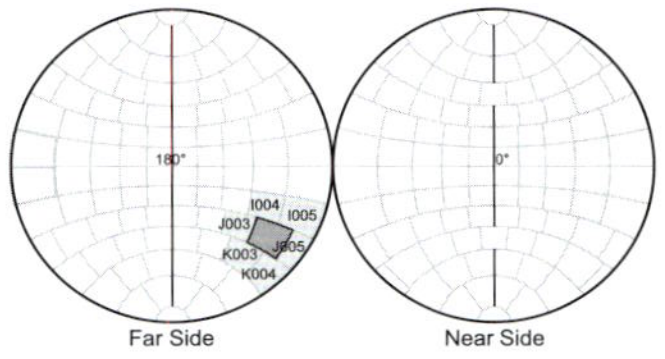

-4000 -3000 -2000 -1000 0 1000 2000 3000 4000 5000m

0 20 40 60 80 100 km

Image Code J005

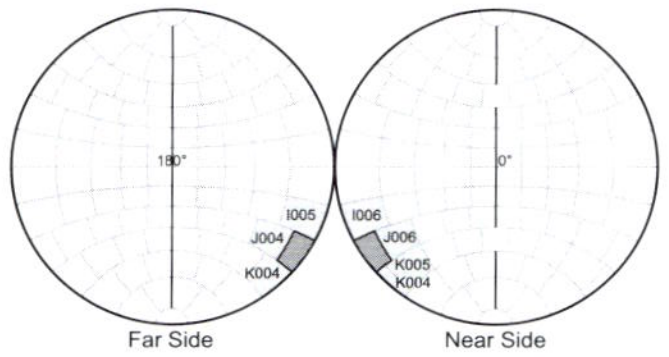

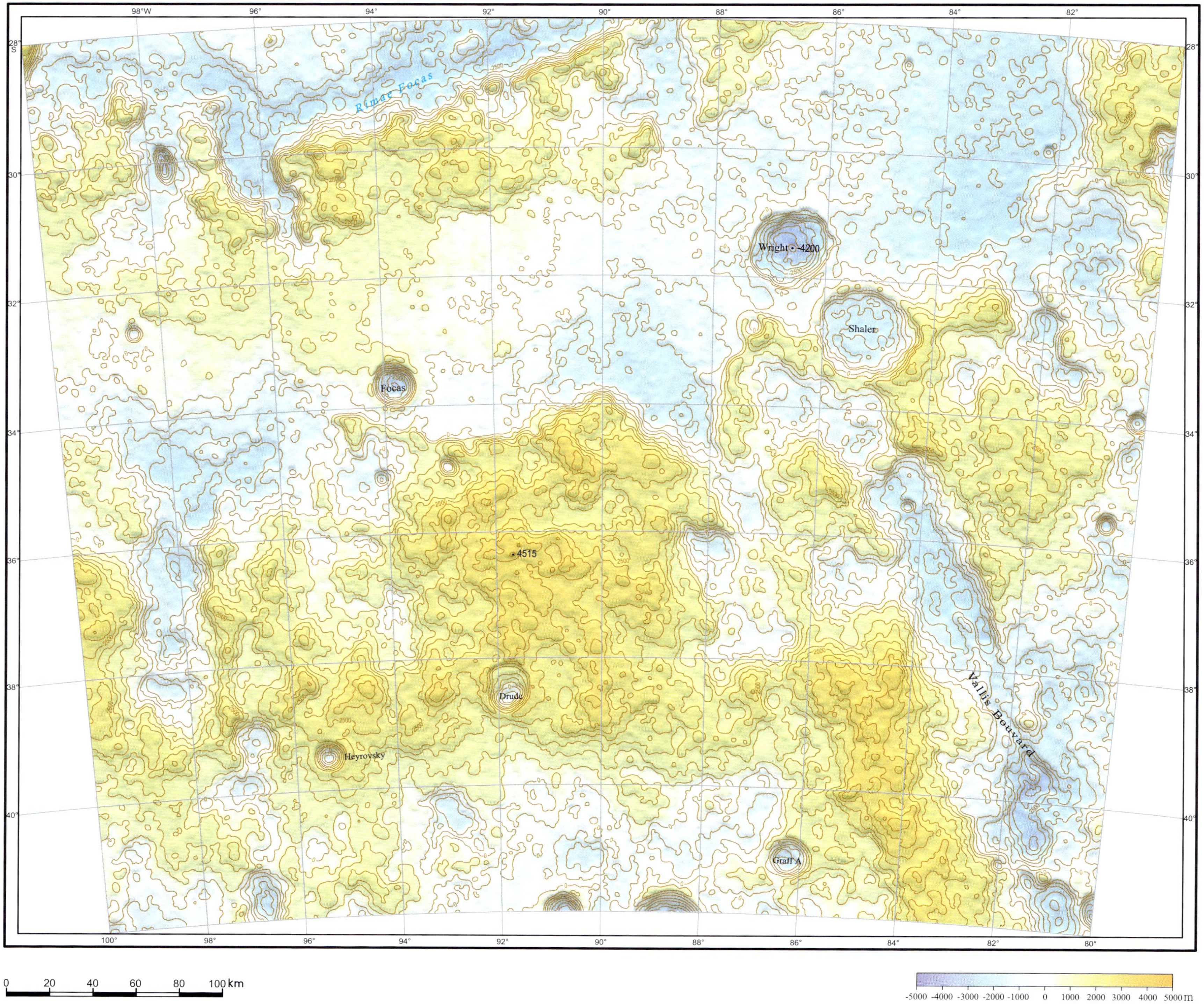

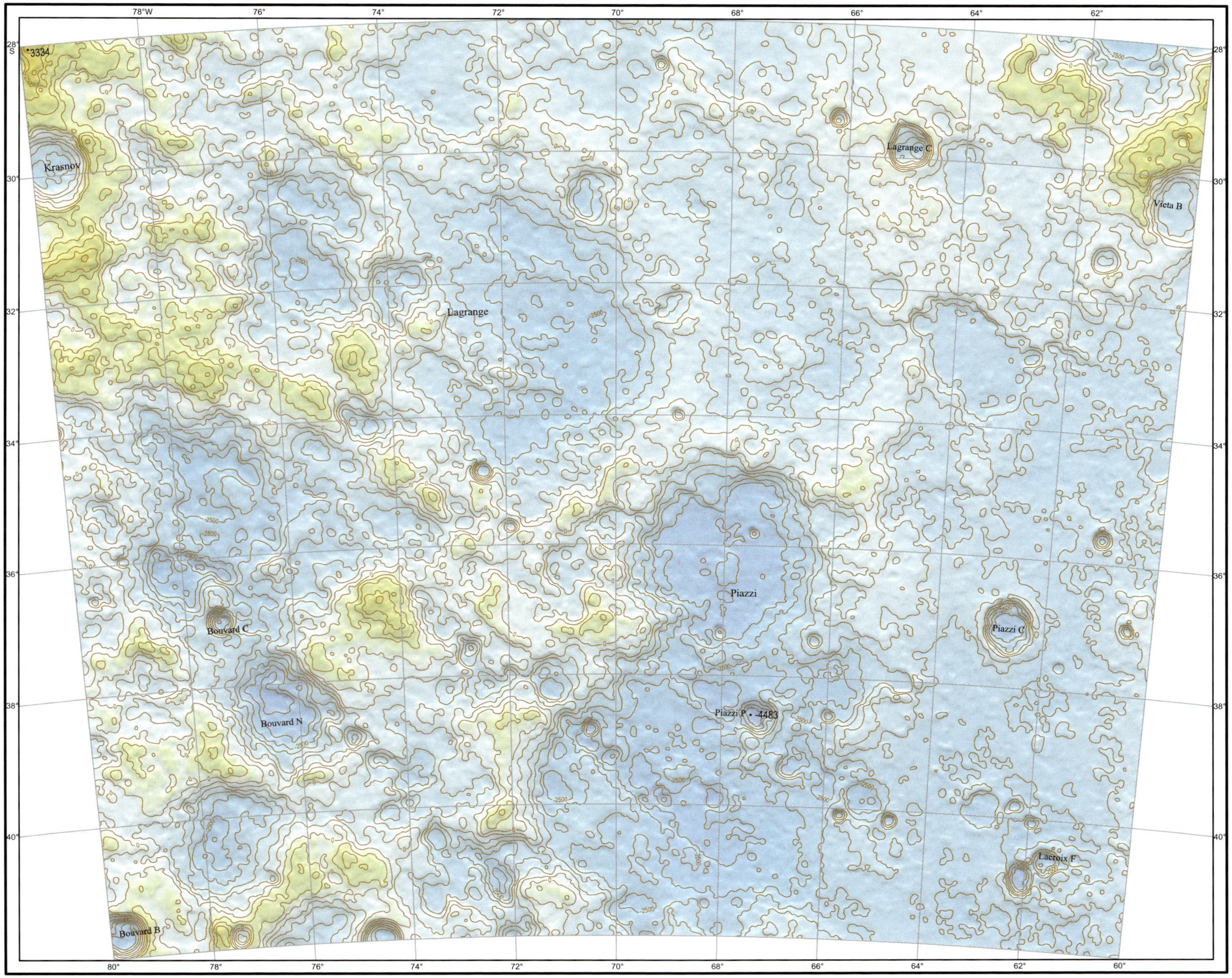

Image Code J006

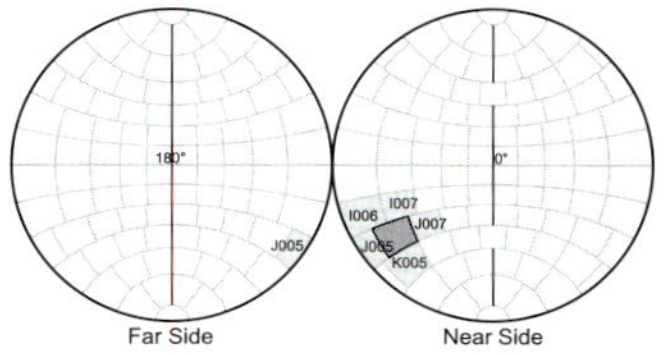

-5000 -4000 -3000 -2000 -1000 0 1000 2000 3000 4000m

0 20 40 60 80 100 km

Image Code K009

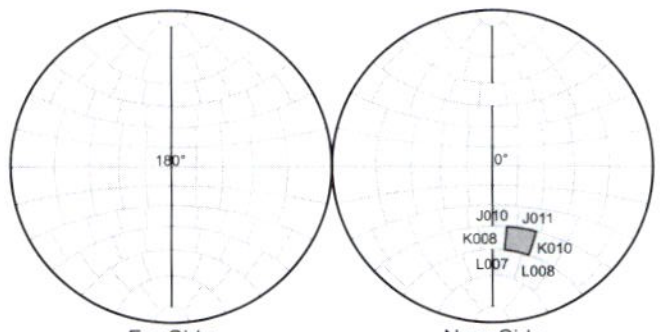

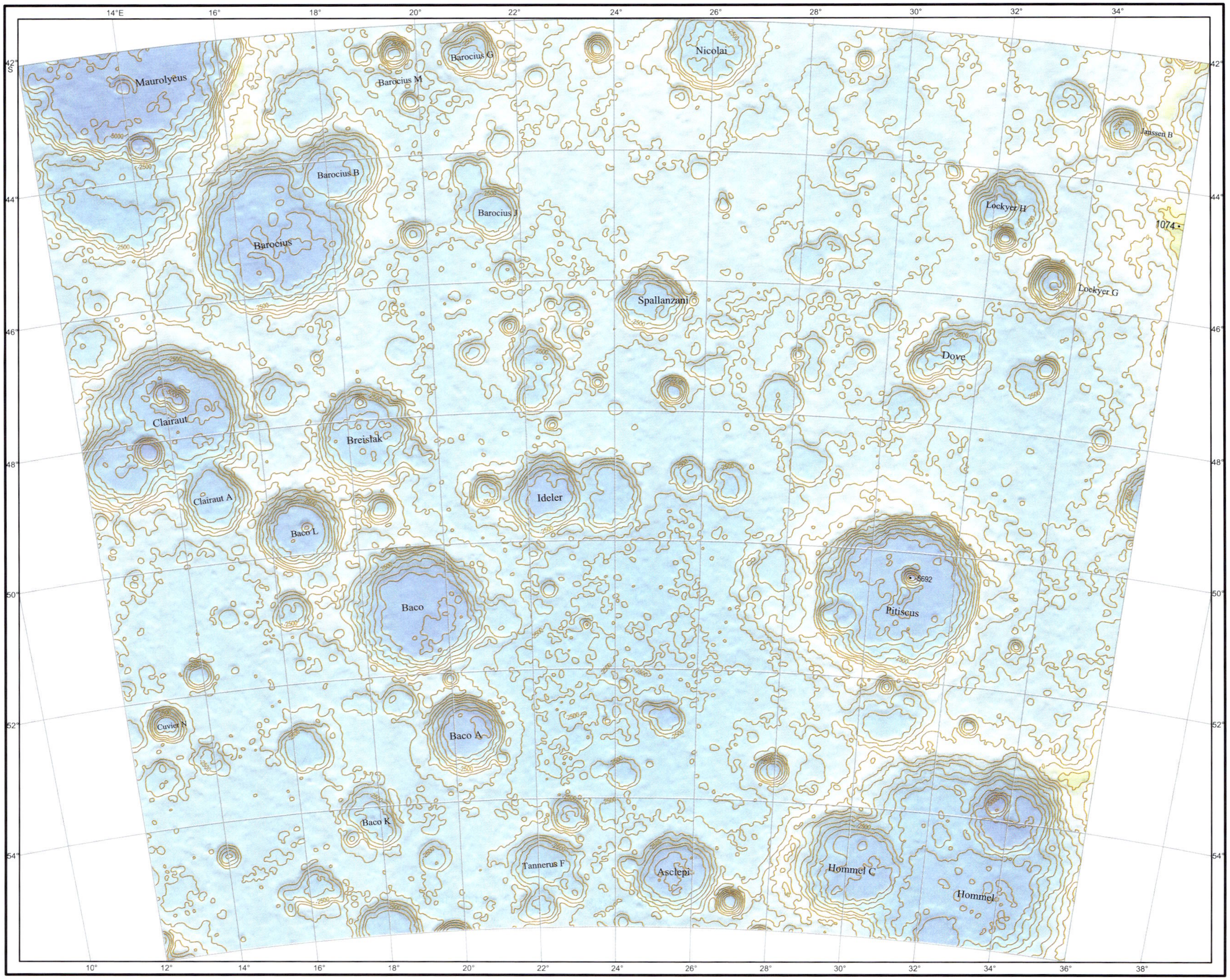

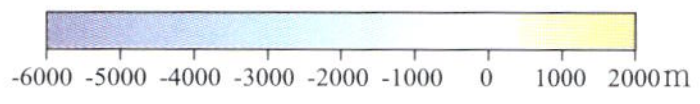

Image Code **K010**

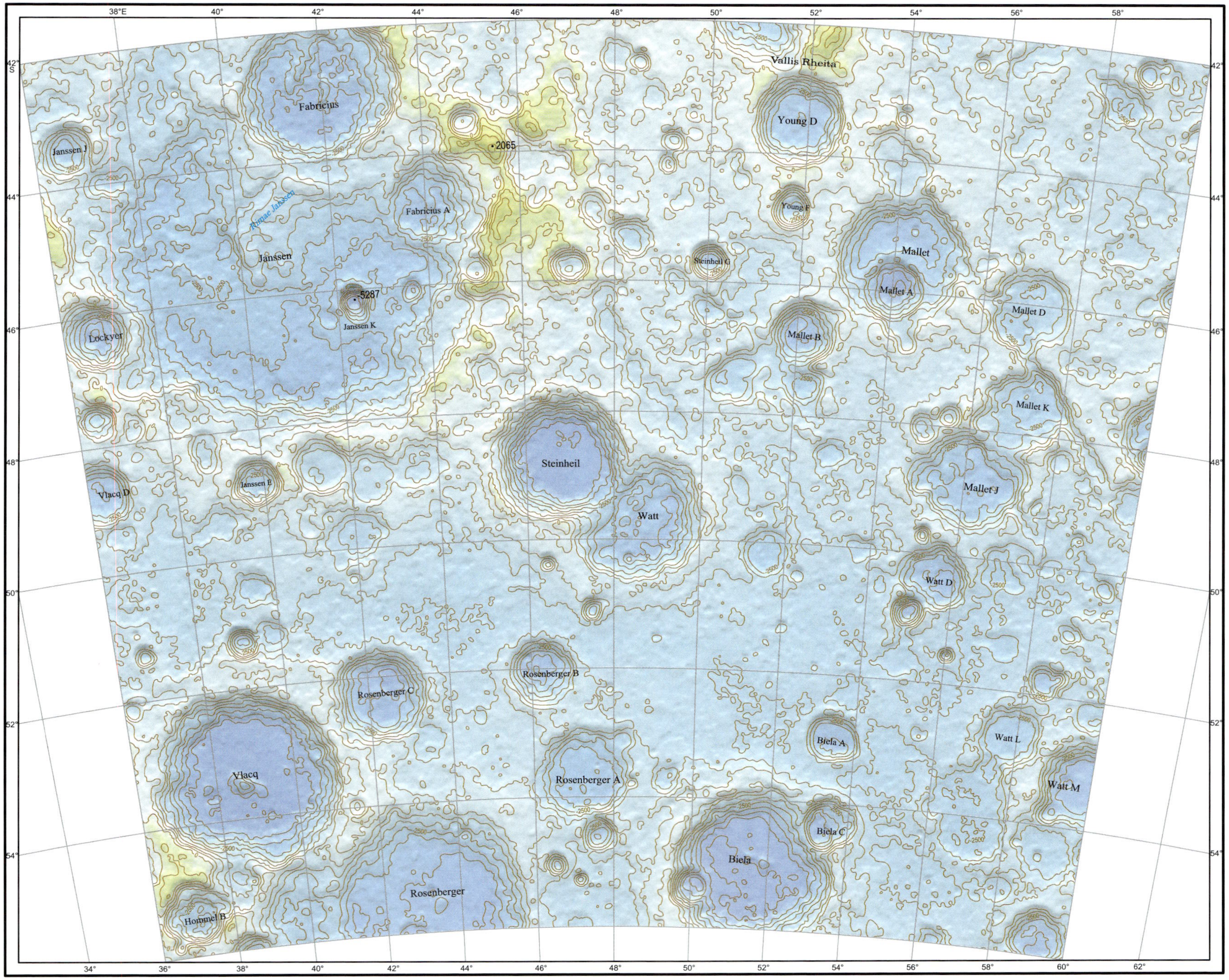

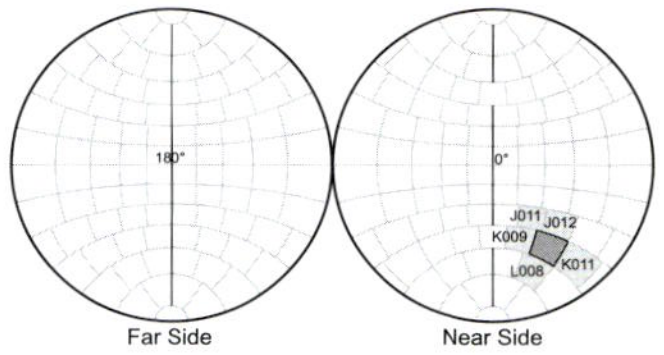

Image Code K011

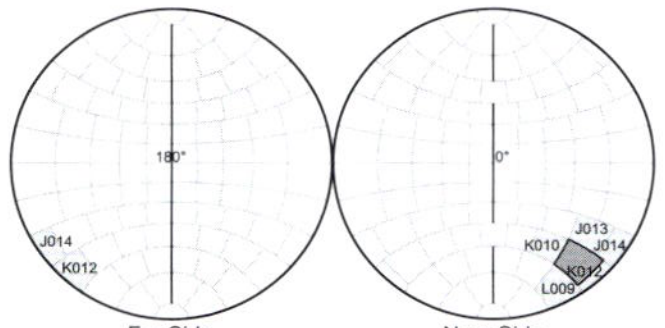

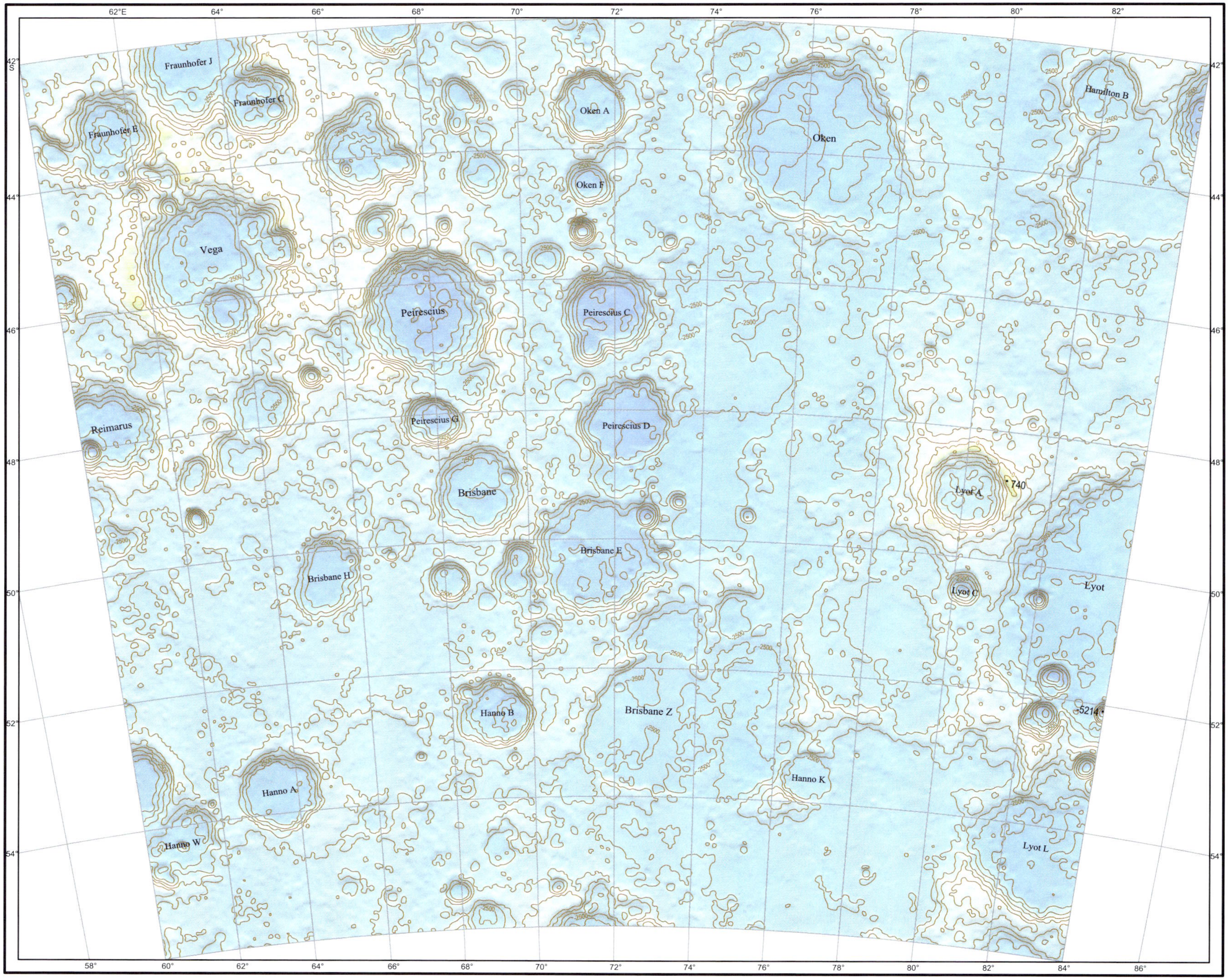

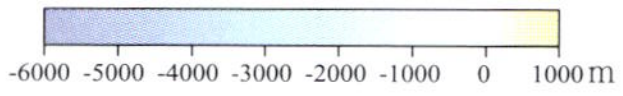

Image Code K012

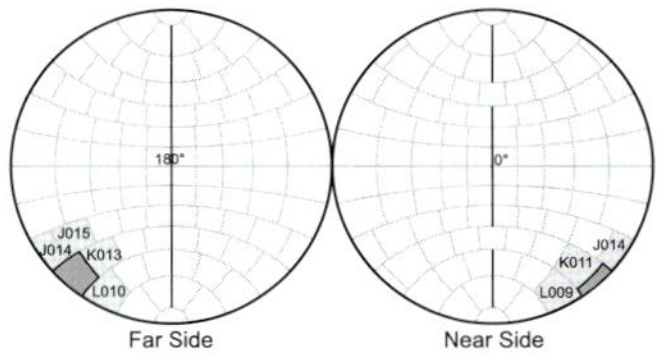

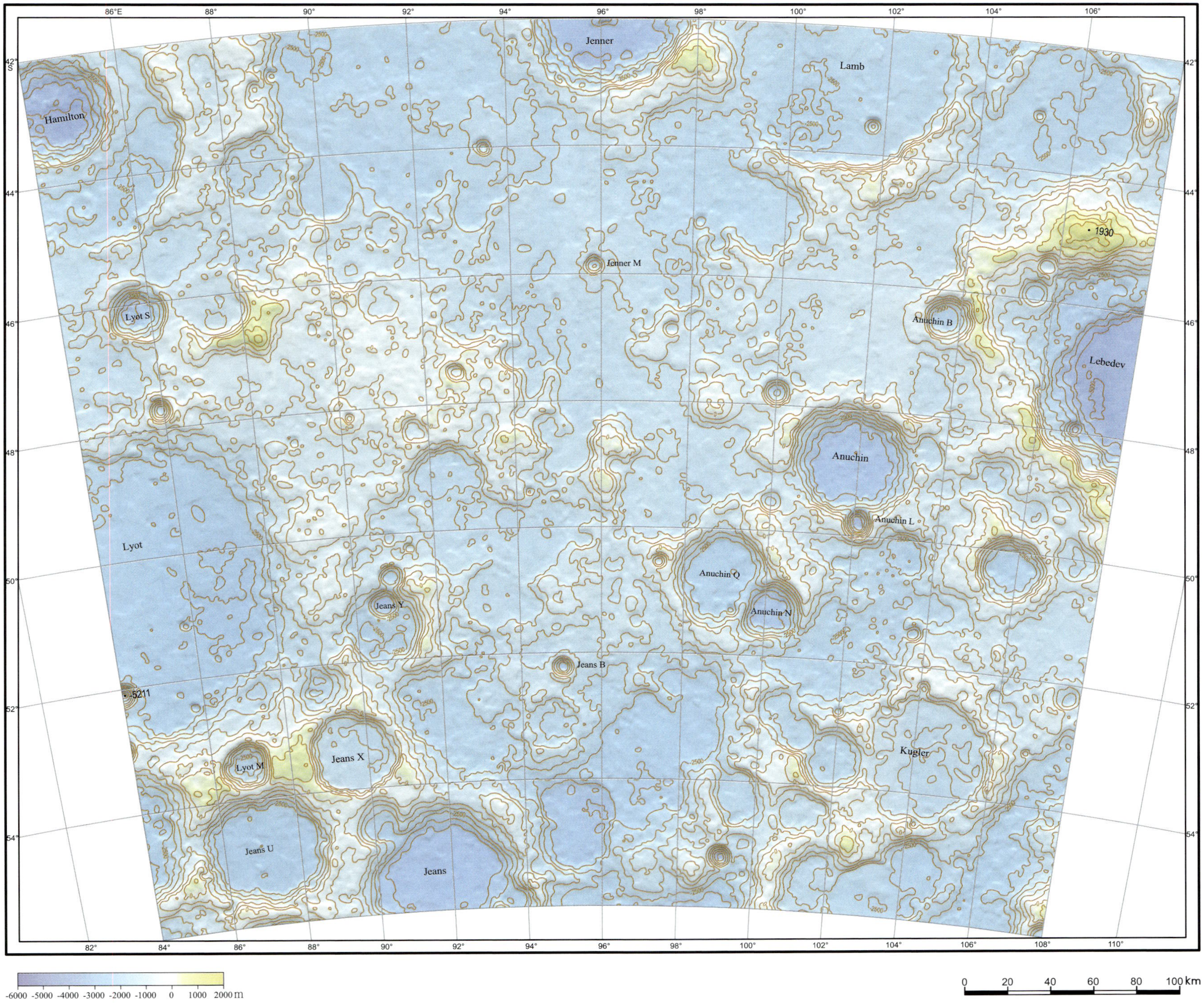

Image Code K013

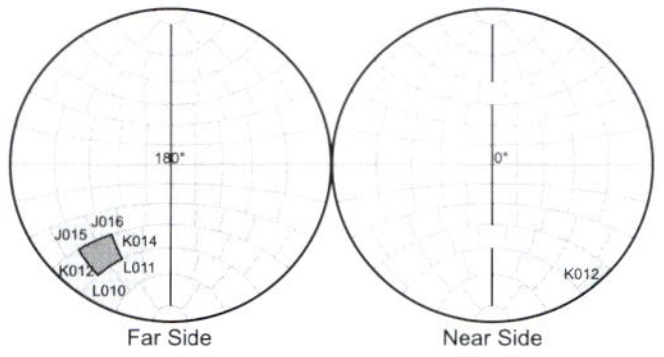

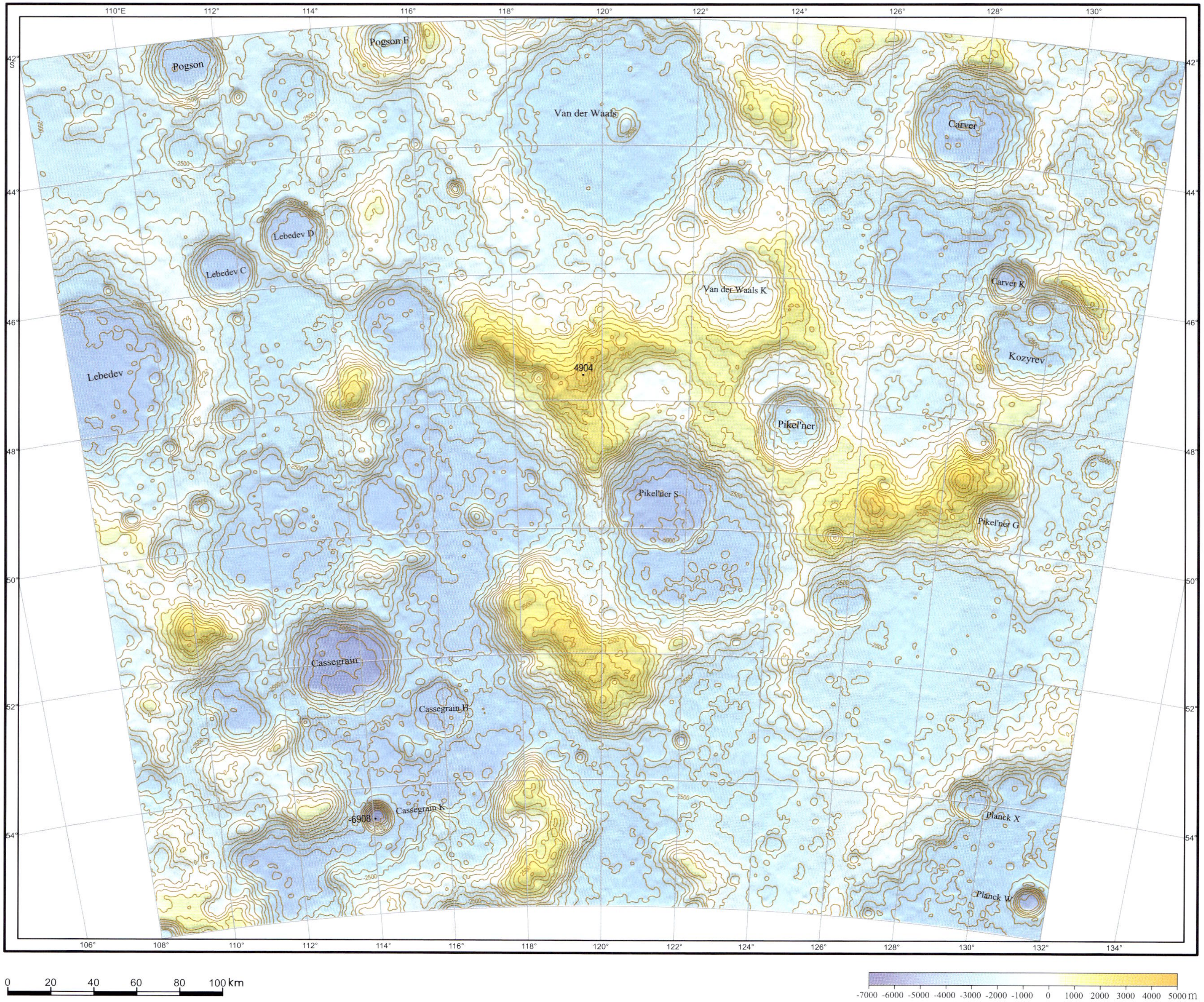

Image Code K014

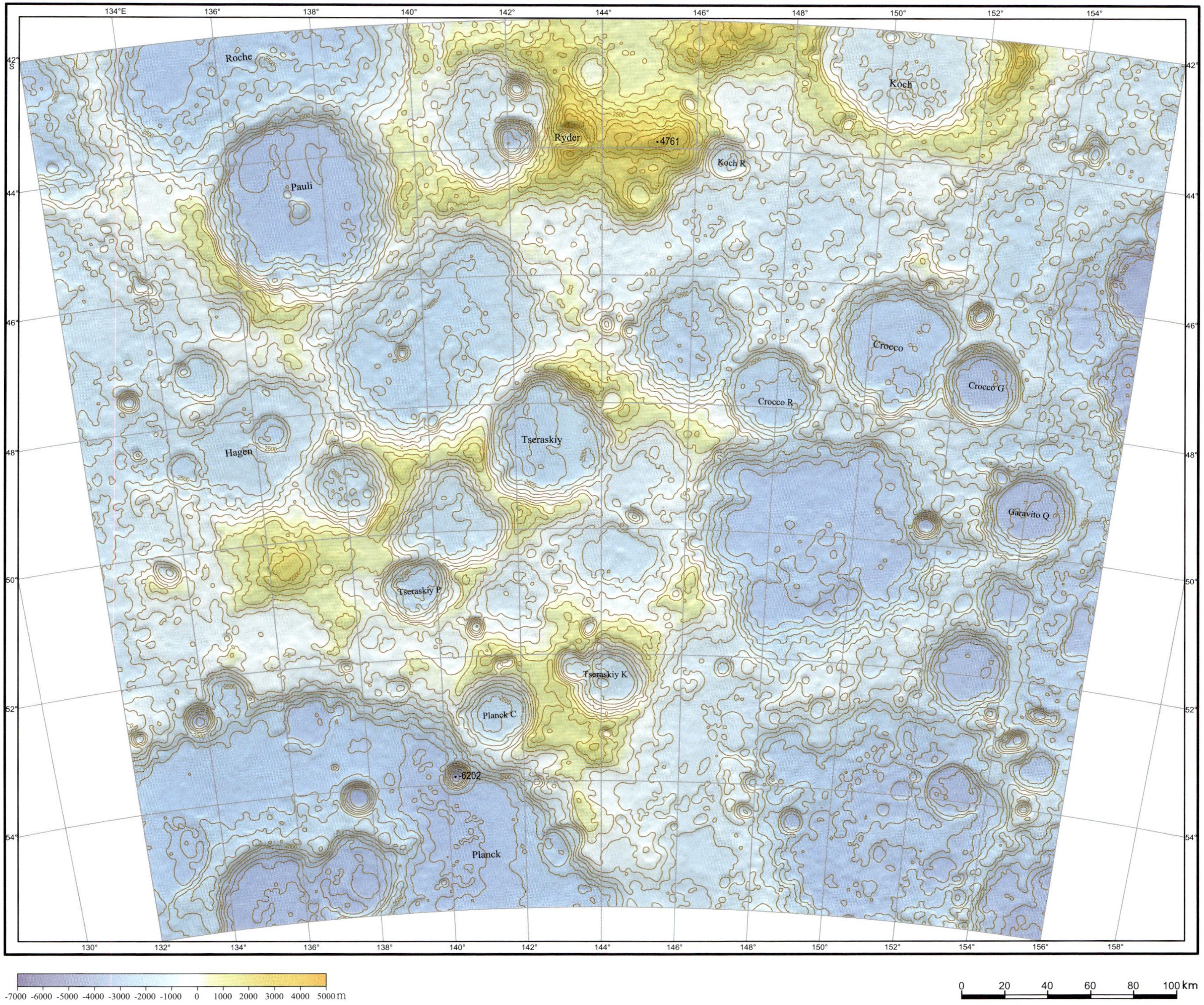

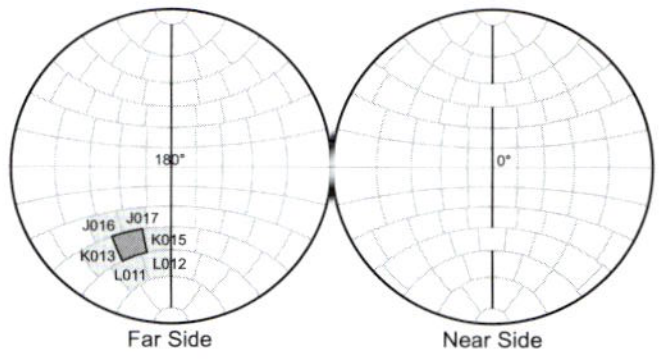

Image Code **K015**

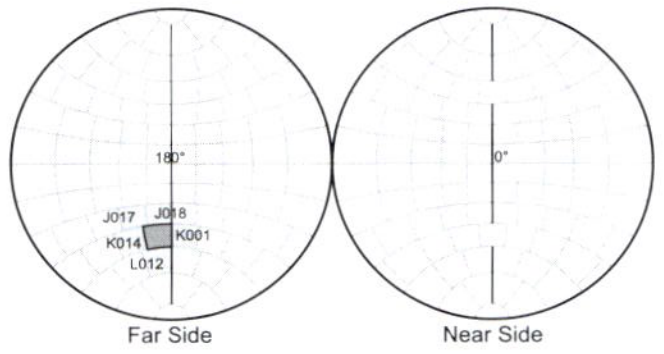

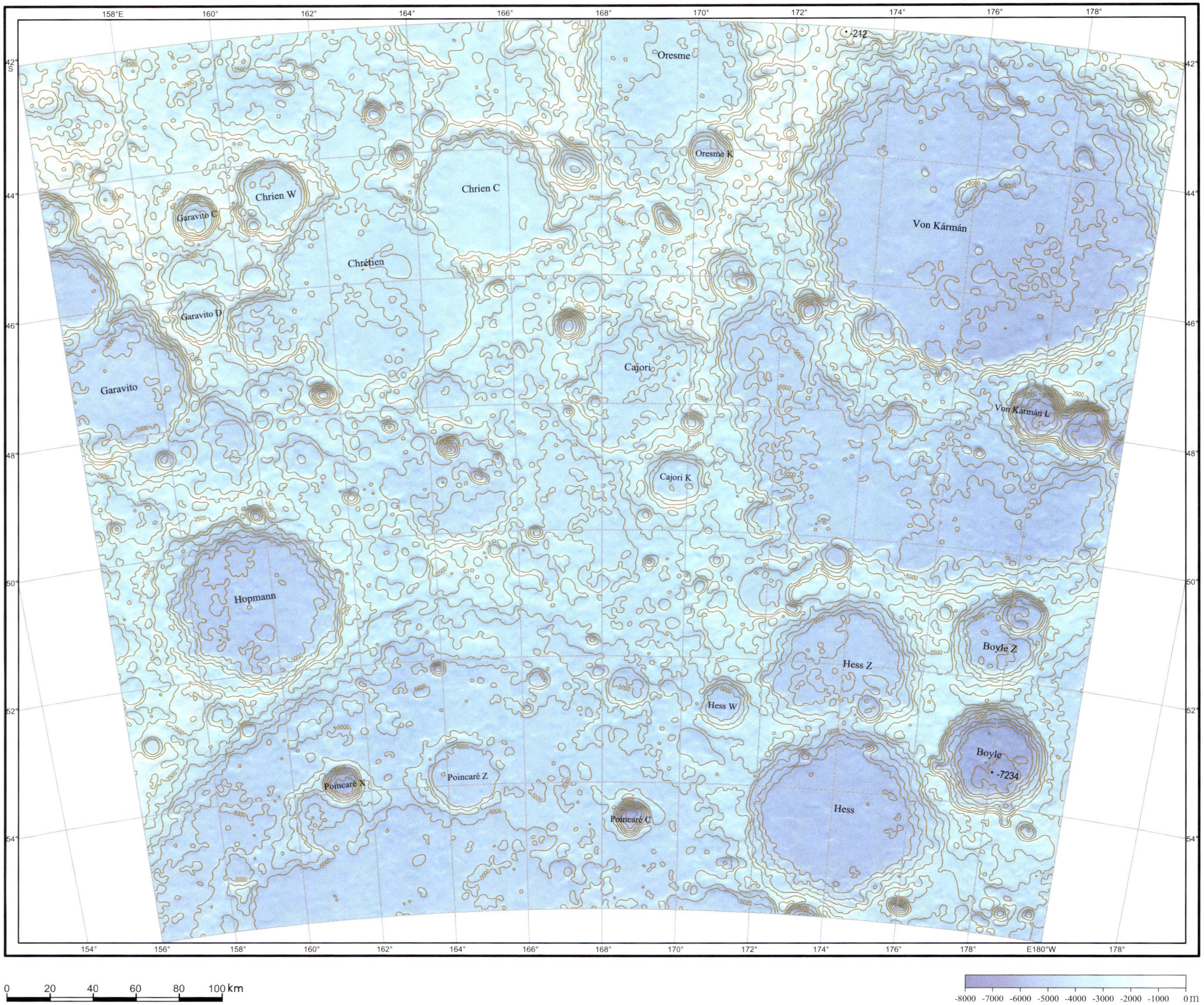

Image Code L001

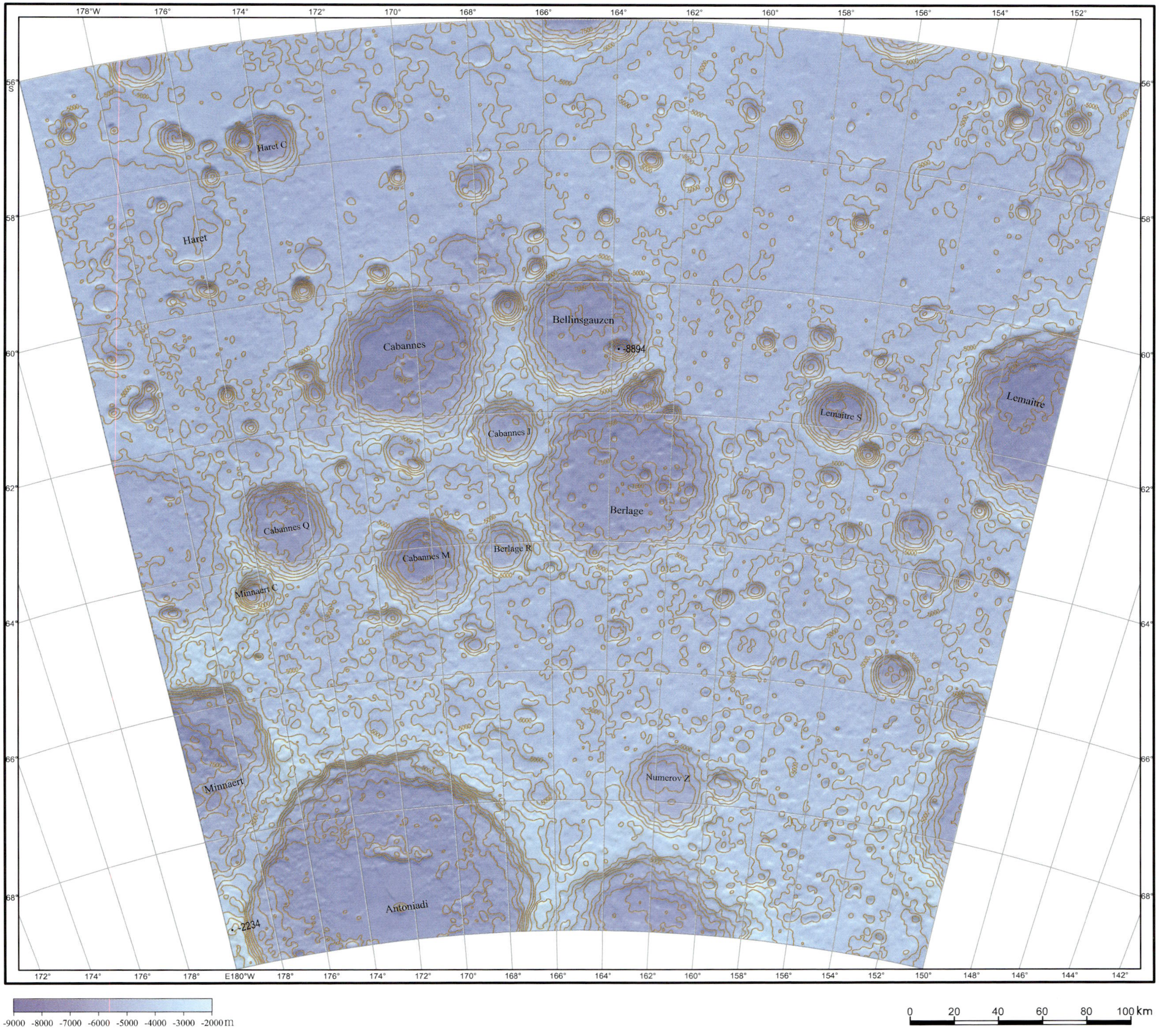

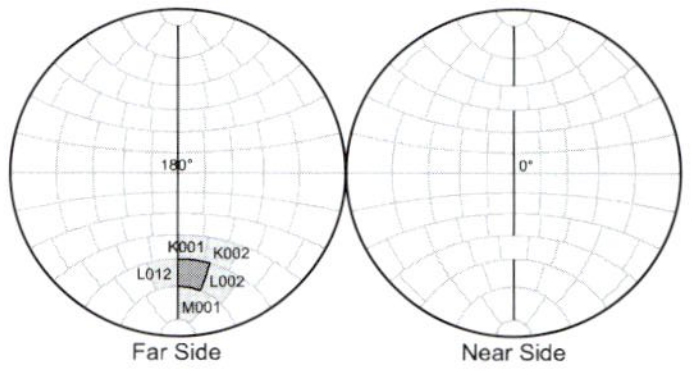

Image Code **L002**

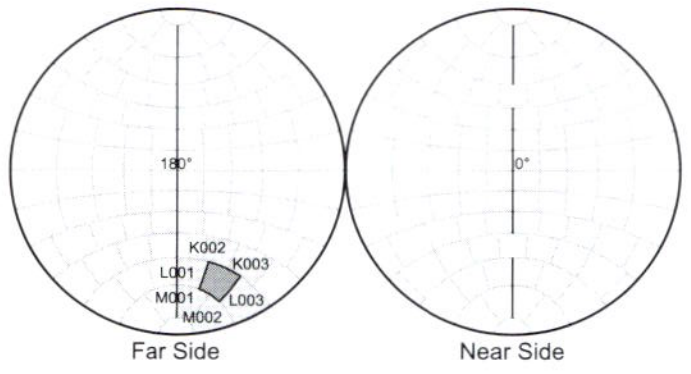

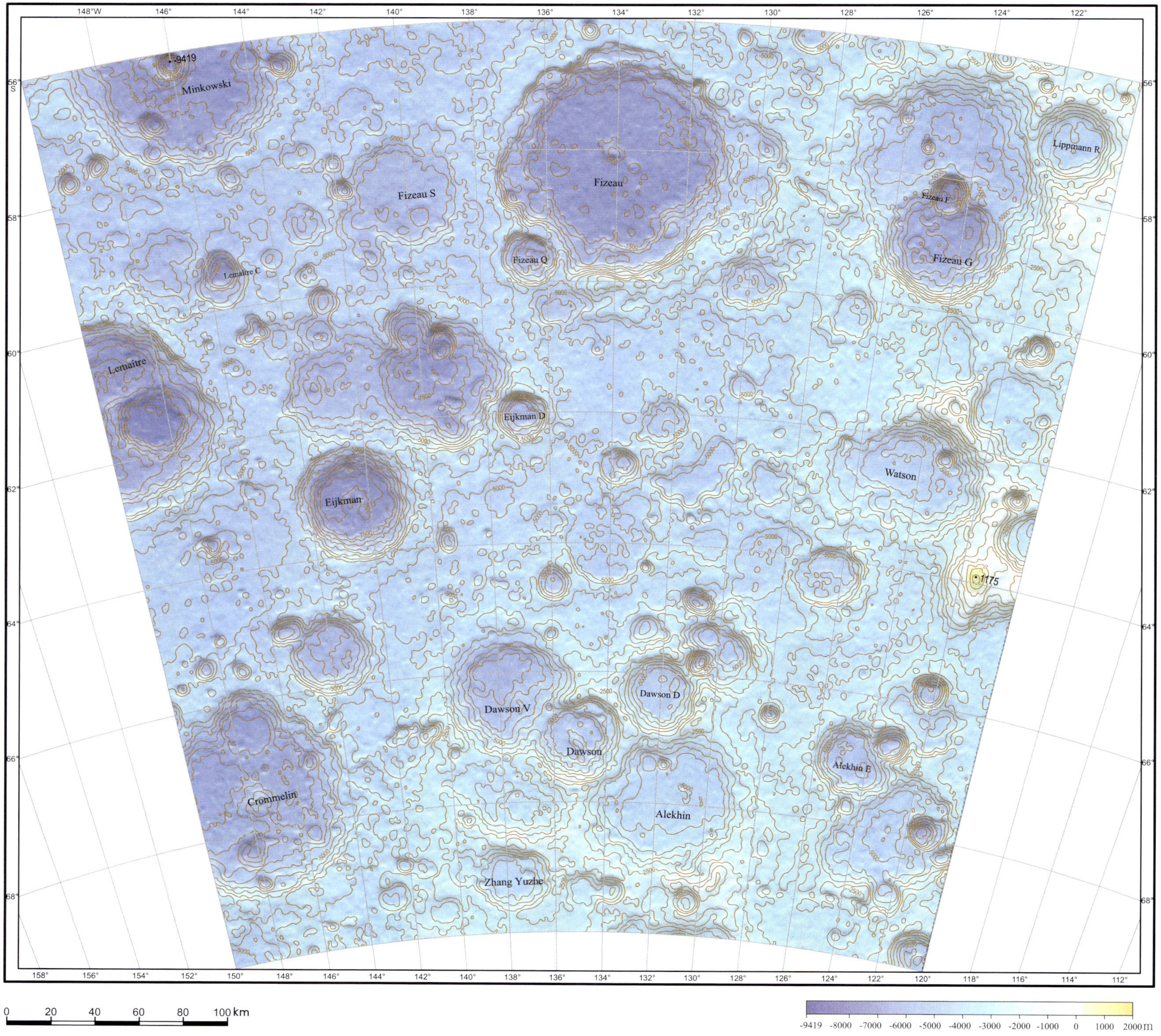

Image Code L003

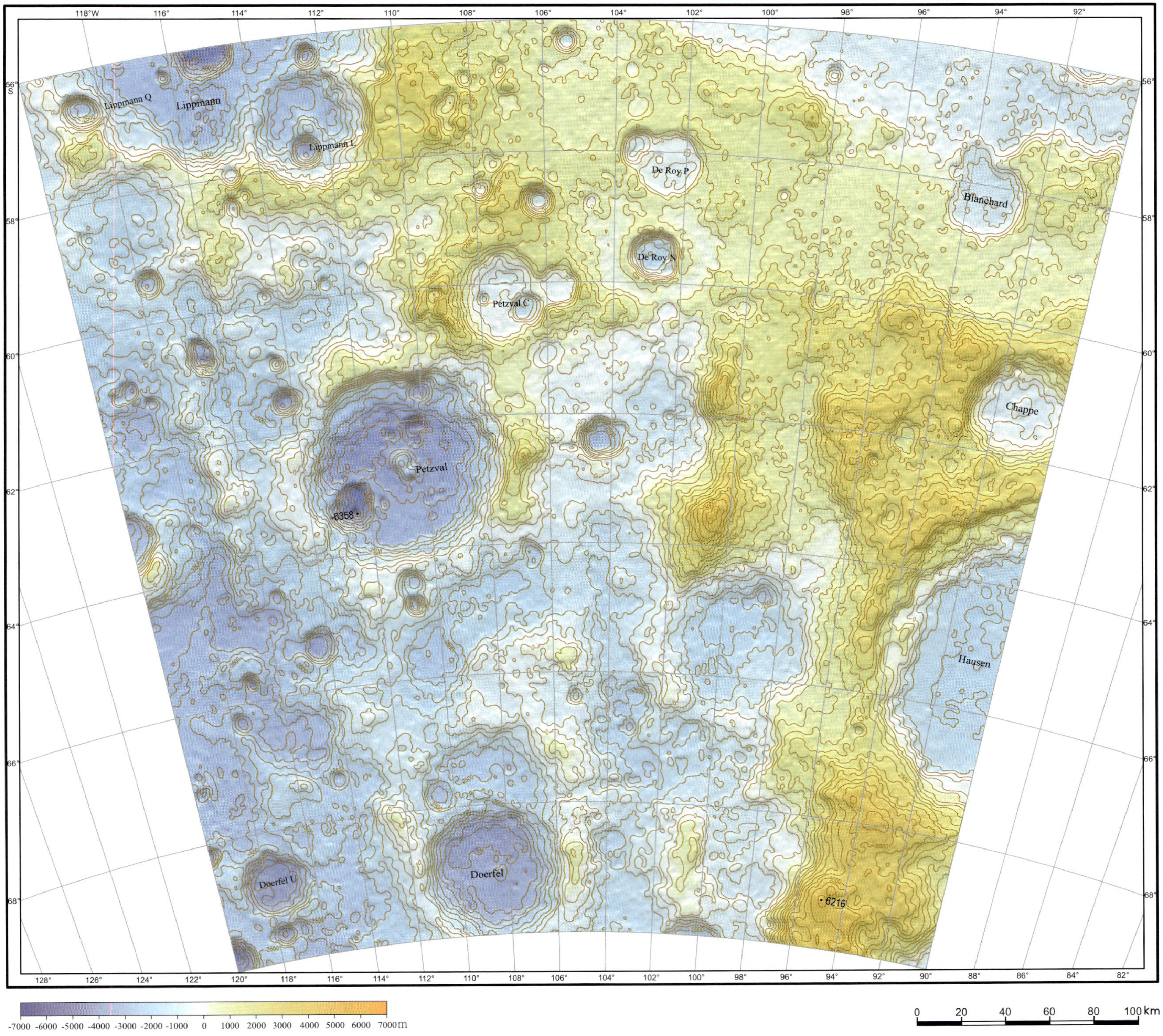

Image Code L004

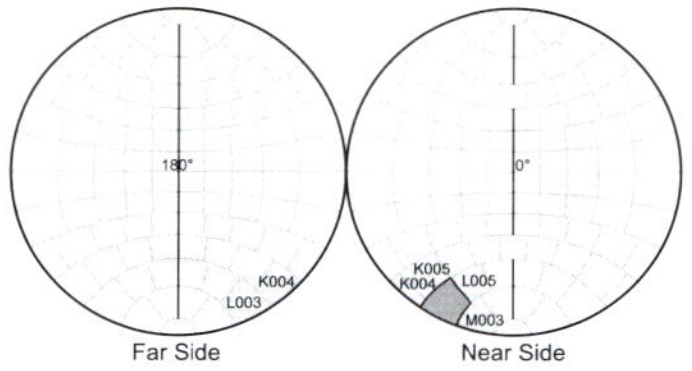

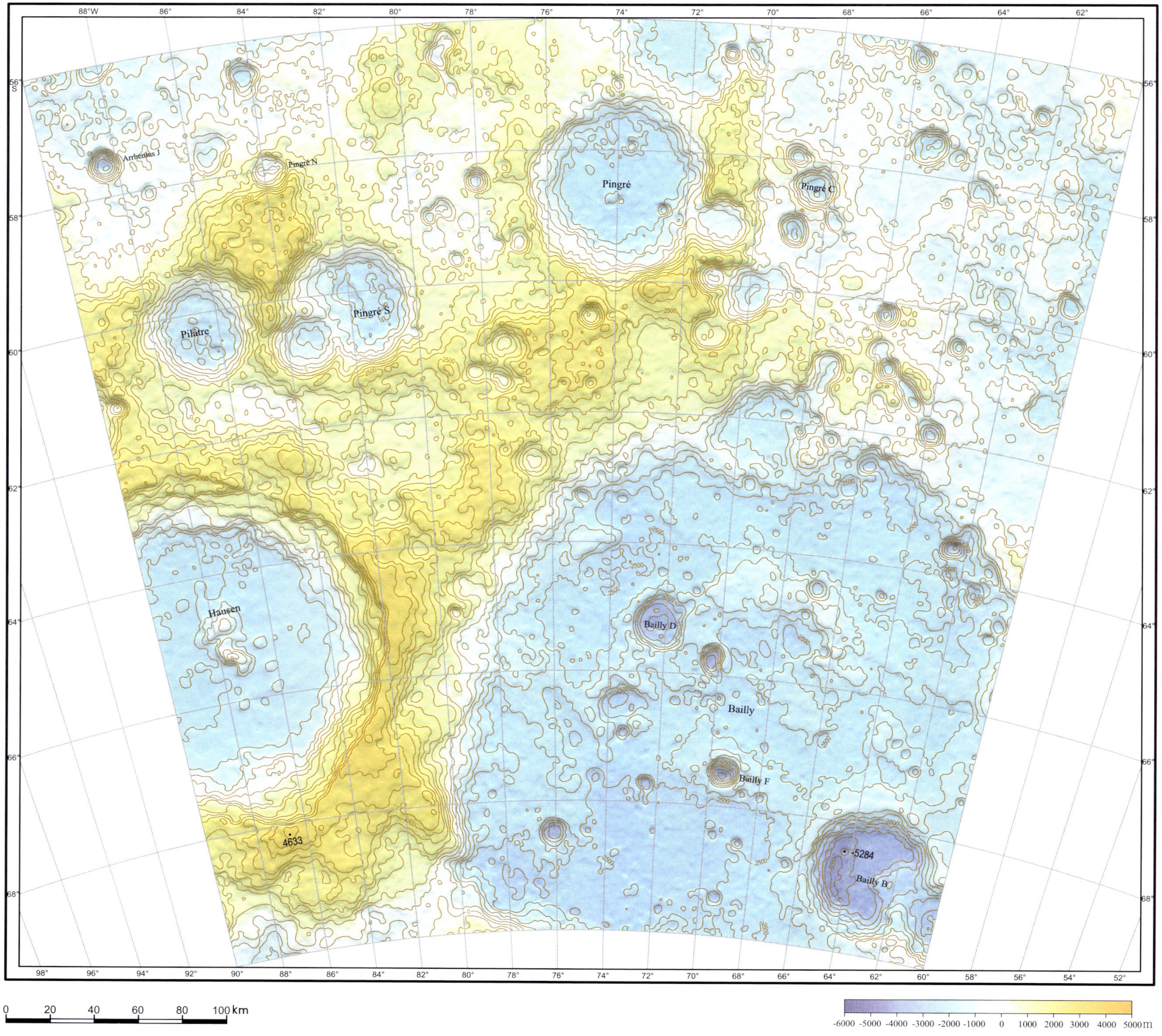

Image Code L005

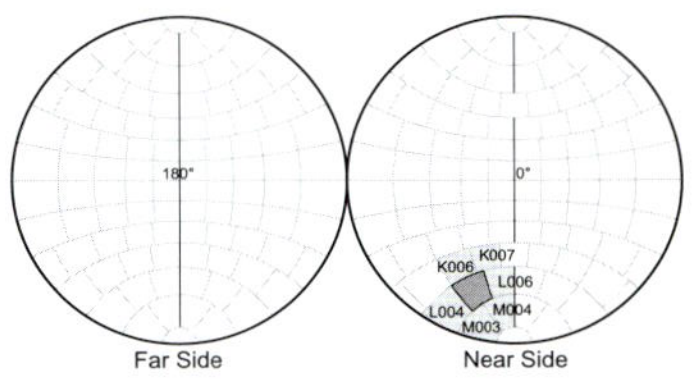

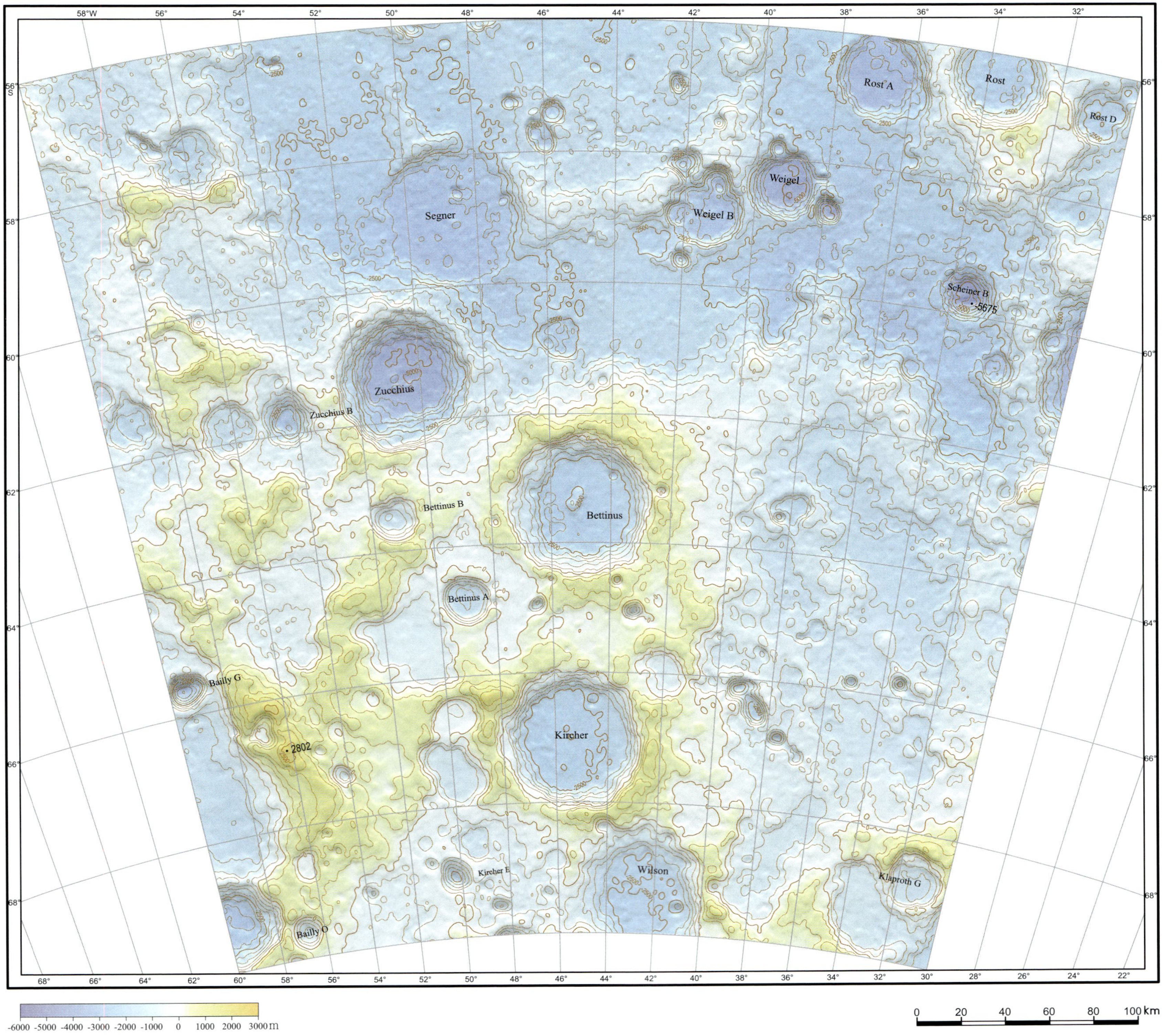

Image Code L006

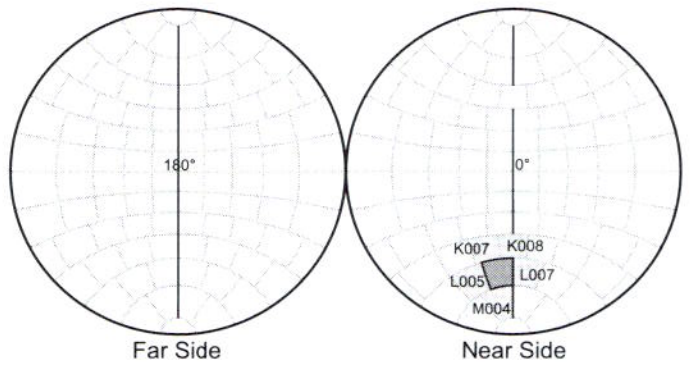

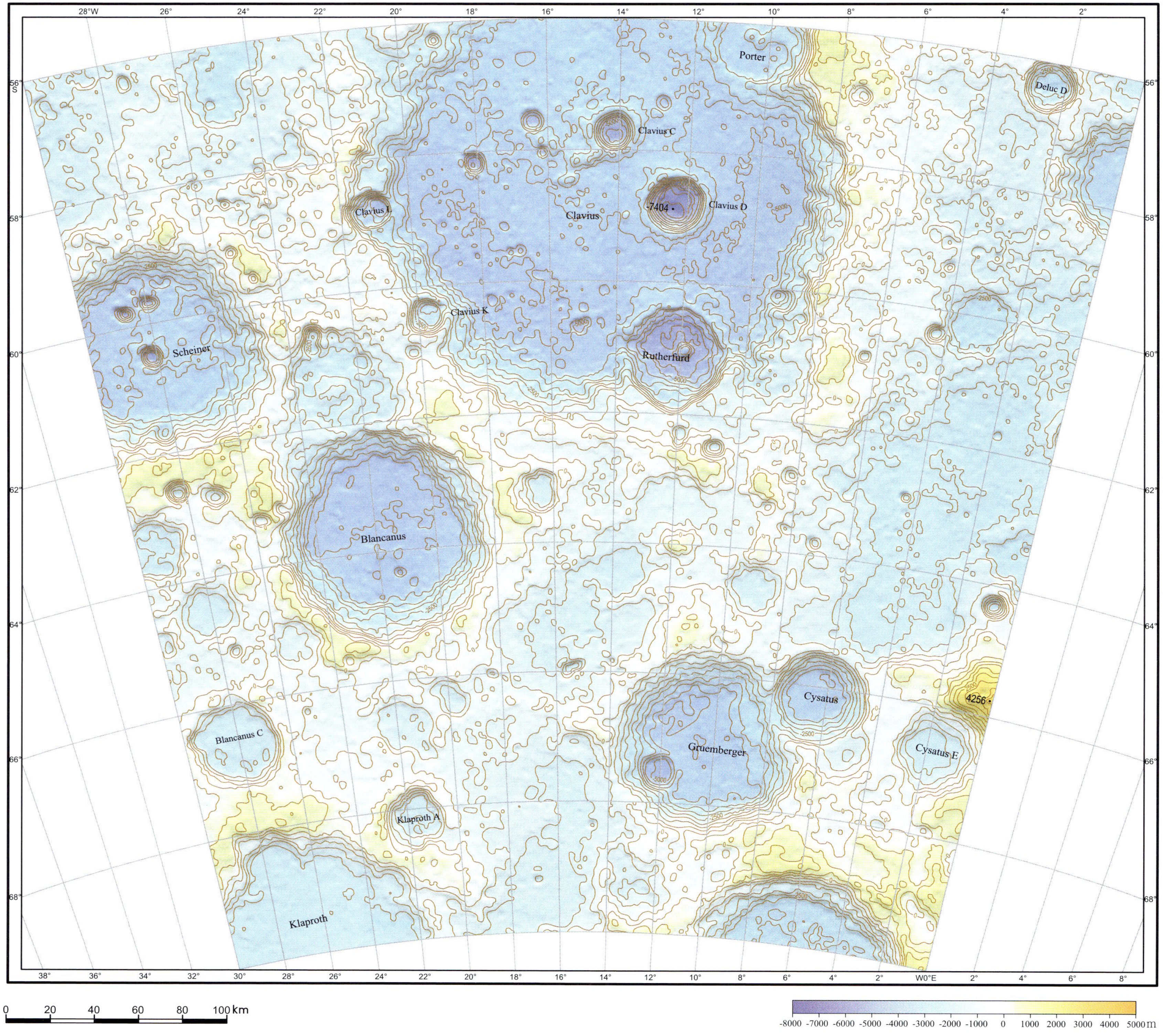

Image Code L007

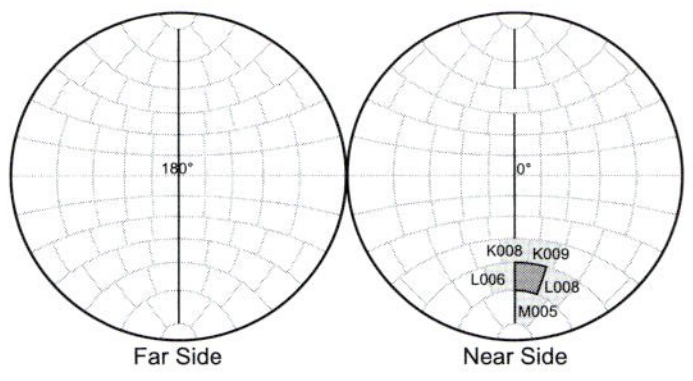

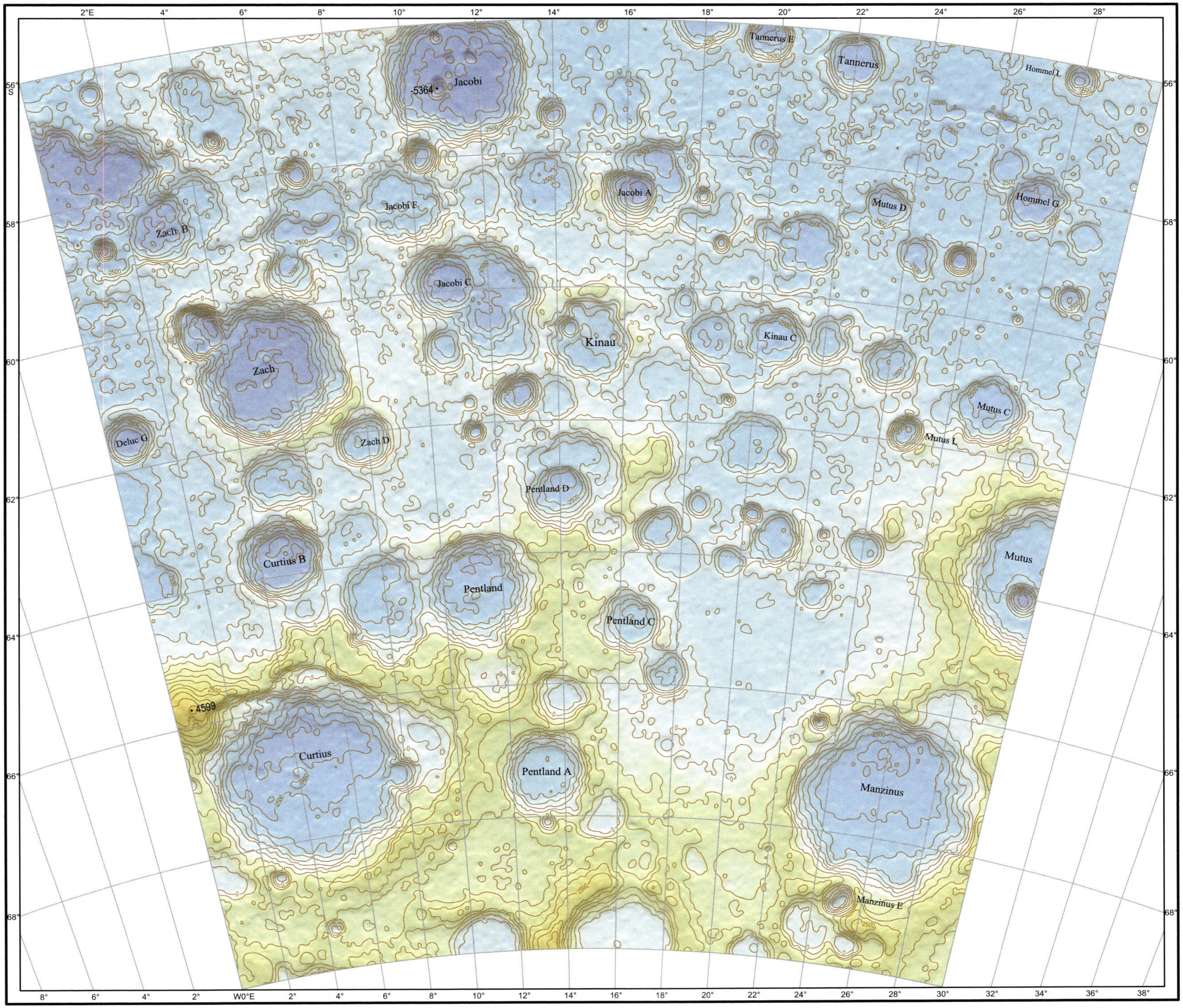

-6000 -5000 -4000 -3000 -2000 -1000 0 1000 2000 3000 4000 5000 m

0 20 40 60 80 100 km

Image Code L008

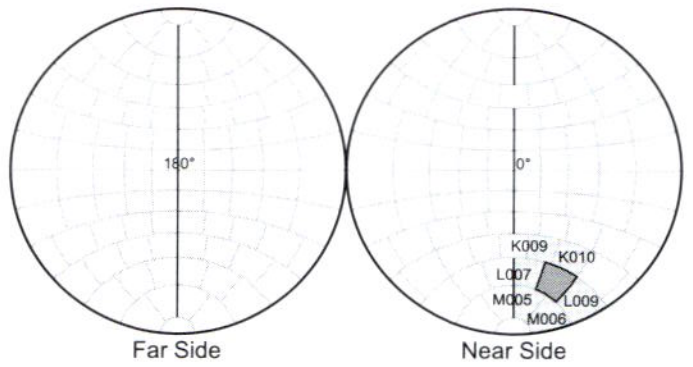

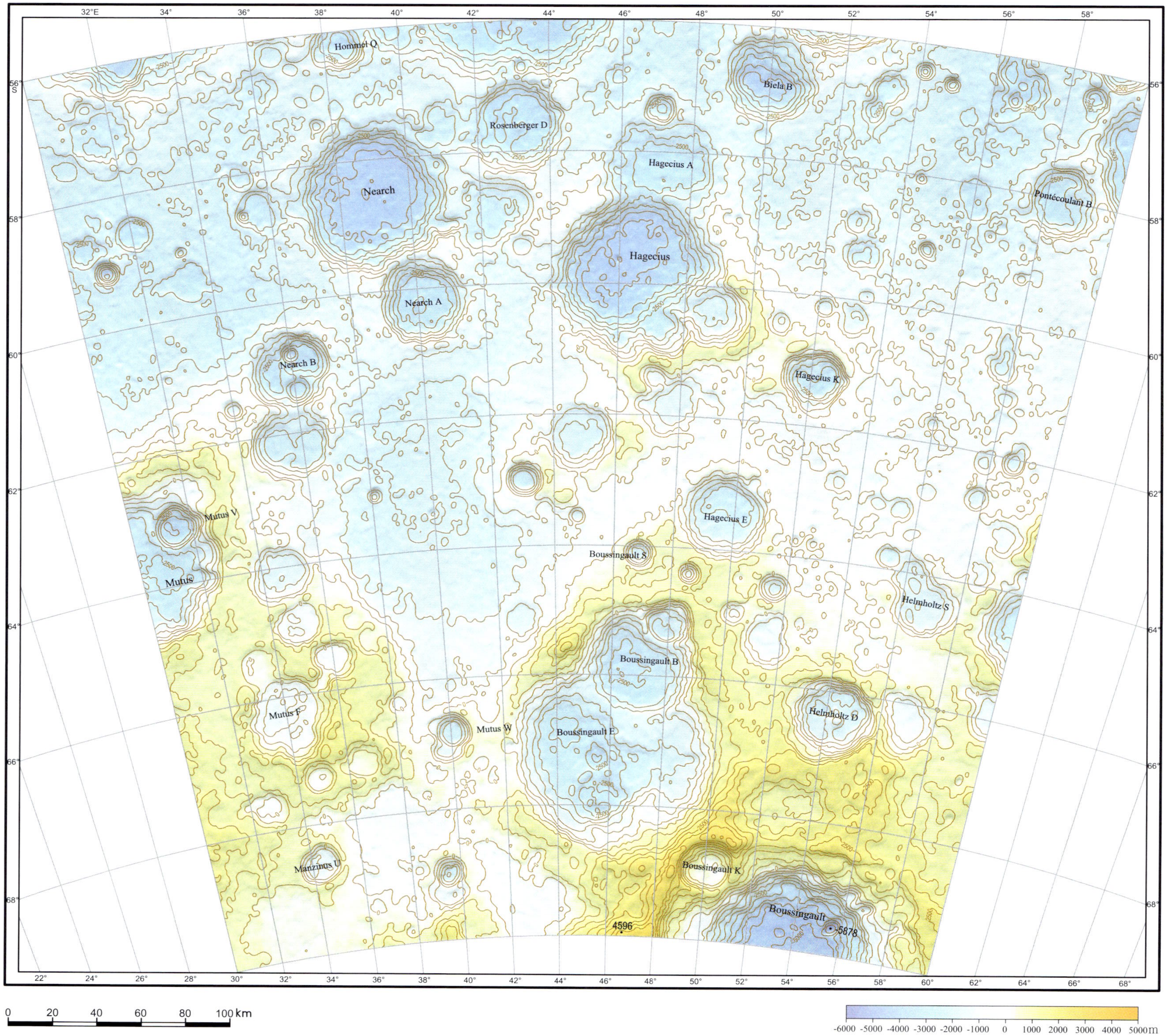

Image Code L009

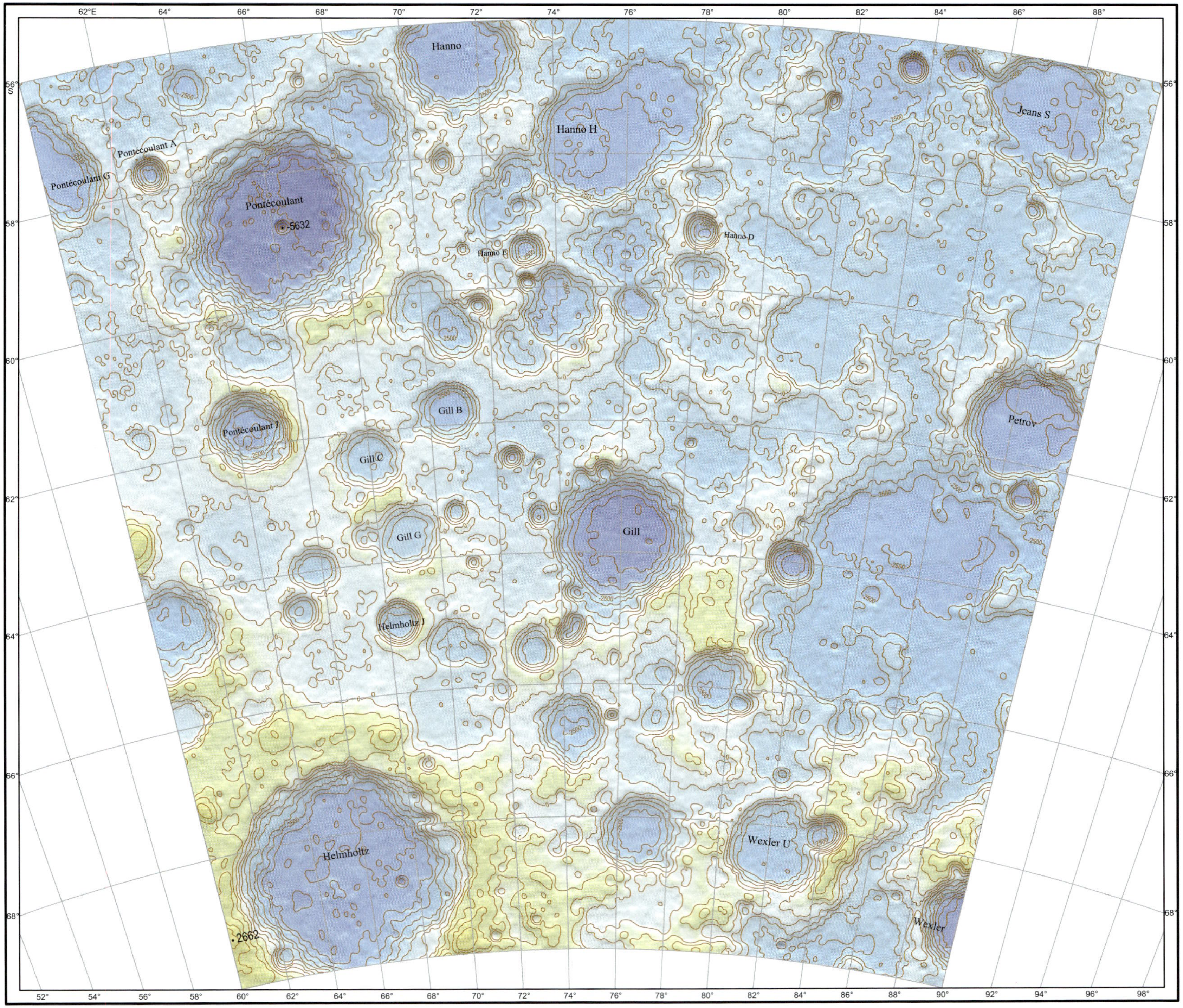

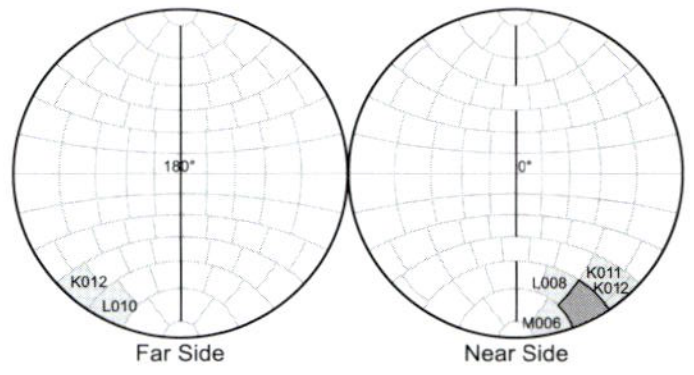

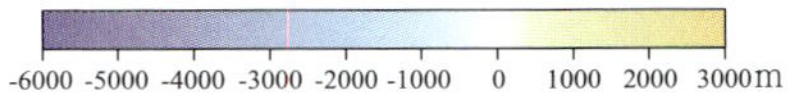

Image Code L010

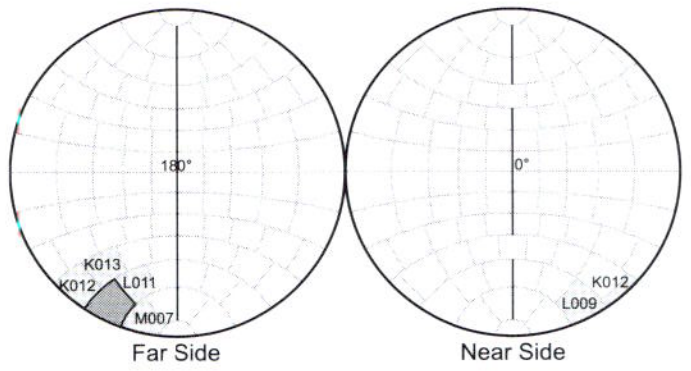

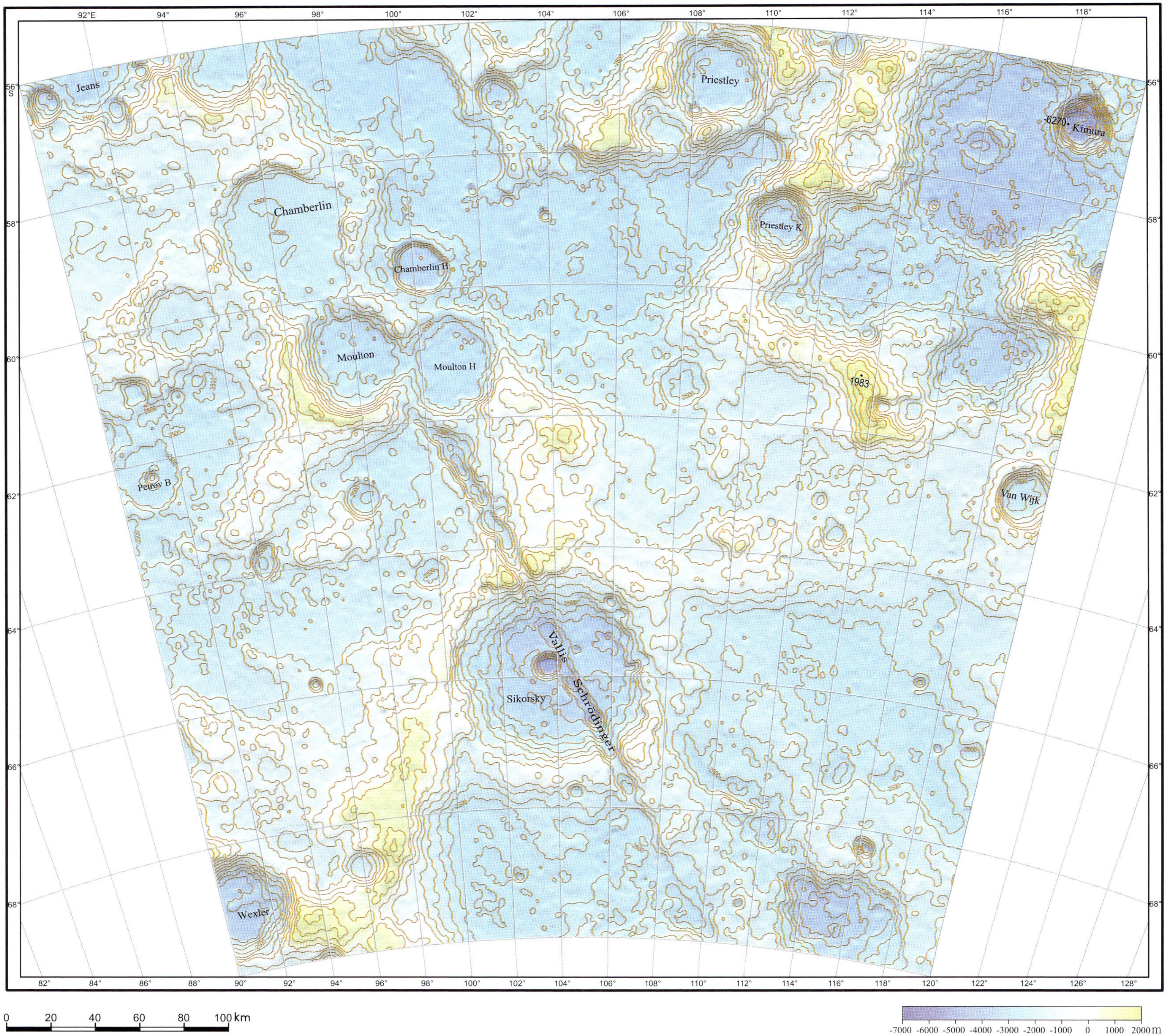

Image Code L011

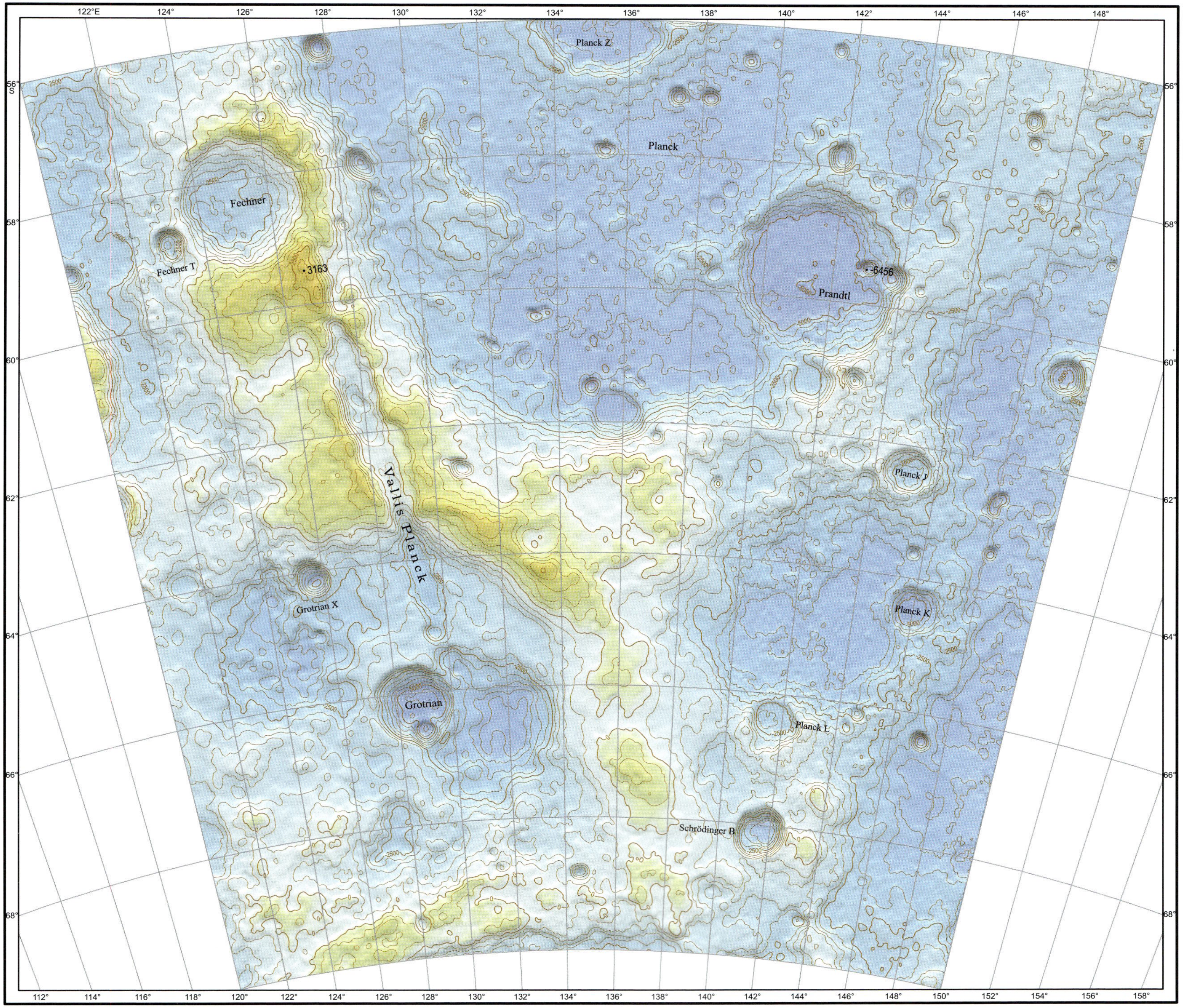

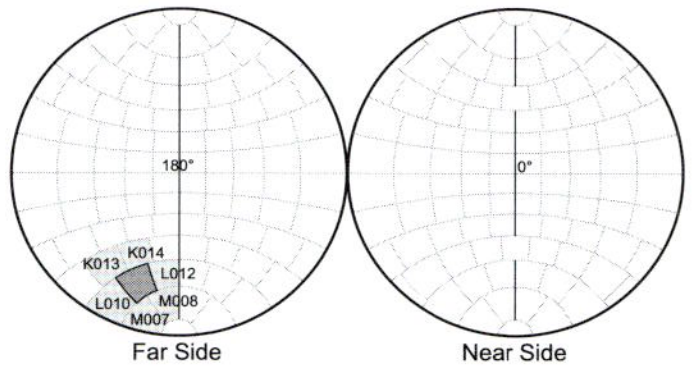

-7000 -6000 -5000 -4000 -3000 -2000 -1000 0 1000 2000 3000 4000 m

0 20 40 60 80 100 km

Image Code L012

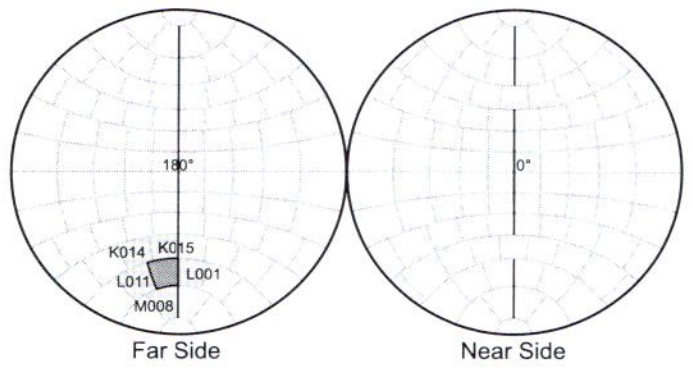

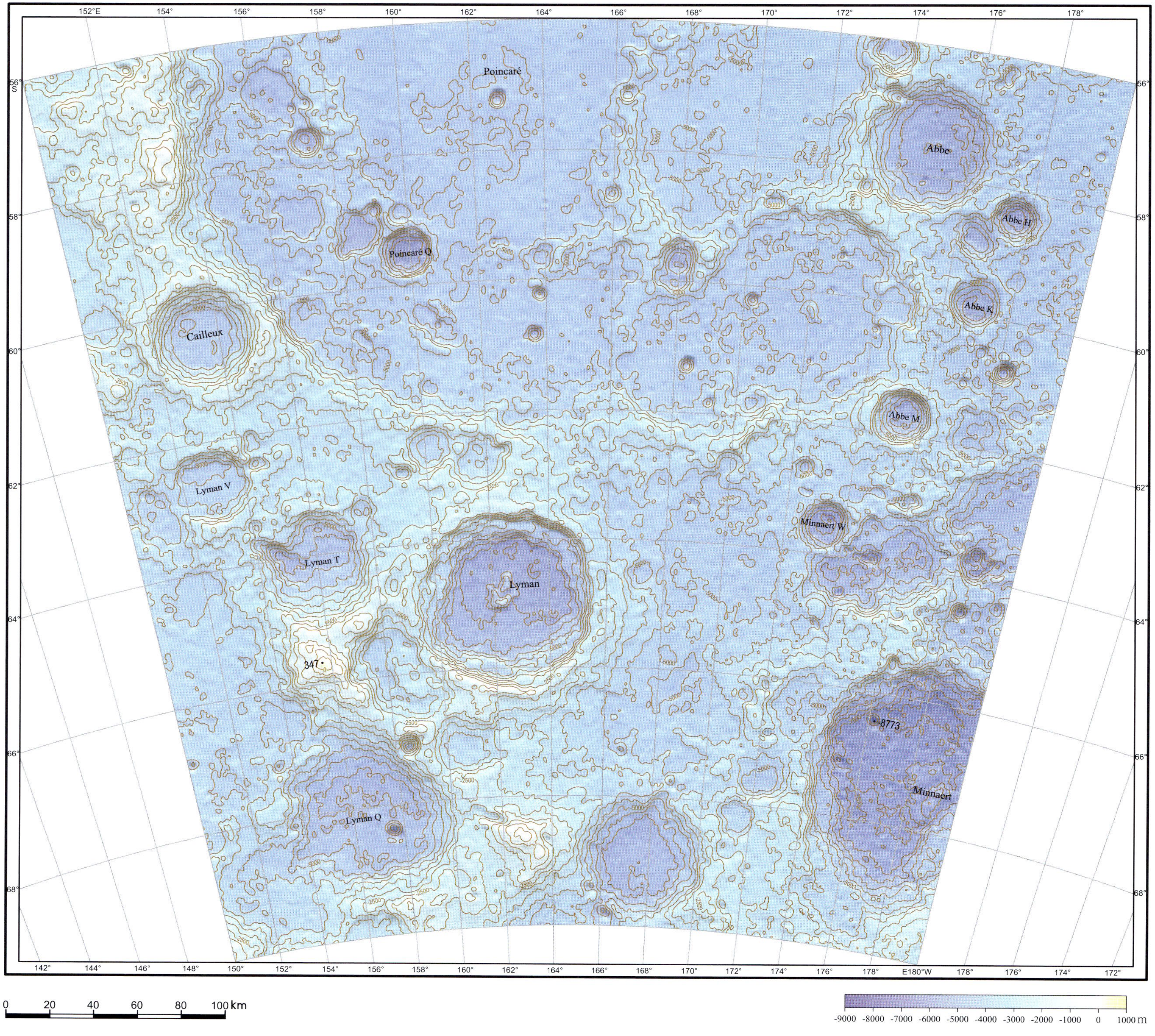

Image Code M001

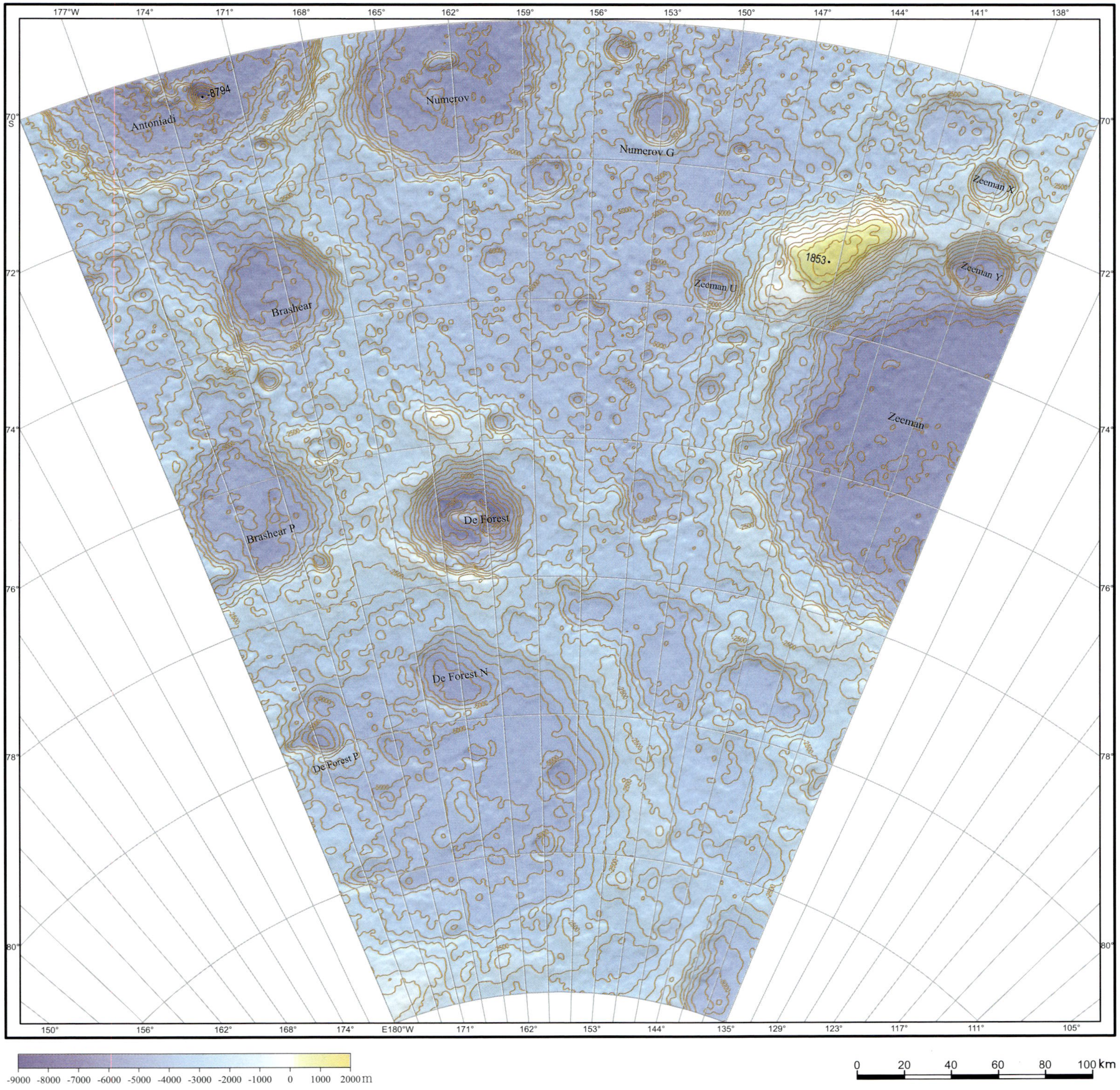

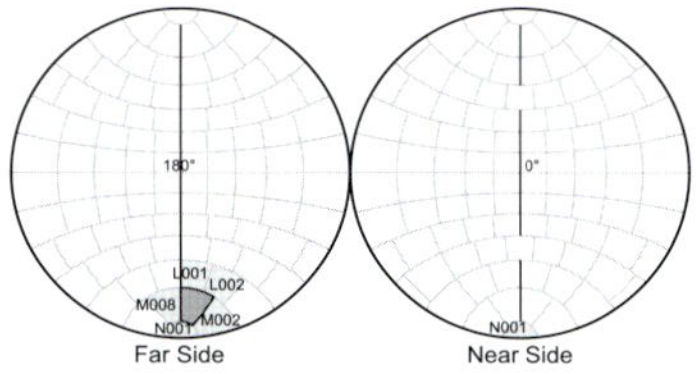

Image Code **M002**

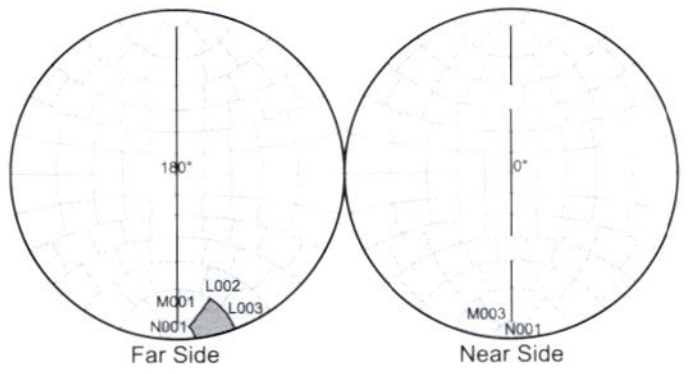

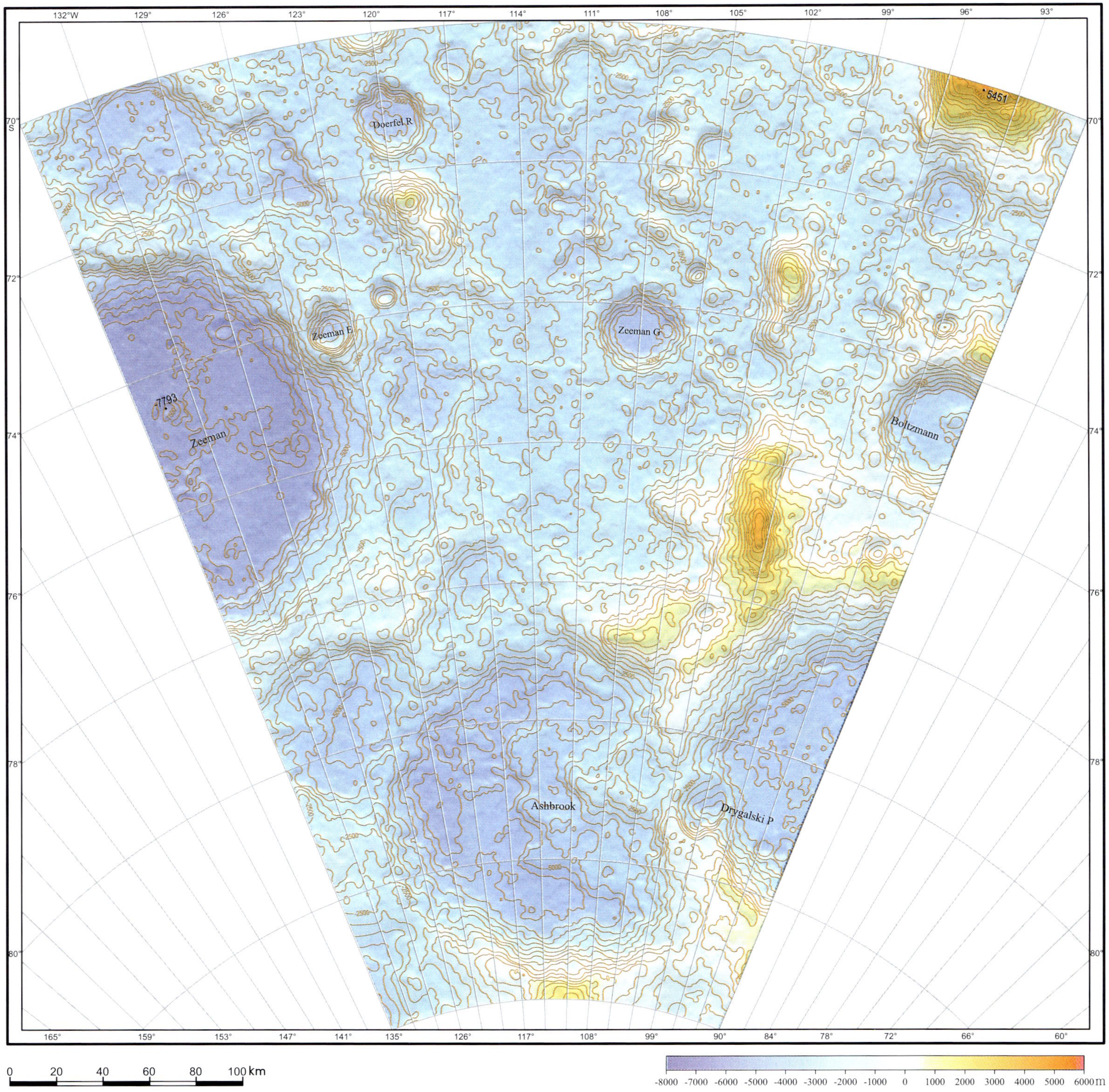

Image Code M003

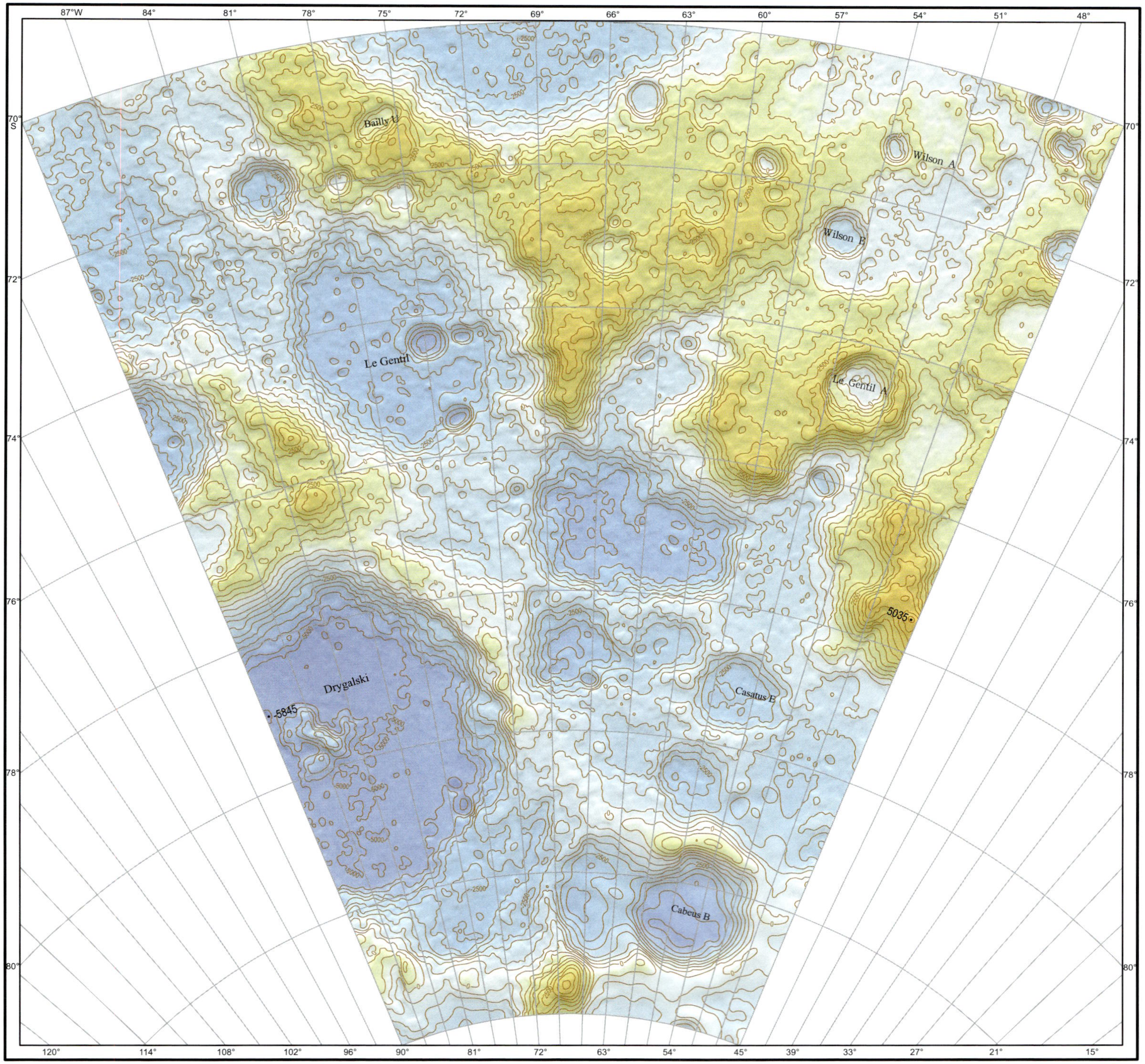

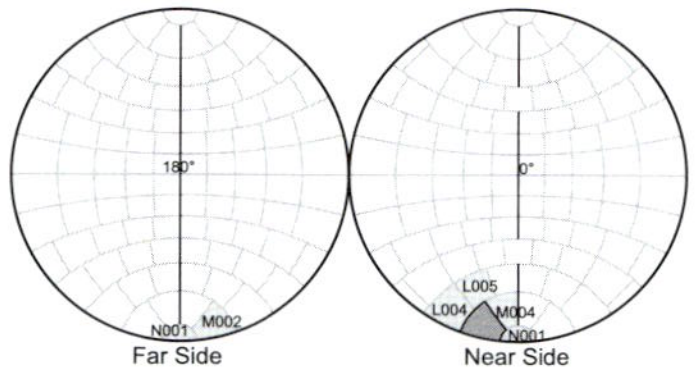

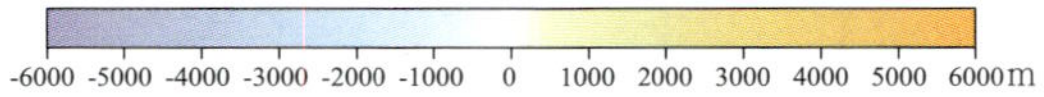

Image Code M004

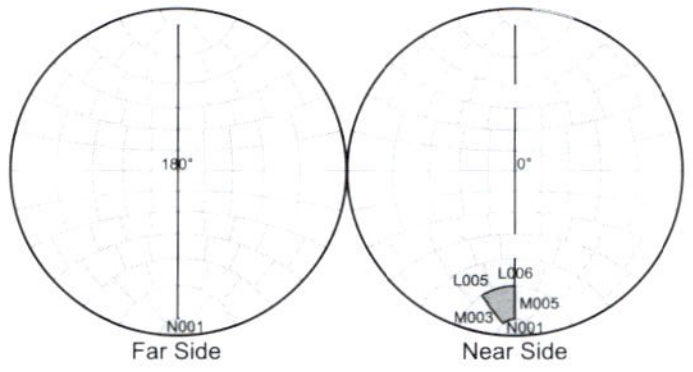

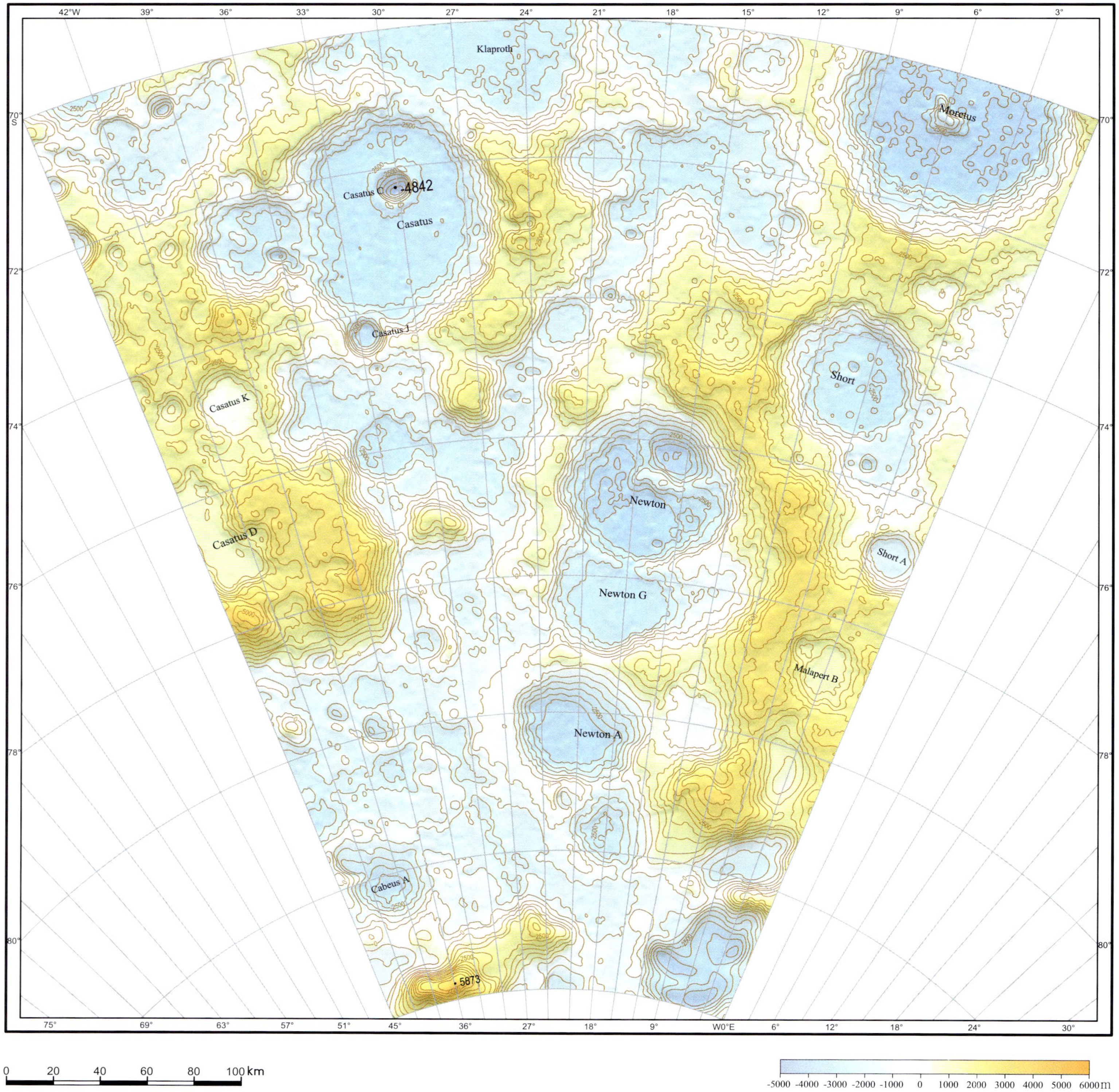

Image Code M005

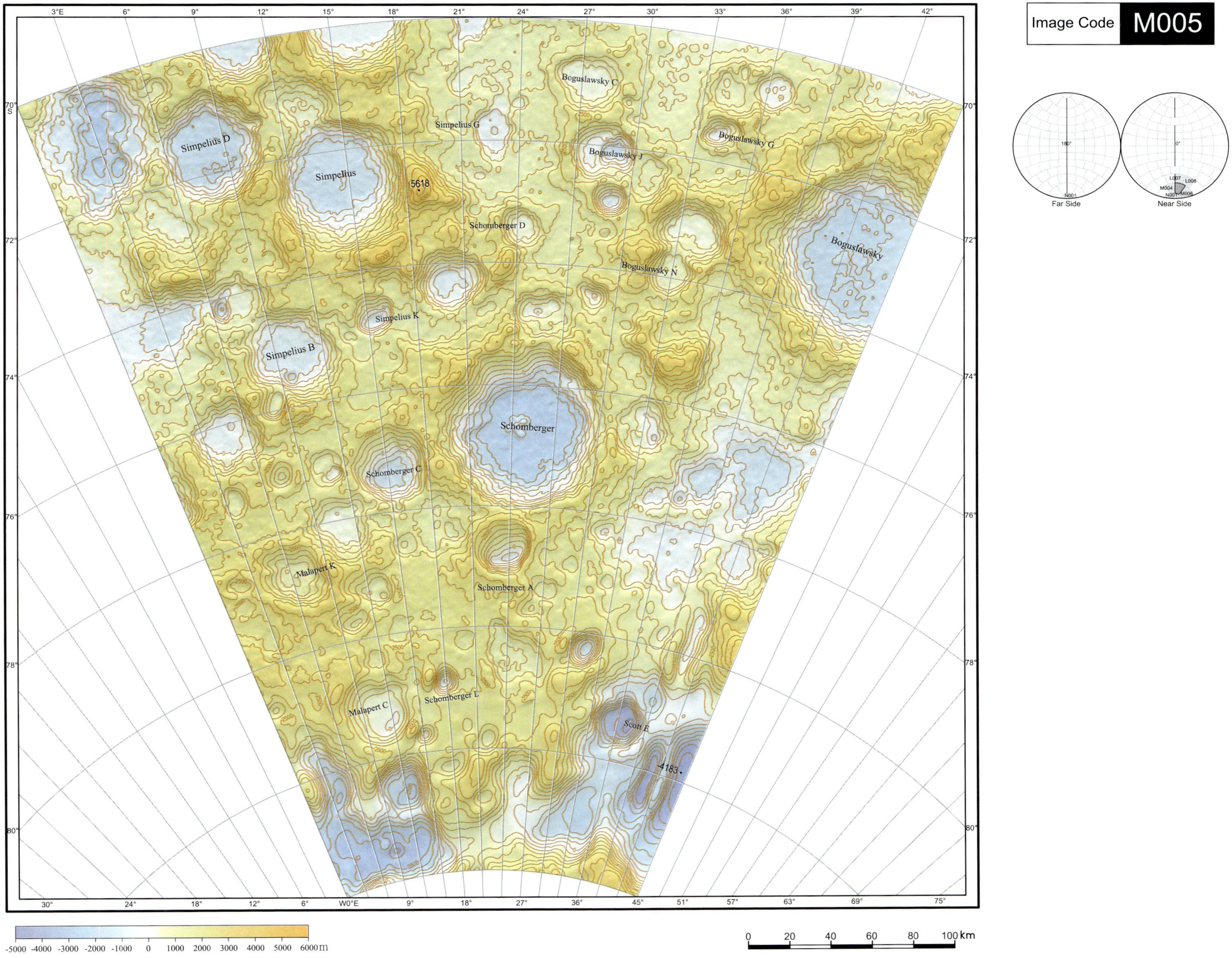

Image Code M006

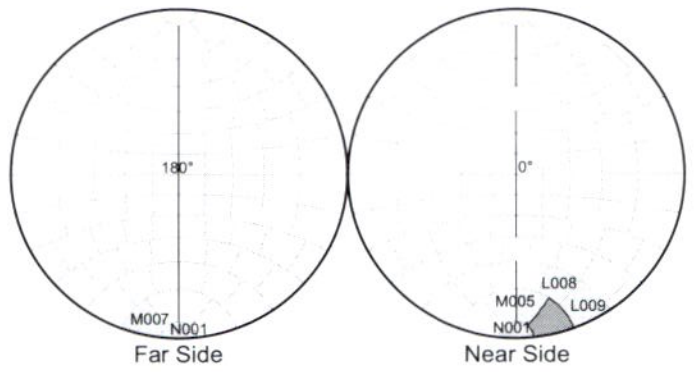

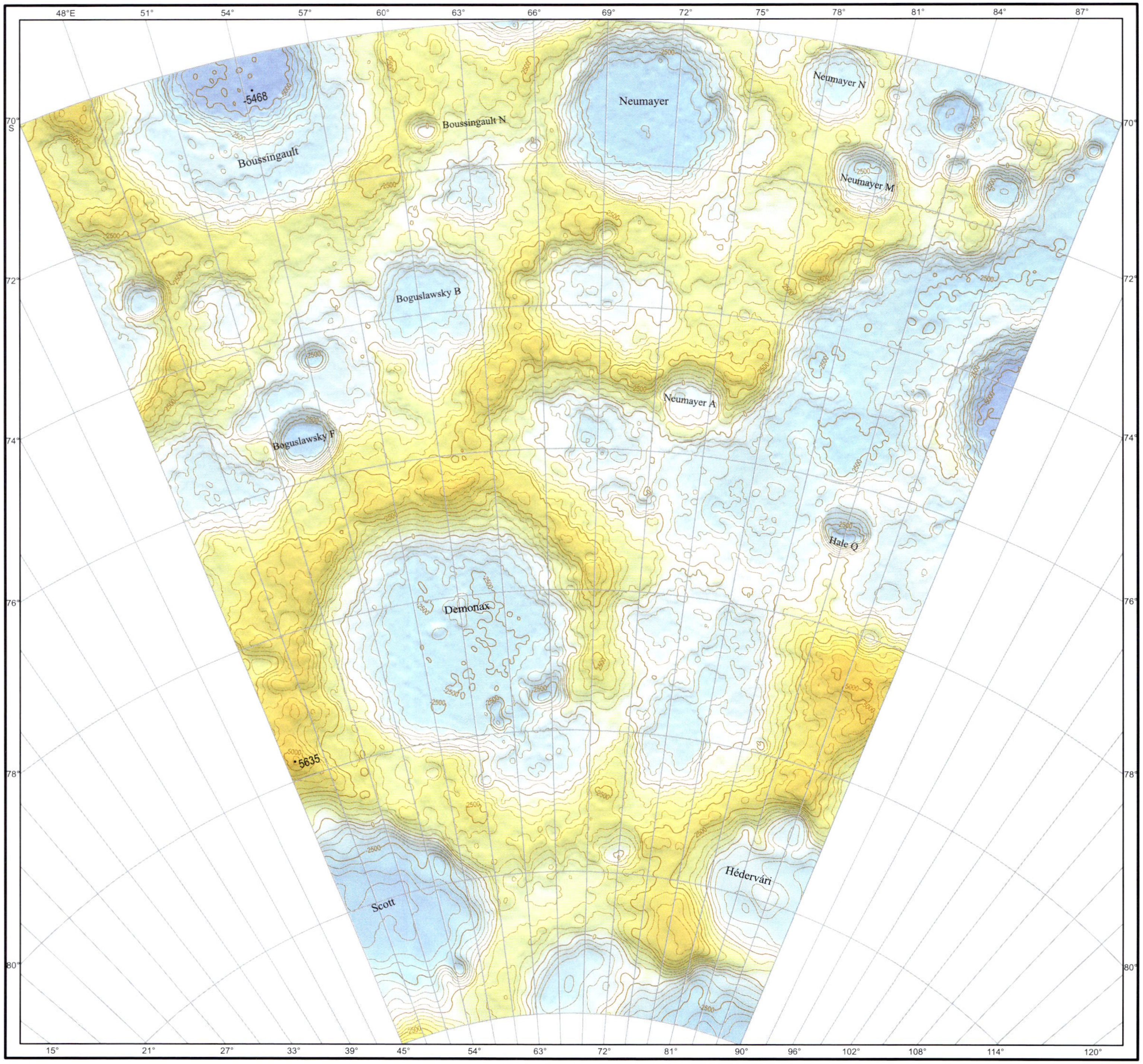

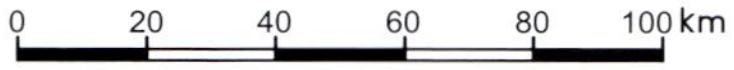

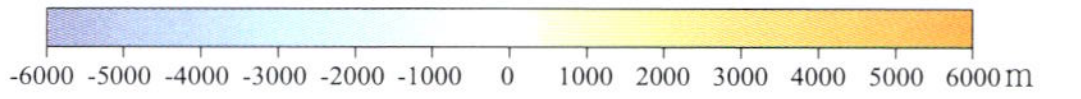

Image Code M007

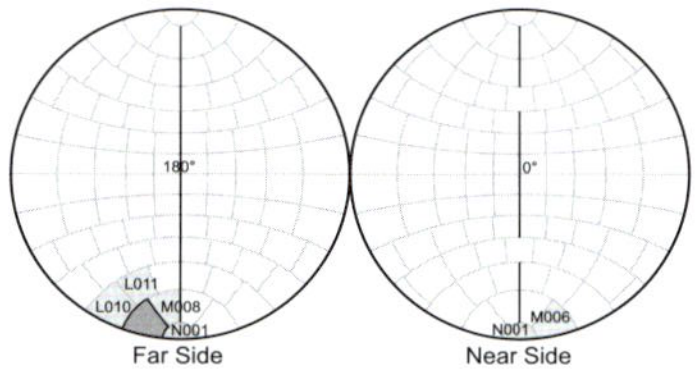

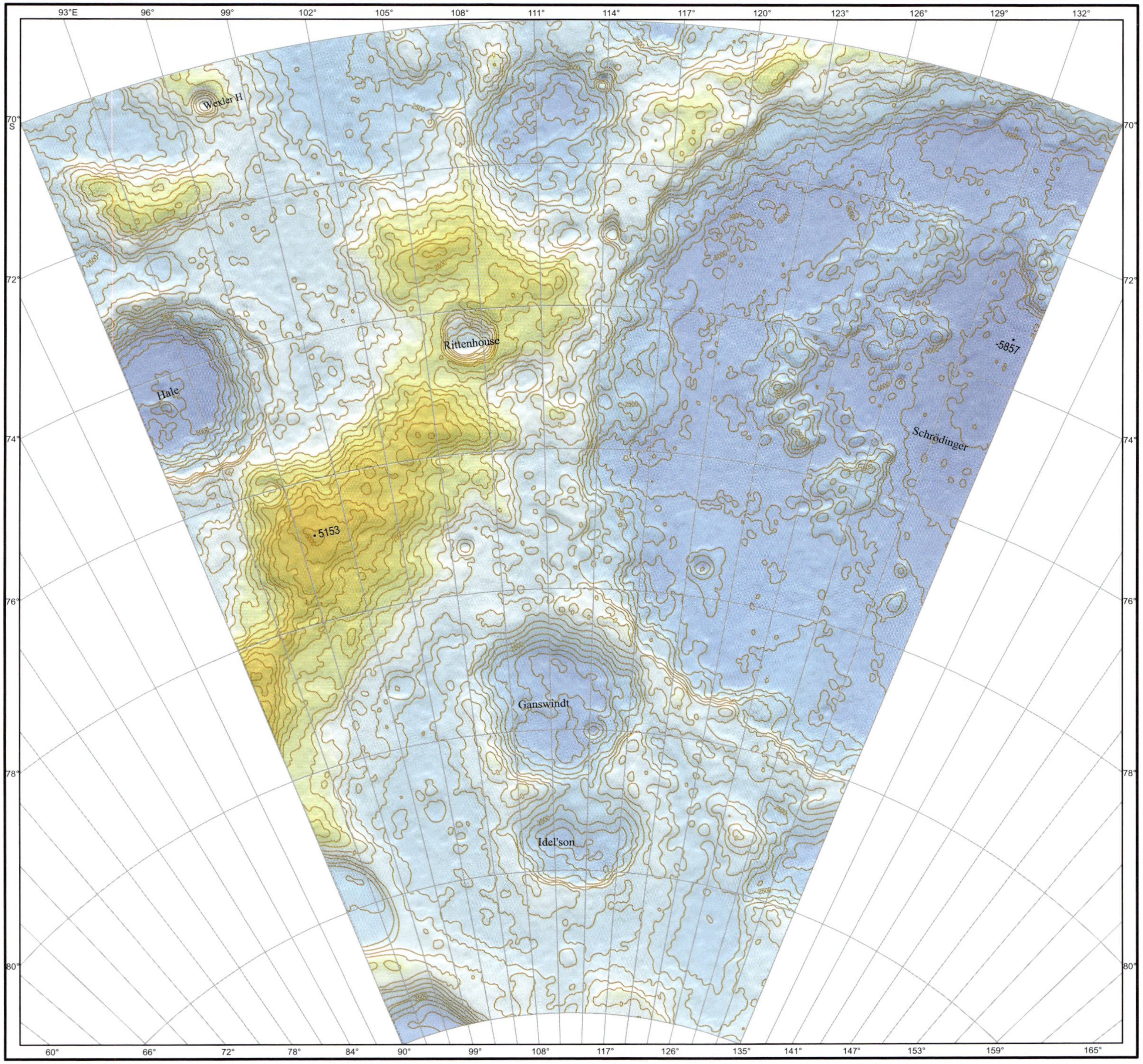

-6000 -5000 -4000 -3000 -2000 -1000 0 1000 2000 3000 4000 5000 6000 m

0 20 40 60 80 100 km

Image Code **M008**

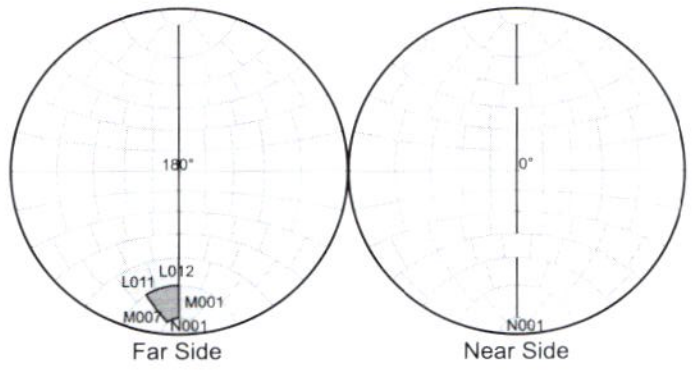

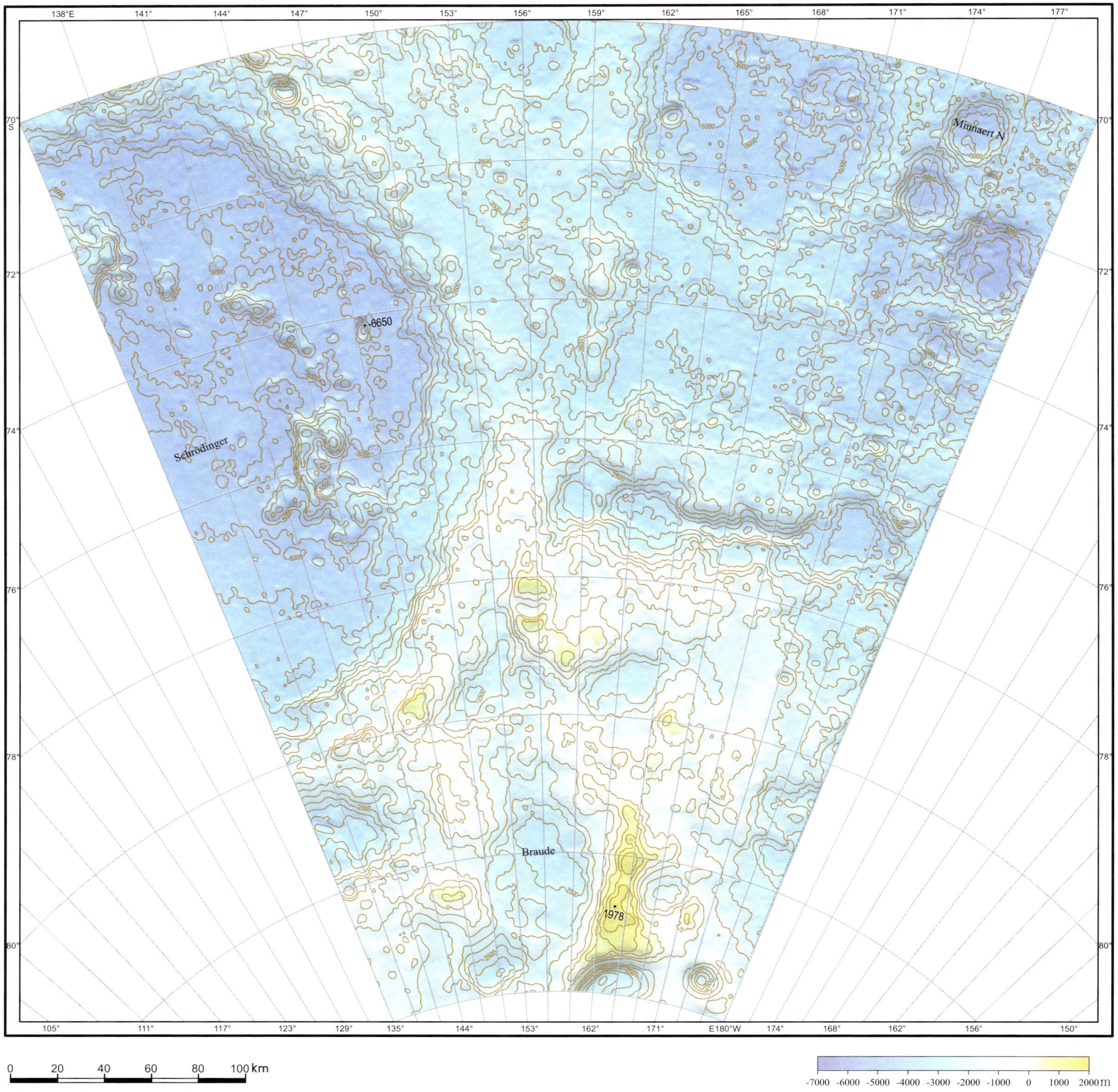

Appendix-Gazetteer

The appendix lists all of the lunar place names marked in this atlas. The lunar place names are from the Gazetteer of Planetary Nomenclature, released by International Astronomical Union (IAU) Working Group for Planetary System Nomenclature (WG-PSN). Before Jan 1,2012, the Gazetteer of Planetary Nomenclature collected 9099 historically used lunar place names, including 8990 valid names approved by IAU and 109 invalid names abandoned or disapproved by IAU. In the naming system, the satellite feature is a feature that shares the name of an associated feature. For example, on the Moon the craters referred to as 'Lettered Craters' are classified in the gazetteer as 'Satellite Feature'. Due to space limitations, this appendix lists only 3698 lunar names marked in this atlas, including 1765 satellite feature names.

Name	Lat.	Long.	Size	Page
Oceanus, Oceani				
Oceanus Procellarum	18°24'N	57°24'W	2568	75
Mare, Maria				
Mare Anguis	22°36'N	67°42'E	150	82
Mare Australe	38°54'S	93°00'E	603	162
Mare Cognitum	10°00'S	23°06'W	376	117
Mare Crisium	17°00'N	59°06'E	418	82
Mare Fecunditatis	7°48'S	51°18'E	909	121
Mare Frigoris	56°00'N	1°24'E	1596	29
Mare Humboldtianum	56°48'N	81°30'E	273	32
Mare Humorum	24°24'S	38°36'W	389	136
Mare Imbrium	32°48'N	15°36'W	1123	59
Mare Ingenii	33°42'S	163°30'E	318	166
Mare Insularum	7°30'N	30°54'W	513	97
Mare Marginis	13°18'N	86°06'E	420	103
Mare Moscoviense	27°18'N	147°54'E	277	87
Mare Nectaris	15°12'S	35°30'E	333	140
Mare Nubium	21°18'S	16°36'W	715	138
Mare Orientale	19°24'S	92°48'W	327	133
Mare Serenitatis	28°00'N	17°30'E	707	60
Mare Smythii	1°18'N	87°30'E	373	103
Mare Spumans	1°06'N	65°06'E	139	102
Mare Tranquillitatis	8°30'N	31°24'E	873	100
Mare Undarum	6°48'N	68°24'E	243	102
Mare Vaporum	13°18'N	3°36'E	245	99
Lacus, Lacūs				
Lacus Aestatis	15°00'S	69°00'W	90	135
Lacus Autumni	9°54'S	83°54'W	183	114
Lacus Bonitatis	23°12'N	43°42'E	92	81
Lacus Doloris	17°06'N	9°00'E	110	79
Lacus Excellentiae	35°24'S	44°00'W	184	155
Lacus Felicitatis	19°00'N	5°00'E	90	79
Lacus Gaudii	16°12'N	12°36'E	113	79
Lacus Hiemalis	15°00'N	14°00'E	50	79
Lacus Lenitatis	14°00'N	12°00'E	80	79
Lacus Luxuriae	19°00'N	176°00'E	50	88
Lacus Mortis	45°00'N	27°12'E	151	44
Lacus Oblivionis	21°00'S	168°00'W	50	129
Lacus Odii	19°00'N	7°00'E	70	79
Lacus Perseverantiae	8°00'N	62°00'E	70	102
Lacus Solitudinis	27°48'S	104°18'E	139	144
Lacus Somniorum	38°00'N	29°12'E	384	61
Lacus Spei	43°00'N	65°00'E	80	46
Lacus Temporis	45°54'N	58°24'E	117	45
Lacus Timoris	38°48'S	27°18'W	117	156
Lacus Veris	16°30'S	86°06'W	396	134
Planitia, Planitiae				
Planitia Descensus	7°06'N	64°24'W		95
Mons, Montes				
Mons Agnes	18°36'N	5°18'E	1	79
Mons Ampère	19°00'N	4°00'W	30	78
Mons André	5°12'N	120°36'E	10	105
Mons Ardeshir	5°00'N	121°00'E	8	105
Mons Argaeus	19°00'N	29°00'E	50	80
Mons Bradley	22°00'N	1°00'E	30	79
Mons Delisle	29°30'N	35°48'W	30	58
Mons Dieter	5°00'N	120°12'E	20	105
Mons Dilip	5°36'N	120°48'E	2	105
Mons Esam	14°36'N	35°42'E	8	80
Mons Ganau	4°48'N	120°36'E	14	105
Mons Gruithuisen Delta	36°00'N	39°30'W	20	58
Mons Gruithuisen Gamma	36°36'N	40°30'W	20	57
Mons Hadley	26°30'N	4°42'E	25	79
Mons Hadley Delta	25°48'N	3°48'E	15	79
Mons Hansteen	12°06'S	50°00'W	30	116
Mons Herodotus	27°30'N	53°00'W	5	76
Mons Huygens	20°00'N	2°54'W	40	78
Mons La Hire	27°48'N	25°30'W	25	77
Mons Maraldi	20°18'N	35°18'E	15	80
Mons Moro	12°00'S	19°42'W	10	117
Mons Penck	10°00'S	21°36'E	30	120
Mons Pico	45°42'N	8°54'W	25	43
Mons Piton	40°36'N	1°06'W	25	59
Mons Rümker	40°48'N	58°06'W	70	57
Mons Usov	12°00'N	63°00'E	15	102
Mons Vinogradov	22°24'N	32°24'W	25	77
Mons Vitruvius	19°24'N	30°48'E	15	80
Mons Wolff	17°00'N	6°48'W	35	78
Mont Blanc	45°00'N	1°00'E	25	43
Montes Agricola	29°06'N	54°12'W	141	57
Montes Alpes	46°24'N	0°48'W	281	43
Montes Apenninus	18°54'N	3°42'W	401	78
Montes Archimedes	25°18'N	4°36'W	163	78
Montes Carpatus	14°30'N	24°24'W	361	77
Montes Caucasus	38°24'N	10°00'E	445	60
Montes Cordillera	17°30'S	81°36'W	574	134
Montes Haemus	19°54'N	9°12'E	560	79
Montes Harbinger	27°00'N	41°00'W	90	76
Montes Jura	47°06'N	34°00'W	422	42

C. Li et al., *The Chang'E-1 Topographic Atlas of the Moon*,
DOI 10.1007/978-3-662-48439-5, © Springer-Verlag Berlin Heidelberg 2016

Name	Lat.	Long.	Size	Page
Montes Pyrenaeus	15°36'S	41°12'E	164	141
Montes Recti	48°00'N	20°00'W	90	42
Montes Riphaeus	7°42'S	28°06'W	189	117
Montes Rook	20°36'S	82°30'W	791	134
Montes Secchi	3°00'N	43°00'E	50	101
Montes Spitzbergen	35°00'N	5°00'W	60	59
Montes Taurus	28°24'N	41°06'E	172	62
Montes Teneriffe	47°06'N	11°48'W	182	42
Dorsum, Dorsa				
Dorsa Aldrovandi	24°00'N	28°30'E	136	80
Dorsa Andrusov	1°00'S	57°00'E	160	122
Dorsa Argand	28°06'N	40°36'W	109	57
Dorsa Barlow	15°00'N	31°00'E	120	80
Dorsa Burnet	28°24'N	57°00'W	194	57
Dorsa Cato	1°00'N	47°00'E	140	101
Dorsa Dana	3°00'N	90°00'E	70	103
Dorsa Ewing	10°12'S	39°24'W	141	116
Dorsa Geikie	4°36'S	52°30'E	228	121
Dorsa Harker	14°30'N	64°00'E	197	82
Dorsa Lister	20°18'N	23°48'E	203	80
Dorsa Mawson	7°00'S	53°00'E	132	121
Dorsa Rubey	10°00'S	42°00'W	100	116
Dorsa Smirnov	27°18'N	25°18'E	156	80
Dorsa Sorby	19°00'N	14°00'E	80	79
Dorsa Stille	27°00'N	19°00'W	80	77
Dorsa Tetyaev	19°54'N	64°12'E	176	82
Dorsa Whiston	29°24'N	56°24'W	85	57
Dorsum Arduino	24°54'N	35°48'W	107	77
Dorsum Azara	26°42'N	19°12'E	105	80
Dorsum Bucher	31°00'N	39°00'W	90	58
Dorsum Buckland	20°24'N	12°48'E	380	79
Dorsum Cayeux	1°36'N	51°12'E	84	101
Dorsum Cloos	1°00'N	91°00'E	100	104
Dorsum Cushman	1°00'N	49°00'E	80	101
Dorsum Gast	24°00'N	9°00'E	60	79

Name	Lat.	Long.	Size	Page
Dorsum Grabau	29°24'N	15°54'W	121	59
Dorsum Guettard	10°00'S	18°00'W	40	117
Dorsum Heim	32°00'N	29°48'W	148	58
Dorsum Higazy	28°00'N	17°00'W	60	59
Dorsum Nicol	18°00'N	23°00'E	50	80
Dorsum Niggli	29°00'N	52°00'W	50	57
Dorsum Oppel	18°42'N	52°36'E	268	81
Dorsum Owen	25°00'N	11°00'E	50	79
Dorsum Scilla	32°48'N	60°24'W	108	56
Dorsum Termier	11°00'N	58°00'E	90	102
Dorsum Thera	24°24'N	31°24'W	7	77
Dorsum Von Cotta	23°12'N	11°54'E	199	79
Dorsum Zirkel	28°06'N	23°30'W	193	58
Crater, Craters				
Abbe	57°18'S	175°12'E	66	193
Abbot	5°36'N	54°48'E	10	102
Abel	34°30'S	87°18'E	122	162
Abenezra	21°00'S	11°54'E	42	139
Abetti	20°06'N	27°48'E	1	80
Abul Wáfa	1°00'N	116°36'E	55	105
Abulfeda	13°48'S	13°54'E	65	119
Acosta	5°36'S	60°06'E	13	122
Adams	31°54'S	68°12'E	66	161
Aepinus	88°01'N	108°18'W	17	14
Agatharchides	19°48'S	30°54'W	48	137
Agrippa	4°06'N	10°30'E	44	99
Airy	18°06'S	5°42'E	36	139
Aitken	16°48'S	173°24'E	135	148
Akis	20°00'N	31°48'W	2	77
Alan	10°54'S	6°06'W	2	118
Al-Bakri	14°18'N	20°12'E	12	80
Albategnius	11°42'S	4°18'E	114	119
Al-Biruni	17°54'N	92°30'E	77	84
Alden	23°36'S	110°48'E	104	145
Alder	48°36'S	177°24'W	77	167

Name	Lat.	Long.	Size	Page
Aldrin	1°24'N	22°06'E	3	100
Alekhin	68°12'S	131°18'W	70	183
Alexander	40°18'N	13°30'E	81	60
Alfraganus	5°24'S	19°00'E	20	120
Alhazen	15°54'N	71°48'E	32	82
Aliacensis	30°36'S	5°12'E	79	158
Al-Khwarizmi	7°06'N	106°24'E	65	104
Almanon	16°48'S	15°12'E	49	139
Al-Marrakushi	10°24'S	55°48'E	8	122
Aloha	29°48'N	53°54'W	3	57
Alpetragius	16°00'S	4°30'W	39	138
Alphonsus	13°42'S	3°12'W	108	118
Alter	18°42'N	107°30'W	64	73
Ameghino	3°18'N	57°00'E	9	102
Amici	9°54'S	172°06'W	54	109
Ammonius	8°30'S	0°48'W	8	118
Amontons	5°18'S	46°48'E	2	121
Amundsen	84°18'S	85°36'E	101	15
Anaxagoras	73°24'N	10°06'W	50	19
Anaximander	66°54'N	51°18'W	67	28
Anaximenes	72°30'N	44°30'W	80	19
Andĕl	10°24'S	12°24'E	35	119
Anders	41°18'S	142°54'W	40	150
Anderson	15°48'N	171°06'E	109	88
Andersson	49°42'S	95°18'W	13	170
Andronov	22°42'S	146°06'E	16	147
Ango	20°30'N	32°18'W	1	77
Angström	29°54'N	41°36'W	9	57
Ann	25°06'N	0°06'W	3	78
Annegrit	29°24'N	25°36'W	1	58
Ansgarius	12°42'S	79°42'E	94	123
Antoniadi	69°42'S	172°00'W	143	182
Anuchin	49°00'S	101°18'E	57	178
Anville	1°54'N	49°30'E	10	101
Apianus	26°54'S	7°54'E	63	139

Name	Lat.	Long.	Size	Page
Apollo	36°06'S	151°48'W	537	150
Apollonius	4°30'N	61°06'E	53	102
Appleton	37°12'N	158°18'E	63	67
Arago	6°12'N	21°24'E	26	100
Aratus	23°36'N	4°30'E	10	79
Archimedes	29°42'N	4°00'W	82	59
Archytas	58°42'N	5°00'E	31	30
Argelander	16°30'S	5°48'E	34	139
Ariadaeus	4°36'N	17°18'E	11	99
Aristarchus	23°42'N	47°24'W	40	76
Aristillus	33°54'N	1°12'E	55	60
Aristoteles	50°12'N	17°24'E	87	44
Armiński	16°24'S	154°12'E	26	147
Armstrong	1°24'N	25°00'E	4	100
Arnold	66°48'N	35°54'E	94	31
Arrhenius	55°36'S	91°18'W	40	170
Artamonov	25°30'N	103°30'E	60	84
Artem'ev	10°48'N	144°24'W	67	90
Artemis	25°00'N	25°24'W	2	77
Artsimovich	27°36'N	36°36'W	8	76
Aryabhata	6°12'N	35°06'E	22	100
Arzachel	18°12'S	1°54'W	96	138
Asada	7°18'N	49°54'E	12	101
Asclepi	55°06'S	25°24'E	42	175
Ashbrook	81°24'S	112°30'W	156	195
Aston	32°54'N	87°42'W	43	55
Atlas	46°42'N	44°24'E	87	45
Atwood	5°48'S	57°42'E	29	122
Autolycus	30°42'N	1°30'E	39	60
Auwers	15°06'N	17°12'E	20	79
Auzout	10°18'N	64°06'E	32	102
Avery	1°24'S	81°24'E	9	123
Avicenna	39°42'N	97°12'W	74	55
Avogadro	63°06'N	164°54'E	139	35
Azophi	22°06'S	12°42'E	47	139

Name	Lat.	Long.	Size	Page
Baade	44°48'S	81°48'W	55	171
Babakin	20°48'S	123°18'E	20	145
Babbage	59°42'N	57°06'W	143	28
Babcock	4°12'N	93°54'E	99	104
Back	1°06'N	80°42'E	35	103
Backlund	16°00'S	103°00'E	75	144
Baco	51°00'S	19°06'E	69	175
Baillaud	74°36'N	37°30'E	89	20
Bailly	66°30'S	69°06'W	287	185
Baily	49°42'N	30°24'E	26	44
Balandin	18°54'S	152°36'E	12	147
Balboa	19°06'N	83°12'W	69	74
Baldet	53°18'S	151°06'W	55	168
Ball	35°54'S	8°24'W	41	157
Balmer	20°18'S	69°48'E	138	142
Banachiewicz	5°12'N	80°06'E	92	103
Bancroft	28°00'N	6°24'W	13	59
Banting	26°36'N	16°24'E	5	79
Barbier	23°48'S	157°54'E	66	147
Barkla	10°42'S	67°12'E	42	122
Barnard	29°30'S	85°36'E	105	162
Barocius	44°54'S	16°48'E	82	175
Barringer	28°00'S	149°42'W	68	130
Barrow	71°18'N	7°42'E	92	20
Bartels	24°30'N	89°48'W	55	74
Bawa	25°18'S	102°36'E	1	144
Bayer	51°36'S	35°00'W	47	173
Beals	37°18'N	86°30'E	48	64
Beaumont	18°00'S	28°48'E	53	140
Becquerel	40°42'N	129°42'E	65	66
Bečvář	1°54'S	125°12'E	67	125
Beer	27°06'N	9°06'W	9	78
Behaim	16°30'S	79°24'E	55	143
Beijerinck	13°30'S	151°48'E	70	127
Beketov	16°18'N	29°12'E	8	80

Name	Lat.	Long.	Size	Page
Béla	24°42'N	2°18'E	11	79
Bel'kovich	61°06'N	90°12'E	214	33
Bell	21°48'N	96°24'W	86	73
Bellinsgauzen	60°36'S	164°36'W	63	182
Bellot	12°24'S	48°12'E	17	121
Belopol'skiy	17°12'S	128°06'W	59	131
Belyaev	23°18'N	143°30'E	54	86
Benedict	4°24'N	141°30'E	14	106
Bergman	7°00'N	137°30'E	21	106
Bergstrand	18°48'S	176°18'E	43	148
Berkner	25°12'N	105°12'W	86	73
Berlage	63°12'S	162°48'W	92	182
Bernoulli	35°00'N	60°42'E	47	63
Berosus	33°30'N	69°54'E	74	63
Berzelius	36°36'N	50°54'E	50	62
Bessarion	14°54'N	37°18'W	10	76
Bessel	21°48'N	17°54'E	15	80
Bettinus	63°24'S	44°48'W	71	186
Bhabha	55°06'S	164°30'W	64	167
Bi Sheng	78°20'N	148°22'E	55	23
Bianchini	48°42'N	34°18'W	38	42
Biela	54°54'S	51°18'E	76	176
Bilharz	5°48'S	56°18'E	43	122
Billy	13°48'S	50°06'W	45	116
Bingham	8°06'N	115°06'E	33	105
Biot	22°36'S	51°06'E	12	141
Birkeland	30°12'S	173°54'E	82	166
Birkhoff	58°42'N	146°06'W	345	25
Birmingham	65°06'N	10°30'W	92	29
Birt	22°24'S	8°30'W	16	138
Bjerknes	38°24'S	113°00'E	48	163
Black	9°12'S	80°24'E	18	123
Blackett	37°30'S	116°06'W	141	152
Blagg	1°18'N	1°30'E	5	99
Blancanus	63°48'S	21°24'W	117	187

Name	Lat.	Long.	Size	Page
Blanchard	58°30'S	94°24'W	40	184
Blanchinus	25°24'S	2°30'E	61	139
Blazhko	31°36'N	148°00'W	54	52
Bliss	53°00'N	13°30'W	20	42
Bobillier	19°36'N	15°30'E	6	79
Bobone	26°54'N	131°48'W	31	71
Bode	6°42'N	2°24'W	18	98
Boethius	5°36'N	72°18'E	10	103
Boguslawsky	72°54'S	43°12'E	97	198
Bohnenberger	16°12'S	40°00'E	33	141
Bohr	12°24'N	86°36'W	71	94
Bok	20°12'S	171°36'W	45	129
Boltzmann	74°54'S	90°42'W	76	195
Bolyai	33°36'S	125°54'E	135	164
Bombelli	5°18'N	56°12'E	10	102
Bondarenko	17°48'S	136°18'E	30	146
Bonpland	8°18'S	17°24'W	60	118
Boole	63°42'N	87°24'W	63	27
Borda	25°06'S	46°36'E	44	141
Borel	22°18'N	26°24'E	4	80
Boris	30°36'N	33°30'W	1	58
Borman	38°48'S	147°42'W	50	150
Born	6°00'S	66°48'E	14	122
Bosch	86°49'N	133°36'E	18	14
Boscovich	9°48'N	11°06'E	46	99
Bose	53°30'S	170°00'W	91	167
Boss	45°48'N	89°12'E	47	47
Bouguer	52°18'N	35°48'W	22	41
Boussingault	70°12'S	54°36'E	142	199
Bowditch	25°00'S	103°06'E	40	144
Bowen	17°36'N	9°06'E	8	79
Boyle	53°06'S	178°06'E	57	181
Brackett	17°54'N	23°36'E	8	80
Bragg	42°30'N	102°54'W	84	39
Brashear	73°48'S	170°42'W	55	194

Name	Lat.	Long.	Size	Page
Braude	81°50'S	157°48'E	10	201
Brayley	20°54'N	36°54'W	14	76
Bredikhin	17°18'N	158°12'W	59	70
Breislak	48°12'S	18°18'E	49	175
Brenner	39°00'S	39°18'E	97	159
Brewster	23°18'N	34°42'E	10	80
Brianchon	75°00'N	86°12'W	134	18
Bridgman	43°30'N	137°06'E	80	49
Briggs	26°30'N	69°06'W	37	75
Brisbane	49°06'S	68°30'E	44	177
Bronk	26°06'N	134°30'W	64	71
Brouwer	36°12'S	126°00'W	158	151
Brown	46°24'S	17°54'W	34	173
Bruce	1°06'N	0°24'E	6	99
Brunner	9°54'S	90°54'E	53	124
Buch	38°48'S	17°42'E	53	158
Buffon	40°24'S	133°24'W	106	151
Buisson	1°24'S	112°30'E	56	125
Bullialdus	20°42'S	22°12'W	60	137
Bunsen	41°24'N	85°18'W	52	55
Burckhardt	31°06'N	56°30'E	56	62
Bürg	45°00'N	28°12'E	39	44
Burnham	13°54'S	7°18'E	24	119
Büsching	38°00'S	20°00'E	52	158
Butlerov	12°30'N	108°42'W	40	92
Buys-Ballot	20°48'N	174°30'E	55	88
Byrd	85°18'N	9°48'E	93	14
Byrgius	24°42'S	65°18'W	87	135
C. Herschel	34°30'N	31°12'W	13	58
C. Mayer	63°12'N	17°18'E	38	30
Cabannes	60°54'S	169°36'W	80	182
Cabeus	84°54'S	35°30'W	98	15
Cai Lun	80°30'N	113°49'E	42	22
Cailleux	60°48'S	153°18'E	50	193
Cajal	12°36'N	31°06'E	9	100

Name	Lat.	Long.	Size	Page
Cajori	47°24'S	168°48'E	70	181
Calippus	38°54'N	10°42'E	32	60
Cameron	6°12'N	45°54'E	10	101
Campanus	28°00'S	27°48'W	48	137
Campbell	45°18'N	151°24'E	219	49
Cannizzaro	55°36'N	99°36'W	56	39
Cannon	19°54'N	81°24'E	56	83
Cantor	38°12'N	118°36'E	81	65
Capella	7°30'S	35°00'E	49	120
Capuanus	34°06'S	26°42'W	59	156
Cardanus	13°12'N	72°30'W	49	94
Carlini	33°42'N	24°06'W	10	58
Carlos	24°54'N	2°18'E	4	79
Carmichael	19°36'N	40°24'E	20	81
Carnot	52°18'N	143°30'W	126	37
Carol	8°30'N	122°18'E	8	105
Carpenter	69°24'N	50°54'W	59	28
Carrel	10°42'N	26°42'E	15	100
Carrillo	2°12'S	80°54'E	16	123
Carrington	44°00'N	62°06'E	30	46
Cartan	4°12'N	59°18'E	15	102
Carver	43°00'S	126°54'E	59	179
Casatus	72°48'S	29°30'W	108	197
Cassegrain	52°24'S	113°30'E	55	179
Cassini	40°12'N	4°36'E	56	60
Catalán	45°42'S	87°18'W	25	170
Catharina	18°06'S	23°24'E	104	140
Cauchy	9°36'N	38°36'E	12	101
Cavalerius	5°06'N	66°48'W	57	95
Cavendish	24°30'S	53°42'W	56	136
Caventou	29°48'N	29°24'W	3	58
Cayley	4°00'N	15°06'E	14	99
Celsius	34°06'S	20°06'E	36	159
Censorinus	0°24'S	32°42'E	3	120
Cepheus	40°48'N	45°48'E	39	62

Name	Lat.	Long.	Size	Page
Chacornac	29°48'N	31°42'E	51	61
Chadwick	52°42'S	101°18'W	30	170
Chaffee	38°48'S	153°54'W	49	150
Challis	79°30'N	9°12'E	55	20
Chalonge	21°12'S	117°18'W	30	132
Chamberlin	58°54'S	95°42'E	58	191
Champollion	37°24'N	175°12'E	58	68
Chandler	43°48'N	171°30'E	85	50
Chang Heng	19°00'N	112°12'E	43	85
Chang-Ngo	12°42'S	2°06'W	3	118
Chant	40°00'S	109°12'W	33	152
Chaplygin	6°12'S	150°18'E	137	127
Chapman	50°24'N	100°42'W	71	39
Chappe	61°12'S	91°30'W	59	184
Chappell	54°42'N	177°00'W	80	36
Charles	29°54'N	26°24'W	1	58
Charlier	36°36'N	131°30'W	99	53
Chaucer	3°42'N	140°00'W	45	91
Chauvenet	11°30'S	137°00'E	81	126
Chawla	42°48'S	147°30'W	15	168
Chebyshev	33°42'S	133°06'W	178	151
Chernyshev	47°18'N	174°12'E	58	50
Chevallier	44°54'N	51°12'E	52	45
Ching-Te	20°00'N	30°00'E	4	80
Chladni	4°00'N	1°06'E	13	99
Chrétien	45°54'S	162°54'E	88	181
Cichus	33°18'S	21°06'W	40	156
Clairaut	47°42'S	13°54'E	75	175
Clark	38°24'S	118°54'E	49	163
Clausius	36°54'S	43°48'W	24	155
Clavius	58°48'S	14°06'W	245	187
Cleomedes	27°42'N	56°00'E	125	82
Cleostratus	60°24'N	77°00'W	62	27
Clerke	21°42'N	29°48'E	6	80
Coblentz	37°54'S	126°06'E	33	164

Name	Lat.	Long.	Size	Page
Cockcroft	31°18'N	162°36'W	93	51
Collins	1°18'N	23°42'E	2	100
Colombo	15°06'S	45°48'E	76	141
Compton	55°18'N	103°48'E	162	47
Comrie	23°18'N	112°42'W	59	72
Comstock	21°48'N	121°30'W	72	72
Condon	1°54'N	60°24'E	34	102
Condorcet	12°06'N	69°36'E	74	102
Congreve	0°12'S	167°18'W	57	109
Conon	21°36'N	2°00'E	21	79
Cook	17°30'S	48°54'E	46	141
Cooper	52°54'N	175°36'E	36	50
Copernicus	9°42'N	20°06'W	93	97
Cori	50°36'S	151°54'W	65	168
Coriolis	0°06'N	171°48'E	78	108
Couder	4°48'S	92°24'W	21	113
Coulomb	54°42'N	114°36'W	89	38
Courtney	25°06'N	30°48'W	1	77
Cremona	67°30'N	90°36'W	85	26
Crile	14°12'N	46°00'E	9	81
Crocco	47°30'S	150°12'E	75	180
Crommelin	68°06'S	146°54'W	94	183
Crookes	10°18'S	164°30'W	49	109
Crozier	13°30'S	50°48'E	22	121
Crüger	16°42'S	66°48'W	45	135
Ctesibius	0°48'N	118°42'E	36	105
Curie	22°54'S	91°00'E	151	144
Curtis	14°36'N	56°36'E	2	82
Curtius	67°12'S	4°24'E	95	188
Cusanus	72°00'N	70°48'E	63	21
Cuvier	50°18'S	9°54'E	75	174
Cyrano	20°30'S	157°42'E	80	147
Cyrillus	13°12'S	24°00'E	98	120
Cysatus	66°12'S	6°06'W	48	187
D. Brown	42°00'S	147°12'W	15	150

Name	Lat.	Long.	Size	Page
da Vinci	9°06'N	45°00'E	37	101
Daedalus	5°54'S	179°24'E	93	128
Dag	18°42'N	5°18'E	0	79
Daguerre	11°54'S	33°36'E	46	120
Dale	9°36'S	82°54'E	22	123
D'Alembert	50°48'N	163°54'E	248	50
Dalton	17°06'N	84°18'W	60	74
Daly	5°42'N	59°36'E	17	102
Damoiseau	4°48'S	61°06'W	36	115
Daniell	35°18'N	31°06'E	29	61
Danjon	11°24'S	124°00'E	71	125
Dante	25°30'N	180°00'E	54	88
Darney	14°30'S	23°30'W	15	137
D'Arrest	2°18'N	14°42'E	30	99
D'Arsonval	10°18'S	124°36'E	28	125
Darwin	20°12'S	69°30'W	120	135
Das	26°36'S	136°48'W	38	131
Daubrée	15°42'N	14°42'E	14	79
Davisson	37°30'S	174°36'W	87	149
Davy	11°48'S	8°06'W	34	118
Dawes	17°12'N	26°24'E	18	80
Dawson	67°24'S	134°42'W	45	183
De Forest	77°18'S	162°06'W	57	194
de Gasparis	25°54'S	50°42'W	30	136
de Gerlache	88°30'S	87°06'W	32	15
De La Rue	59°06'N	52°18'E	134	31
De Moraes	49°30'N	143°12'E	53	49
De Morgan	3°18'N	14°54'E	10	99
De Roy	55°18'S	99°06'W	43	170
De Sitter	80°06'N	39°36'E	64	20
De Vico	19°42'S	60°12'W	20	135
De Vries	19°54'S	176°42'W	59	129
Debes	29°30'N	51°42'E	30	62
Debus	10°30'S	99°36'E	20	124
Debye	49°36'N	176°12'W	142	36

Name	Lat.	Long.	Size	Page
Dechen	46°06'N	68°12'W	12	40
Delambre	1°54'S	17°30'E	51	119
Delaunay	22°12'S	2°30'E	46	139
Delia	10°54'S	6°06'W	2	118
Delisle	29°54'N	34°36'W	25	58
Dellinger	6°48'S	140°36'E	81	126
Delmotte	27°06'N	60°12'E	32	82
Delporte	16°00'S	121°36'E	45	145
Deluc	55°00'S	2°48'W	46	174
Dembowski	2°54'N	7°12'E	26	99
Democritus	62°18'N	35°00'E	39	31
Demonax	77°54'S	60°48'E	128	199
Denning	16°24'S	142°36'E	44	146
Desargues	70°12'N	73°18'W	85	18
Descartes	11°42'S	15°42'E	48	119
Deseilligny	21°06'N	20°36'E	6	80
Deslandres	33°06'S	4°48'W	256	157
Deutsch	24°06'N	110°30'E	66	85
Dewar	2°42'S	165°30'E	50	128
Diana	14°18'N	35°42'E	2	80
Diderot	20°24'S	121°30'E	20	145
Dionysius	2°48'N	17°18'E	18	99
Diophantus	27°36'N	34°18'W	17	77
Dirichlet	11°06'N	151°24'W	47	90
Dobrovol'skiy	12°48'S	129°42'E	38	126
Doerfel	69°06'S	107°54'W	68	184
Dollond	10°24'S	14°24'E	11	119
Donati	20°42'S	5°12'E	36	139
Donna	7°12'N	38°18'E	2	101
Donner	31°24'S	98°00'E	58	162
Doppelmayer	28°30'S	41°24'W	63	155
Doppler	12°36'S	159°36'W	110	110
Douglass	35°54'N	122°24'W	49	53
Dove	46°42'S	31°30'E	30	175
Draper	17°36'N	21°42'W	8	77

Name	Lat.	Long.	Size	Page
Drebbel	40°54'S	49°00'W	30	155
Dreyer	10°00'N	96°54'E	61	104
Drude	38°30'S	91°48'W	24	153
Dryden	33°00'S	155°12'W	51	150
Drygalski	79°18'S	84°54'W	149	196
Dubyago	4°24'N	70°00'E	51	102
Dufay	5°30'N	169°30'E	39	108
Dugan	64°12'N	103°18'E	50	33
Dunér	44°48'N	179°30'E	62	50
Dunthorne	30°06'S	31°36'W	15	156
Dyson	61°18'N	121°12'W	63	25
Dziewulski	21°12'N	98°54'E	63	84
Eckert	17°18'N	58°18'E	2	82
Eddington	21°18'N	72°12'W	118	74
Edison	25°00'N	99°06'E	62	84
Edith	25°48'S	102°18'E	8	144
Egede	48°42'N	10°36'E	37	43
Ehrlich	40°54'N	172°24'W	30	51
Eichstadt	22°36'S	78°18'W	49	134
Eijkman	63°06'S	141°30'W	54	183
Eimmart	24°00'N	64°48'E	46	82
Einstein	16°18'N	88°42'W	198	74
Einthoven	4°54'S	109°36'E	69	125
Elger	35°18'S	29°48'W	21	156
Ellerman	25°18'S	120°06'W	47	132
Ellison	55°06'N	107°30'W	36	39
Elmer	10°06'S	84°06'E	16	123
Elvey	8°48'N	100°30'W	74	93
Emden	63°18'N	177°18'W	111	24
Encke	4°36'N	36°36'W	28	96
Endymion	53°54'N	57°00'E	123	45
Engel'gardt	5°42'N	159°00'W	43	90
Eötvös	35°30'S	133°48'E	99	164
Epigenes	67°30'N	4°36'W	55	29
Epimenides	40°54'S	30°12'W	27	156

Name	Lat.	Long.	Size	Page
Eratosthenes	14°30'N	11°18'W	58	78
Erlanger	86°56'N	28°37'E	9	14
Erro	5°42'N	98°30'E	61	104
Esclangon	21°30'N	42°06'E	15	81
Esnault-Pelterie	47°42'N	141°24'W	79	37
Espin	28°06'N	109°06'E	75	65
Euclides	7°24'S	29°30'W	11	117
Euctemon	76°24'N	31°18'E	62	20
Eudoxus	44°18'N	16°18'E	67	44
Euler	23°18'N	29°12'W	27	77
Evans	9°30'S	133°30'W	67	111
Evdokimov	34°48'N	153°00'W	50	52
Evershed	35°42'N	159°30'W	66	52
Ewen	7°42'N	121°24'E	3	105
Fabbroni	18°42'N	29°12'E	10	80
Fabricius	42°54'S	42°00'E	78	176
Fabry	42°54'N	100°42'E	184	47
Fahrenheit	13°06'N	61°42'E	6	102
Fairouz	26°06'S	102°54'E	3	144
Faraday	42°24'S	8°42'E	69	174
Faustini	87°18'S	77°00'E	39	15
Fauth	6°18'N	20°06'W	12	97
Faye	21°24'S	3°54'E	36	139
Fechner	59°00'S	124°54'E	63	192
Fedorov	28°12'N	37°00'W	6	58
Felix	25°06'N	25°24'W	1	77
Fényi	44°54'S	105°06'W	38	170
Feoktistov	30°54'N	140°42'E	23	67
Fermat	22°36'S	19°48'E	38	140
Fermi	19°18'S	122°36'E	183	145
Fernelius	38°06'S	4°54'E	65	158
Fersman	18°42'N	126°00'W	151	71
Fesenkov	23°12'S	135°06'E	35	146
Feuillée	27°24'N	9°24'W	9	78
Fibiger	86°05'N	37°18'E	17	14

Name	Lat.	Long.	Size	Page
Finsch	23°36'N	21°18'E	4	80
Finsen	42°00'S	177°54'W	72	149
Firmicus	7°18'N	63°24'E	56	102
Firsov	4°30'N	112°12'E	51	105
Fischer	8°00'N	142°24'E	30	106
Fitzgerald	27°30'N	171°42'W	110	69
Fizeau	58°36'S	133°54'W	111	183
Flammarion	3°24'S	3°42'W	74	118
Flamsteed	4°30'S	44°18'W	20	116
Fleming	15°00'N	109°36'E	106	85
Florensky	25°18'N	131°30'E	71	86
Florey	86°49'N	17°50'W	54	14
Focas	33°42'S	93°48'W	22	153
Fontana	16°06'S	56°36'W	31	135
Fontenelle	63°24'N	18°54'W	38	29
Foster	23°42'N	141°30'W	33	71
Foucault	50°24'N	39°42'W	23	41
Fourier	30°18'S	53°00'W	51	155
Fowler	42°18'N	145°00'W	146	37
Fox	0°30'N	98°12'E	24	104
Fra Mauro	6°06'S	17°00'W	101	118
Fracastorius	21°30'S	33°12'E	112	140
Franck	22°36'N	35°30'E	12	80
Franklin	38°48'N	47°42'E	56	62
Franz	16°36'N	40°12'E	25	81
Fraunhofer	39°30'S	59°06'E	56	160
Fredholm	18°24'N	46°30'E	14	81
Freud	25°48'N	52°18'W	2	76
Freundlich	25°00'N	171°00'E	85	88
Fridman	12°36'S	126°00'W	102	111
Froelich	80°18'N	109°42'W	58	17
Frost	37°42'N	118°24'W	75	54
Fryxell	21°18'S	101°24'W	18	133
Furnerius	36°00'S	60°36'E	135	161
G. Bond	32°24'N	36°12'E	20	61

Name	Lat.	Long.	Size	Page
Gadomski	36°24'N	147°18'W	65	52
Gagarin	20°12'S	149°12'E	265	147
Galen	21°54'N	5°00'E	10	79
Galilaei	10°30'N	62°42'W	15	95
Galle	55°54'N	22°18'E	21	44
Galois	14°12'S	151°54'W	222	130
Galvani	49°36'N	84°36'W	80	39
Gambart	1°00'N	15°12'W	25	98
Gamow	65°18'N	145°18'E	129	34
Ganskiy	9°42'S	97°00'E	43	124
Ganswindt	79°36'S	110°18'E	74	200
Garavito	47°30'S	156°42'E	74	181
Gardner	17°42'N	33°48'E	18	80
Gärtner	59°06'N	34°36'E	115	31
Gassendi	17°36'S	40°06'W	101	136
Gaston	30°54'N	34°00'W	2	58
Gaudibert	10°54'S	37°48'E	34	121
Gauricus	33°48'S	12°36'W	79	157
Gauss	35°42'N	79°00'E	177	63
Gavrilov	17°24'N	130°54'E	60	86
Gay-Lussac	13°54'N	20°48'W	26	97
Geber	19°24'S	13°54'E	44	139
Geiger	14°36'S	158°30'E	34	147
Geissler	2°36'S	76°30'E	16	123
Geminus	34°30'N	56°42'E	85	62
Gemma Frisius	34°12'S	13°18'E	87	158
Gerard	44°30'N	80°00'W	90	40
Gerasimovich	22°54'S	122°36'W	86	132
Gernsback	36°30'S	99°42'E	48	162
Gibbs	18°24'S	84°18'E	76	143
Gilbert	3°12'S	76°00'E	112	123
Gill	63°54'S	75°54'E	66	190
Ginzel	14°18'N	97°24'E	55	84
Gioja	83°18'N	2°00'E	41	20
Giordano Bruno	35°54'N	102°48'E	22	65

Name	Lat.	Long.	Size	Page
Glaisher	13°12'N	49°30'E	15	101
Glauber	11°30'N	142°36'E	15	106
Glazenap	1°36'S	137°36'E	43	126
Glushko	8°24'N	77°36'W	43	94
Goclenius	10°00'S	45°00'E	72	121
Goddard	14°48'N	89°00'E	89	83
Godin	1°48'N	10°12'E	34	99
Goldschmidt	73°12'N	3°48'W	113	19
Golgi	27°48'N	60°00'W	5	75
Golitsyn	25°06'S	105°00'W	36	133
Golovin	39°54'N	161°06'E	37	68
Goodacre	32°42'S	14°06'E	46	158
Gore	86°010'N	61°50'W	8	14
Gould	19°12'S	17°12'W	34	138
Grace	14°12'N	35°54'E	1	80
Grachev	3°42'S	108°12'W	35	112
Graff	42°24'S	88°36'W	36	170
Grave	17°06'S	150°18'E	40	147
Greaves	13°12'N	52°42'E	13	101
Green	4°06'N	132°54'E	65	106
Gregory	2°12'N	127°12'E	67	106
Grigg	12°54'N	129°24'W	36	91
Grignard	84°29'N	75°37'W	12	14
Grimaldi	5°30'S	68°18'W	172	115
Grissom	47°00'S	147°24'W	58	168
Grotrian	66°30'S	128°18'E	37	192
Grove	40°18'N	32°54'E	28	61
Gruemberger	66°54'S	10°00'W	93	187
Gruithuisen	32°54'N	39°42'W	15	58
Guericke	11°30'S	14°06'W	63	118
Guillaume	45°24'N	173°24'W	57	36
Gullstrand	45°12'N	129°18'W	43	38
Gum	40°24'S	88°36'E	54	162
Gutenberg	8°36'S	41°12'E	74	121
Guthnick	47°42'S	93°54'W	36	170

Name	Lat.	Long.	Size	Page
Guyot	11°24'N	117°30'E	92	105
Gyldén	5°18'S	0°18'E	47	119
H. G. Wells	40°42'N	122°48'E	114	66
Haber	83°25'N	95°19'W	55	17
Hagecius	59°48'S	46°36'E	76	189
Hagen	48°18'S	135°06'E	55	180
Hahn	31°18'N	73°36'E	84	63
Haidinger	39°12'S	25°00'W	22	156
Hainzel	41°18'S	33°30'W	70	156
Haldane	1°42'S	84°06'E	37	123
Hale	74°12'S	90°48'E	83	200
Hall	33°42'N	37°00'E	35	61
Halley	8°00'S	5°42'E	36	119
Hamilton	42°48'S	84°42'E	57	178
Hanno	56°18'S	71°12'E	56	190
Hansen	14°00'N	72°30'E	39	83
Hansteen	11°30'S	52°00'W	44	116
Harden	5°30'N	143°30'E	15	106
Harding	43°30'N	71°42'W	22	40
Haret	59°00'S	176°30'W	29	182
Hargreaves	2°12'S	64°00'E	16	122
Harkhebi	39°36'N	98°18'E	237	64
Harlan	38°30'S	79°30'E	65	161
Harold	10°54'S	6°00'W	2	118
Harpalus	52°36'N	43°24'W	39	41
Harriot	33°06'N	114°18'E	56	65
Hartmann	3°12'N	135°18'E	61	106
Hartwig	6°06'S	80°30'W	79	114
Harvey	19°30'N	146°30'W	60	70
Hase	29°24'S	62°30'E	83	161
Haskin	81°31'N	133°58'E	58	22
Hatanaka	29°42'N	121°30'W	26	53
Hausen	65°00'S	88°06'W	167	185
Haworth	86°54'S	4°00'W	35	15
Hayford	12°42'N	176°24'W	27	89

Name	Lat.	Long.	Size	Page
Hayn	64°42'N	85°12'E	87	32
Healy	32°48'N	110°30'W	38	54
Heaviside	10°24'S	167°06'E	165	128
Hecataeus	21°48'S	79°24'E	167	143
Hédervári	81°48'S	84°00'E	69	199
Hedin	2°00'N	76°30'W	150	94
Heinrich	24°48'N	15°18'W	6	78
Heinsius	39°30'S	17°42'W	64	157
Heis	32°24'N	31°54'W	14	58
Helberg	22°30'N	102°12'W	62	73
Helicon	40°24'N	23°06'W	24	58
Hell	32°24'S	7°48'W	33	157
Helmert	7°36'S	87°36'E	26	123
Helmholtz	68°06'S	64°06'E	94	190
Henderson	4°48'N	152°06'E	47	107
Hendrix	46°36'S	159°12'W	18	167
Henry	24°00'S	56°48'W	41	135
Henry Frères	23°30'S	58°54'W	42	135
Henyey	13°30'N	151°36'W	63	90
Heraclitus	49°12'S	6°12'E	90	174
Hercules	46°42'N	39°06'E	69	45
Herigonius	13°18'S	33°54'W	15	117
Hermann	0°54'S	57°00'W	15	115
Hermite	86°00'N	89°54'W	104	14
Herodotus	23°12'N	49°42'W	34	76
Heron	0°42'N	119°48'E	24	105
Herschel	5°42'S	2°06'W	40	118
Hertz	13°24'N	104°30'E	90	104
Hertzsprung	2°36'N	129°12'W	591	91
Hesiodus	29°24'S	16°18'W	42	157
Hess	54°18'S	174°36'E	88	181
Hevelius	2°12'N	67°36'W	115	95
Hevesy	83°08'N	150°06'E	49	23
Heymans	75°18'N	144°06'W	50	16
Heyrovsky	39°36'S	95°18'W	16	153

Name	Lat.	Long.	Size	Page
Hilbert	17°54'S	108°12'E	151	145
Hill	20°54'N	40°48'E	16	81
Hind	7°54'S	7°24'E	29	119
Hinshelwood	89°22'N	46°22'W	14	14
Hippalus	24°48'S	30°12'W	57	137
Hipparchus	5°06'S	5°12'E	138	119
Hippocrates	70°42'N	145°54'W	60	16
Hirayama	6°06'S	93°30'E	132	124
Hoffmeister	15°12'N	136°54'E	45	86
Hogg	33°36'N	121°54'E	38	66
Hohmann	17°54'S	94°06'W	16	133
Holden	19°06'S	62°30'E	47	142
Holetschek	27°36'S	150°54'E	38	147
Hommel	54°42'S	33°48'E	126	175
Hooke	41°12'N	54°54'E	36	62
Hopmann	50°48'S	160°18'E	88	181
Hornsby	23°48'N	12°30'E	3	79
Horrebow	58°42'N	40°48'W	24	28
Horrocks	4°00'S	5°54'E	30	119
Hortensius	6°30'N	28°00'W	14	97
Houssay	83°05'N	98°31'E	24	22
Houtermans	9°24'S	87°12'E	29	123
Houzeau	17°06'S	123°30'W	71	132
Hubble	22°06'N	86°54'E	80	83
Huggins	41°06'S	1°24'W	65	157
Humason	30°42'N	56°36'W	4	57
Humboldt	27°00'S	80°54'E	189	143
Hume	4°42'S	90°24'E	23	124
Husband	40°48'S	147°54'W	29	150
Hutton	37°18'N	168°42'E	50	68
Huxley	20°12'N	4°30'W	4	78
Hyginus	7°48'N	6°18'E	9	99
Hypatia	4°18'S	22°36'E	40	120
Ian	25°42'N	0°24'W	1	78
Ibn Bajja	86°24'S	72°49'W	12	15

Name	Lat.	Long.	Size	Page
Ibn Battuta	6°54'S	50°24'E	11	121
Ibn Firnas	6°48'N	122°18'E	89	105
Ibn Yunus	14°06'N	91°06'E	58	84
Ibn-Rushd	11°42'S	21°42'E	32	120
Icarus	5°18'S	173°12'W	96	109
Ideler	49°12'S	22°18'E	38	175
Idel'son	81°30'S	110°54'E	60	200
Il'in	17°48'S	97°30'W	13	133
Ina	18°36'N	5°18'E	3	79
Ingalls	26°24'N	153°06'W	37	70
Inghirami	47°30'S	68°48'W	91	171
Innes	27°48'N	119°12'E	42	85
Ioffe	14°24'S	129°12'W	86	131
Isabel	28°12'N	34°06'W	1	58
Isaev	17°30'S	147°30'E	90	147
Isidorus	8°00'S	33°30'E	42	120
Isis	18°54'N	27°30'E	1	80
Ivan	26°54'N	43°18'W	4	76
Izsak	23°18'S	117°06'E	30	145
J. Herschel	62°00'N	42°00'W	165	28
Jackson	22°24'N	163°06'W	71	69
Jacobi	56°42'S	11°24'E	68	188
Jansen	13°30'N	28°42'E	23	100
Jansky	8°30'N	89°30'E	72	103
Janssen	45°24'S	40°18'E	199	176
Jarvis	34°54'S	147°30'W	38	150
Jeans	55°48'S	91°24'E	79	178
Jehan	20°42'N	31°54'W	5	77
Jenkins	0°18'N	78°06'E	38	103
Jenner	42°06'S	95°54'E	71	178
Jerik	18°30'N	27°36'E	1	80
Joliot	25°48'N	93°06'E	164	84
Jomo	24°24'N	2°24'E	7	79
José	12°42'S	1°36'W	2	118
Joule	27°18'N	144°12'W	96	70

Name	Lat.	Long.	Size	Page
Joy	25°00'N	6°36'E	5	79
Jules Verne	35°00'S	147°00'E	143	165
Julienne	26°00'N	3°12'E	2	79
Julius Caesar	9°00'N	15°24'E	90	99
Kaiser	36°30'S	6°30'E	52	158
Kamerlingh Onnes	15°00'N	115°48'W	66	72
Kane	63°06'N	26°06'E	54	30
Kant	10°36'S	20°06'E	33	120
Kao	6°42'S	87°36'E	34	123
Kapteyn	10°48'S	70°36'E	49	122
Karima	25°54'S	103°00'E	3	144
Karpinskiy	73°18'N	166°18'E	92	23
Karrer	52°06'S	141°48'W	51	168
Kasper	8°18'N	122°06'E	12	105
Kästner	6°48'S	78°30'E	108	123
Katchalsky	5°54'N	116°06'E	32	105
Kathleen	25°24'N	0°42'W	5	78
Kearons	11°24'S	112°36'W	23	112
Keeler	10°12'S	161°54'E	160	127
Kekulé	16°24'N	138°06'W	94	71
Keldysh	51°12'N	43°36'E	33	45
Kepínski	28°48'N	126°36'E	31	66
Kepler	8°06'N	38°00'W	31	96
Khvol'son	13°48'S	111°24'E	54	125
Kibal'chich	3°00'N	146°30'W	92	90
Kidinnu	35°54'N	122°54'E	56	66
Kies	26°18'S	22°30'W	45	137
Kiess	6°24'S	84°00'E	63	123
Kimura	57°06'S	118°24'E	28	191
Kinau	60°48'S	15°06'E	41	188
King	5°00'N	120°30'E	76	105
Kira	17°36'S	132°48'E	3	146
Kirch	39°12'N	5°36'W	11	59
Kircher	67°06'S	45°18'W	72	186
Kirchhoff	30°18'N	38°48'E	24	61

Name	Lat.	Long.	Size	Page
Kirkwood	68°48'N	156°06'W	67	24
Klaproth	69°48'S	26°00'W	119	187
Klein	12°00'S	2°36'E	44	119
Kleymenov	32°24'S	140°12'W	55	150
Klute	37°12'N	141°18'W	75	52
Knox-Shaw	5°18'N	80°12'E	12	103
Koch	42°48'S	150°06'E	95	180
Kocher	84°36'S	134°30'W	21	15
Kohlschütter	14°24'N	154°00'E	53	87
Kolhörster	11°12'N	114°36'W	97	92
Komarov	24°42'N	152°30'E	78	87
Kondratyuk	14°54'S	115°30'E	108	145
König	24°06'S	24°36'W	23	137
Konoplev	28°30'S	125°30'W	25	151
Konstantinov	19°48'N	158°24'E	66	87
Kopff	17°24'S	89°36'W	41	134
Korolev	4°00'S	157°24'W	437	110
Kosberg	20°12'S	149°36'E	15	147
Kostinskiy	14°42'N	118°48'E	75	85
Kovalevskaya	30°48'N	129°36'W	115	53
Koval'skiy	21°54'S	101°00'E	49	144
Kozyrev	46°48'S	129°18'E	65	179
Krafft	16°36'N	72°36'W	51	74
Kramarov	2°18'S	98°48'W	20	113
Kramers	53°36'N	127°36'W	61	38
Krasnov	29°54'S	79°36'W	40	154
Krasovskiy	3°54'N	175°30'W	59	89
Kreiken	9°00'S	84°36'E	23	123
Krieger	29°00'N	45°36'W	22	57
Krogh	9°24'N	65°42'E	19	102
Krusenstern	26°12'S	5°54'E	47	139
Krylov	35°36'N	165°48'W	49	51
Kugler	53°48'S	103°42'E	65	178
Kuhn	84°36'S	153°38'W	16	15
Kuiper	9°48'S	22°42'W	6	117

Name	Lat.	Long.	Size	Page
Kulik	42°24'N	154°30'W	58	37
Kundt	11°30'S	11°30'W	10	118
Kunowsky	3°12'N	32°30'W	18	97
Kuo Shou Ching	8°24'N	133°42'W	34	91
Kurchatov	38°18'N	142°06'E	106	67
L. Clark	43°42'S	147°42'W	16	168
La Caille	23°48'S	1°06'E	67	139
La Condamine	53°24'N	28°12'W	37	42
La Pérouse	10°42'S	76°18'E	77	123
Lacchini	41°42'N	107°30'W	58	54
Lacroix	37°54'S	59°00'W	37	155
Lade	1°18'S	10°06'E	55	119
Lagalla	44°36'S	22°30'W	85	173
Lagrange	32°18'S	72°48'W	225	154
Lalande	4°24'S	8°36'W	24	118
Lallemand	14°18'S	84°06'W	18	134
Lamarck	22°54'S	69°48'W	100	135
Lamb	42°54'S	100°06'E	106	178
Lambert	25°48'N	21°00'W	30	77
Lamé	14°42'S	64°30'E	84	142
Lamèch	42°42'N	13°06'E	13	44
Lamont	4°24'N	23°42'E	106	100
Lampland	31°00'S	131°00'E	65	164
Landau	41°36'N	118°06'W	214	54
Lander	15°18'S	131°48'E	40	146
Landsteiner	31°18'N	14°48'W	6	59
Lane	9°30'S	132°00'E	55	126
Langemak	10°18'S	118°42'E	97	125
Langevin	44°18'N	162°42'E	58	50
Langley	51°06'N	86°18'W	59	39
Langmuir	35°42'S	128°24'W	91	151
Langrenus	8°54'S	61°06'E	127	122
Lansberg	0°18'S	26°36'W	38	117
Larmor	32°06'N	179°42'W	97	51
Lassell	15°30'S	7°54'W	23	138

Name	Lat.	Long.	Size	Page
Laue	28°00'N	96°42'W	87	55
Lauritsen	27°36'S	96°06'E	52	144
Laveran	81°56'S	160°38'W	12	194
Lavoisier	38°12'N	81°12'W	70	55
Lawrence	7°24'N	43°12'E	24	101
Le Gentil	74°36'S	75°42'W	128	196
Le Monnier	26°36'N	30°36'E	60	80
Le Verrier	40°18'N	20°36'W	20	58
Leakey	3°12'S	37°24'E	12	121
Leavitt	44°48'S	139°18'W	66	168
Lebedev	47°18'S	107°48'E	102	178
Lebedinskiy	8°18'N	164°18'W	62	89
Lebesgue	5°06'S	89°00'E	11	123
Lee	30°42'S	40°42'W	41	155
Leeuwenhoek	29°18'S	178°42'W	125	149
Legendre	28°54'S	70°12'E	78	161
Lehmann	40°00'S	56°00'W	53	155
Leibnitz	38°18'S	179°12'E	245	166
Lemaître	61°12'S	149°36'W	91	183
Lenard	85°12'N	105°00'W	48	14
Lents	2°48'N	102°06'W	21	93
Leonov	19°00'N	148°12'E	33	87
Lepaute	33°18'S	33°36'W	16	156
Letronne	10°48'S	42°30'W	116	116
Leucippus	29°06'N	116°00'W	56	54
Leuschner	1°48'N	108°48'W	49	92
Levi-Civita	23°42'S	143°24'E	121	146
Lewis	18°30'S	113°48'W	42	132
Lexell	35°48'S	4°12'W	62	157
Ley	42°12'N	154°54'E	79	49
Licetus	47°06'S	6°42'E	74	174
Lichtenberg	31°48'N	67°42'W	20	56
Lick	12°24'N	52°42'E	31	101
Liebig	24°18'S	48°12'W	37	136
Lilius	54°30'S	6°12'E	61	174

Name	Lat.	Long.	Size	Page
Linda	30°42'N	33°24'W	1	58
Lindbergh	5°24'S	52°54'E	12	121
Lindblad	70°24'N	98°48'W	66	17
Lindenau	32°18'S	24°54'E	53	159
Lindsay	7°00'S	13°00'E	32	119
Linné	27°42'N	11°48'E	2	79
Liouville	2°36'N	73°30'E	16	103
Lippershey	25°54'S	10°18'W	6	138
Lippmann	56°00'S	114°54'W	160	169
Lipskiy	2°12'S	179°30'W	80	109
Litke	16°48'S	123°06'E	39	145
Littrow	21°30'N	31°24'E	30	80
Lobachevskiy	9°54'N	112°36'E	84	105
Lockyer	46°12'S	36°42'E	34	176
Lodygin	17°42'S	146°48'W	62	130
Loewy	22°42'S	32°48'W	24	137
Lohrmann	0°30'S	67°12'W	30	115
Lohse	13°42'S	60°12'E	41	122
Lomonosov	27°18'N	98°00'E	92	84
Longomontanus	49°36'S	21°48'W	157	173
Lorentz	32°36'N	95°18'W	312	55
Louise	28°30'N	34°12'W	0	58
Louville	44°00'N	46°00'W	36	41
Love	6°18'S	129°00'E	84	126
Lovelace	82°18'N	106°24'W	54	17
Lovell	36°48'S	141°54'W	34	150
Lowell	12°54'S	103°06'W	66	113
Lubbock	3°54'S	41°48'E	13	121
Lubiniezky	17°48'S	23°48'W	43	137
Lucian	14°18'N	36°42'E	7	81
Lucretius	8°12'S	120°48'W	63	112
Ludwig	7°42'S	97°24'E	23	124
Lundmark	39°42'S	152°30'E	106	165
Luther	33°12'N	24°06'E	9	61
Lyapunov	26°18'N	89°18'E	66	83

Name	Lat.	Long.	Size	Page
Lyell	13°36'N	40°36'E	32	101
Lyman	64°48'S	163°36'E	84	193
Lyot	49°48'S	84°30'E	132	178
M. Anderson	41°36'S	149°00'W	17	150
Mach	18°30'N	149°18'W	180	70
Maclaurin	1°54'S	68°00'E	50	122
Maclear	10°30'N	20°06'E	20	100
MacMillan	24°12'N	7°48'W	7	78
Macrobius	21°18'N	46°00'E	64	81
Mädler	11°00'S	29°48'E	27	120
Maestlin	4°54'N	40°36'W	7	96
Magelhaens	11°54'S	44°06'E	40	121
Maginus	50°30'S	6°18'W	194	174
Main	80°48'N	10°06'E	46	20
Mairan	41°36'N	43°24'W	40	57
Maksutov	40°30'S	168°42'W	83	149
Malapert	84°54'S	12°54'E	69	15
Mallet	45°24'S	54°12'E	58	176
Malyy	21°54'N	105°18'E	41	84
Mandel'shtam	5°24'N	162°24'E	197	108
Manilius	14°30'N	9°06'E	38	79
Manners	4°36'N	20°00'E	15	100
Manuel	24°30'N	11°18'E	0	79
Manzinus	67°42'S	26°48'E	98	188
Maraldi	19°24'N	34°54'E	39	80
Marci	22°36'N	167°00'W	25	69
Marco Polo	15°24'N	2°00'W	28	78
Marconi	9°36'S	145°06'E	73	127
Marinus	39°24'S	76°30'E	58	161
Mariotte	28°30'S	139°06'W	65	151
Marius	11°54'N	50°48'W	41	96
Markov	53°24'N	62°42'W	40	40
Marth	31°06'S	29°18'W	6	156
Mary	18°54'N	27°24'E	1	80
Maskelyne	2°12'N	30°06'E	23	100

Name	Lat.	Long.	Size	Page
Mason	42°36'N	30°30'E	33	44
Maunder	14°36'S	93°48'W	55	133
Maupertuis	49°36'N	27°18'W	45	42
Maurolycus	42°00'S	14°00'E	114	158
Maury	37°06'N	39°36'E	17	61
Mavis	29°48'N	26°24'W	1	58
Maxwell	30°12'N	98°54'E	107	64
McAdie	2°06'N	92°06'E	45	104
McAuliffe	33°00'S	148°54'W	19	150
McClure	15°18'S	50°18'E	23	141
McCool	41°42'S	146°18'W	21	150
McDonald	30°24'N	20°54'W	7	58
McKellar	15°42'S	170°48'W	51	129
McLaughlin	47°06'N	92°54'W	79	39
McMath	17°18'N	165°36'W	86	69
McNair	35°42'S	147°18'W	29	150
McNally	22°36'N	127°12'W	47	71
Mechnikov	11°00'S	149°00'W	60	110
Mee	43°42'S	35°18'W	126	173
Mees	13°36'N	96°06'W	50	93
Meggers	24°18'N	123°00'E	52	85
Meitner	10°30'S	112°42'E	87	125
Melissa	8°06'N	121°48'E	18	105
Mendel	48°48'S	109°24'W	138	169
Mendeleev	5°42'N	140°54'E	313	106
Menelaus	16°18'N	16°00'E	26	79
Menzel	3°24'N	36°54'E	3	101
Mercator	29°18'S	26°06'W	46	156
Mercurius	46°36'N	66°12'E	67	46
Merrill	75°12'N	116°18'W	57	17
Mersenius	21°30'S	49°12'W	84	136
Meshcherskiy	12°12'N	125°30'E	65	105
Messala	39°12'N	60°30'E	125	63
Messier	1°54'S	47°36'E	11	121
Metius	40°18'S	43°18'E	87	160

Name	Lat.	Long.	Size	Page
Meton	73°36'N	18°48'E	130	20
Mezentsev	72°06'N	128°42'W	89	17
Michael	25°06'N	0°12'E	4	79
Michelson	7°12'N	120°42'W	123	92
Milanković	77°12'N	168°48'E	101	23
Milichius	10°00'N	30°12'W	12	97
Miller	39°18'S	0°48'E	61	158
Millikan	46°48'N	121°30'E	98	48
Mills	8°36'N	156°00'E	32	107
Milne	31°24'S	112°12'E	272	163
Mineur	25°00'N	161°18'W	73	70
Minkowski	56°30'S	146°00'W	113	183
Minnaert	67°48'S	179°36'E	125	193
Mitchell	49°42'N	20°12'E	30	44
Mitra	18°00'N	154°42'W	92	70
Möbius	15°48'N	101°12'E	50	84
Mohorovičić	19°00'S	165°00'W	51	129
Moigno	66°24'N	28°54'E	36	30
Moiseev	9°30'N	103°18'E	59	104
Moissan	4°48'N	137°24'E	21	106
Moltke	0°36'S	24°12'E	6	120
Monge	19°12'S	47°36'E	36	141
Monira	12°36'S	1°42'W	2	118
Montanari	45°48'S	20°36'W	76	173
Montgolfier	47°18'N	159°48'W	88	36
Moore	37°24'N	177°30'W	54	51
Moretus	70°36'S	5°48'W	111	197
Morley	2°48'S	64°36'E	14	122
Morozov	5°00'N	127°24'E	42	106
Morse	22°06'N	175°06'W	77	69
Moseley	20°54'N	90°06'W	90	73
Mösting	0°42'S	5°54'W	24	118
Mouchez	78°18'N	26°36'W	81	19
Moulton	61°06'S	97°12'E	49	191
Müller	7°36'S	2°06'E	22	119

Name	Lat.	Long.	Size	Page
Murakami	23°18'S	140°30'W	45	131
Murchison	5°06'N	0°06'W	57	98
Mutus	63°36'S	30°06'E	77	189
Nagaoka	19°24'N	154°00'E	46	87
Nansen	80°54'N	95°18'E	104	22
Naonobu	4°36'S	57°48'E	34	122
Nasireddin	41°00'S	0°12'E	52	158
Nasmyth	50°30'S	56°12'W	76	172
Nassau	24°54'S	177°24'E	76	148
Natasha	20°00'N	31°18'W	12	77
Naumann	35°24'N	62°00'W	9	56
Neander	31°18'S	39°54'E	50	159
Nearch	58°30'S	39°06'E	75	189
Necho	5°00'S	123°06'E	30	125
Nefed'ev	81°08'S	135°09'E	40	200
Neison	68°18'N	25°06'E	53	30
Neper	8°30'N	84°36'E	137	103
Nernst	35°18'N	94°48'W	116	55
Neujmin	27°00'S	125°00'E	101	145
Neumayer	71°06'S	70°42'E	76	199
Newcomb	29°54'N	43°48'E	41	62
Newton	76°42'S	16°54'W	78	197
Nicholson	26°12'S	85°06'W	38	134
Nicolai	42°24'S	25°54'E	42	175
Nicollet	21°54'S	12°30'W	15	138
Nielsen	31°48'N	51°48'W	9	57
Niepce	72°42'N	119°06'W	57	17
Nijland	33°00'N	134°06'E	35	66
Nikolaev	35°12'N	151°18'E	41	67
Nishina	44°36'S	170°24'W	65	167
Nobel	15°00'N	101°18'W	48	73
Nobile	85°12'S	53°30'E	73	15
Nobili	0°12'N	75°54'E	42	103
Nöggerath	48°48'S	45°42'W	30	172
Nonius	34°48'S	3°48'E	69	158

Name	Lat.	Long.	Size	Page
Norman	11°48'S	30°24'W	10	117
Nöther	66°36'N	113°30'W	67	26
Numerov	70°42'S	160°42'W	113	194
Nunn	4°36'N	91°06'E	19	104
Nušl	32°18'N	167°36'E	61	68
Oberth	62°24'N	155°24'E	60	35
Obruchev	38°54'S	162°06'E	71	166
O'Day	30°36'S	157°30'E	71	165
Oenopides	57°00'N	64°06'W	67	27
Oersted	43°06'N	47°12'E	42	45
Ohm	18°24'N	113°30'W	64	72
Oken	43°42'S	75°54'E	71	177
Olbers	7°24'N	75°54'W	74	94
Olcott	20°36'N	117°48'E	81	85
Olivier	59°06'N	138°30'E	69	34
Omar Khayyam	58°00'N	102°06'W	70	26
Onizuka	36°12'S	148°54'W	29	150
Opelt	16°18'S	17°30'W	48	138
Oppenheimer	35°12'S	166°18'W	208	149
Oppolzer	1°30'S	0°30'W	40	118
Oresme	42°24'S	169°12'E	76	181
Orlov	25°42'S	175°00'W	81	129
Orontius	40°36'S	4°36'W	105	157
Osama	18°36'N	5°12'E	0	79
Osiris	18°36'N	27°36'E	1	80
Osman	11°00'S	6°12'W	2	118
Ostwald	10°24'N	121°54'E	104	105
Palisa	9°24'S	7°12'W	33	118
Palitzsch	28°00'S	64°30'E	41	142
Pallas	5°30'N	1°36'W	46	98
Palmieri	28°36'S	47°42'W	40	155
Paneth	63°00'N	94°48'W	65	26
Pannekoek	4°12'S	140°30'E	71	126
Papaleksi	10°12'N	164°00'E	97	108
Paracelsus	23°00'S	163°06'E	83	148

Name	Lat.	Long.	Size	Page
Paraskevopoulos	50°24'N	149°54'W	94	37
Parenago	25°54'N	108°30'W	93	72
Parkhurst	33°24'S	103°36'E	96	163
Parrot	14°30'S	3°18'E	70	139
Parry	7°54'S	15°48'W	47	118
Parsons	37°18'N	171°12'W	40	51
Pascal	74°36'N	70°18'W	115	18
Paschen	13°30'S	139°48'W	124	111
Pasteur	11°54'S	104°36'E	224	124
Patricia	25°00'N	0°18'E	5	79
Patsaev	16°42'S	133°24'E	55	146
Pauli	44°30'S	137°30'E	84	180
Pavlov	28°48'S	142°30'E	148	165
Pawsey	44°30'N	145°00'E	60	49
Peary	88°36'N	33°00'E	73	14
Pease	12°30'N	106°06'W	38	93
Peek	2°36'N	86°54'E	12	103
Peirce	18°18'N	53°30'E	18	81
Peirescius	46°30'S	67°36'E	61	177
Pentland	64°36'S	11°30'E	56	188
Perel'man	24°00'S	106°00'E	46	144
Perepelkin	10°00'S	129°00'E	97	126
Perkin	47°12'N	175°54'W	62	36
Perrine	42°30'N	127°48'W	86	38
Petavius	25°06'S	60°24'E	188	142
Petermann	74°12'N	66°18'E	73	21
Peters	68°06'N	29°30'E	15	30
Petit	2°18'N	63°30'E	5	102
Petrie	45°18'N	108°24'E	33	48
Petropavlovskiy	37°12'N	114°48'W	63	54
Petrov	61°24'S	88°00'E	49	190
Pettit	27°30'S	86°36'W	35	134
Petzval	62°42'S	110°24'W	90	184
Phillips	26°36'S	75°18'E	122	143
Philolaus	72°06'N	32°24'W	70	19

Name	Lat.	Long.	Size	Page
Phocylides	52°42'S	57°00'W	121	172
Piazzi	36°36'S	67°54'W	134	154
Piazzi Smyth	41°54'N	3°12'W	13	59
Picard	14°36'N	54°42'E	22	82
Piccolomini	29°42'S	32°12'E	87	159
Pickering	2°54'S	7°00'E	15	119
Pictet	43°36'S	7°24'W	62	174
Pikel'ner	47°54'S	123°18'E	47	179
Pilâtre	60°12'S	86°54'W	50	185
Pingré	58°42'S	73°42'W	88	185
Pirquet	20°18'S	139°36'E	65	146
Pitatus	29°54'S	13°30'W	106	157
Pitiscus	50°24'S	30°54'E	82	175
Pizzetti	34°54'S	118°48'E	44	163
Plana	42°12'N	28°12'E	44	44
Planck	57°54'S	136°48'E	314	192
Planté	10°12'S	163°18'E	37	128
Plaskett	82°06'N	174°18'E	109	23
Plato	51°36'N	9°24'W	109	43
Playfair	23°30'S	8°24'E	47	139
Plinius	15°24'N	23°42'E	43	80
Plummer	25°00'S	155°00'W	73	130
Plutarch	24°06'N	79°00'E	68	83
Poczobutt	57°06'N	98°48'W	195	26
Pogson	42°12'S	110°30'E	50	179
Poincaré	56°42'S	163°36'E	319	193
Poinsot	79°30'N	145°42'W	68	16
Poisson	30°24'S	10°36'E	42	158
Polybius	22°24'S	25°36'E	41	140
Polzunov	25°18'N	114°36'E	67	85
Pomortsev	0°42'N	66°54'E	23	102
Poncelet	75°48'N	54°06'W	69	18
Pons	25°18'S	21°30'E	41	140
Pontanus	28°24'S	14°24'E	57	158
Pontécoulant	58°42'S	66°00'E	91	190

Name	Lat.	Long.	Size	Page
Popov	17°12'N	99°42'E	65	84
Porter	56°06'S	10°06'W	51	187
Posidonius	31°48'N	29°54'E	95	61
Poynting	18°06'N	133°24'W	128	71
Prager	3°54'S	130°30'E	60	126
Prandtl	60°06'S	141°48'E	91	192
Priestley	57°18'S	108°24'E	52	191
Prinz	25°30'N	44°06'W	46	76
Priscilla	10°54'S	6°12'W	1	118
Proclus	16°06'N	46°48'E	28	81
Proctor	46°24'S	5°06'W	52	174
Protagoras	56°00'N	7°18'E	21	30
Ptolemaeus	9°18'S	1°54'W	164	118
Puiseux	27°48'S	39°00'W	24	136
Pupin	23°48'N	11°00'W	2	78
Purbach	25°30'S	2°18'W	115	138
Purkyně	1°36'S	94°54'E	48	124
Pythagoras	63°30'N	63°00'W	142	27
Pytheas	20°30'N	20°36'W	20	77
Quételet	43°06'N	134°54'W	55	37
Rabbi Levi	34°42'S	23°36'E	81	159
Racah	13°48'S	179°48'W	63	109
Raimond	14°36'N	159°18'W	70	70
Raman	27°00'N	55°06'W	10	75
Ramon	41°36'S	148°06'W	17	150
Ramsay	40°12'S	144°30'E	81	165
Ramsden	32°54'S	31°48'W	24	156
Rankine	3°54'S	71°30'E	8	122
Raspletin	22°30'S	151°48'E	48	147
Ravi	12°30'S	1°54'W	2	118
Rayet	44°42'N	114°30'E	27	48
Rayleigh	29°18'N	89°36'E	114	64
Razumov	39°06'N	114°18'W	70	54
Réaumur	2°24'S	0°42'E	52	119
Recht	9°48'N	124°00'E	20	105

Name	Lat.	Long.	Size	Page
Regiomontanus	28°18'S	1°00'W	108	157
Regnault	54°06'N	88°00'W	46	39
Reichenbach	30°18'S	48°00'E	71	160
Reimarus	47°42'S	60°18'E	48	177
Reiner	7°00'N	54°54'W	29	95
Reinhold	3°18'N	22°48'W	42	97
Repsold	51°18'N	78°36'W	109	40
Resnik	33°48'S	150°06'W	20	150
Respighi	2°48'N	71°54'E	18	102
Rhaeticus	0°00'N	4°54'E	45	99
Rheita	37°06'S	47°12'E	70	160
Riccioli	3°18'S	74°36'W	139	114
Riccius	36°54'S	26°30'E	71	159
Ricco	75°36'N	176°18'E	65	23
Richards	7°42'N	140°06'E	16	106
Richardson	31°06'N	100°30'E	141	65
Riedel	48°54'S	139°36'W	47	168
Riemann	38°54'N	86°48'E	163	64
Ritchey	11°06'S	8°30'E	24	119
Rittenhouse	74°30'S	106°30'E	26	200
Ritter	2°00'N	19°12'E	29	100
Ritz	15°06'S	92°12'E	51	144
Robert	19°00'N	27°24'E	1	80
Roberts	71°06'N	174°30'W	89	16
Robertson	21°48'N	105°12'W	88	73
Robinson	59°00'N	45°54'W	24	28
Rocca	12°42'S	72°48'W	89	114
Rocco	28°54'N	45°00'W	4	57
Roche	42°18'S	136°30'E	160	180
Romeo	7°30'N	122°36'E	8	105
Römer	25°24'N	36°24'E	39	81
Röntgen	33°00'N	91°24'W	126	55
Rosa	20°18'N	32°18'W	1	77
Rosenberger	55°24'S	43°06'E	95	176
Ross	11°42'N	21°42'E	24	100

Name	Lat.	Long.	Size	Page
Rosse	17°54'S	35°00'E	11	140
Rosseland	41°00'S	131°00'E	75	164
Rost	56°24'S	33°42'W	48	186
Rothmann	30°48'S	27°42'E	42	159
Rowland	57°24'N	162°30'W	171	24
Rozhdestvenskiy	85°12'N	155°24'W	177	14
Rumford	28°48'S	169°48'W	61	149
Runge	2°30'S	86°42'E	38	123
Russell	26°30'N	75°24'W	103	74
Ruth	28°42'N	45°06'W	3	57
Rutherford	10°42'N	137°00'E	13	106
Rutherfurd	60°54'S	12°06'W	48	187
Rydberg	46°30'S	96°18'W	49	170
Ryder	44°30'S	143°12'E	17	180
Rynin	47°00'N	103°30'W	75	39
Sabatier	13°12'N	79°00'E	10	103
Sabine	1°24'N	20°06'E	30	100
Sacrobosco	23°42'S	16°42'E	98	139
Saenger	4°18'N	102°24'E	75	104
Šafařík	10°36'N	176°54'E	27	108
Saha	1°36'S	102°42'E	99	124
Samir	28°30'N	34°18'W	2	58
Sampson	29°42'N	16°30'W	1	59
Sanford	32°36'N	138°54'W	55	53
Santbech	20°54'S	44°00'E	64	141
Santos-Dumont	27°42'N	4°48'E	8	79
Sarabhai	24°42'N	21°00'E	7	80
Sarton	49°18'N	121°06'W	69	38
Sasserides	39°06'S	9°18'W	90	157
Saunder	4°12'S	8°48'E	44	119
Saussure	43°24'S	3°48'W	54	174
Scaliger	27°06'S	108°54'E	84	145
Schaeberle	26°12'S	117°12'E	62	145
Scheele	9°24'S	37°48'W	4	116
Scheiner	60°30'S	27°30'W	110	187

Name	Lat.	Long.	Size	Page
Schiaparelli	23°24'N	58°48'W	24	75
Schickard	44°18'S	55°18'W	206	172
Schiller	51°54'S	39°00'W	180	172
Schjellerup	69°42'N	157°06'E	62	35
Schlesinger	47°24'N	138°36'W	97	37
Schliemann	2°06'S	155°12'E	80	127
Schlüter	5°54'S	83°18'W	89	114
Schmidt	1°00'N	18°48'E	11	100
Schneller	41°48'N	163°36'W	54	51
Schomberger	76°42'S	24°54'E	85	198
Schönfeld	44°48'N	98°06'W	25	39
Schorr	19°30'S	89°42'E	53	143
Schrödinger	75°00'S	132°24'E	312	200
Schröter	2°36'N	7°00'W	35	98
Schubert	2°48'N	81°00'E	54	103
Schumacher	42°24'N	60°42'E	60	46
Schuster	4°12'N	146°30'E	108	107
Schwabe	65°06'N	45°36'E	25	31
Schwarzschild	70°06'N	121°12'E	212	22
Scobee	31°06'S	148°54'W	40	150
Scoresby	77°42'N	14°06'E	55	20
Scott	82°06'S	48°30'E	103	199
Seares	73°30'N	145°48'E	110	23
Secchi	2°24'N	43°30'E	22	101
Sechenov	7°06'S	142°36'W	62	111
Seeliger	2°12'S	3°00'E	8	119
Segers	47°06'N	127°42'E	17	48
Segner	58°54'S	48°18'W	67	186
Seidel	32°48'S	152°12'E	62	165
Seleucus	21°00'N	66°36'W	43	75
Seneca	26°36'N	80°12'E	46	83
Seyfert	29°06'N	114°36'E	110	65
Shackleton	89°54'S	0°00'E	19	15
Shahinaz	7°30'N	122°24'E	15	105
Shaler	32°54'S	85°12'W	48	153

Name	Lat.	Long.	Size	Page
Shapley	9°24'N	56°54'E	23	102
Sharonov	12°24'N	173°18'E	74	108
Sharp	45°42'N	40°12'W	39	41
Shatalov	24°18'N	141°30'E	21	86
Shayn	32°36'N	172°30'E	93	68
Sheepshanks	59°12'N	16°54'E	25	30
Sherrington	11°06'S	118°00'E	18	125
Shi Shen	76°00'N	104°06'E	43	22
Shirakatsi	12°06'S	128°36'E	51	126
Shoemaker		88°06'W	44	15
Short	74°36'S	7°18'W	70	197
Shternberg	19°30'N	116°18'W	70	72
Shuckburgh	42°36'N	52°48'E	38	45
Shuleykin	27°06'S	92°30'W	15	133
Siedentopf	22°00'N	135°30'E	61	86
Sierpinski	27°12'S	154°30'E	69	147
Sikorsky	66°06'S	103°12'E	98	191
Silberschlag	6°12'N	12°30'E	13	99
Simpelius	73°00'S	15°12'E	70	198
Sinas	8°48'N	31°36'E	11	100
Sirsalis	12°30'S	60°24'W	42	115
Sisakyan	41°12'N	109°00'E	34	65
Sita	4°36'N	120°48'E	2	105
Sklodowska	18°12'S	95°30'E	127	144
Slipher	49°30'N	160°06'E	69	50
Slocum	3°00'S	89°00'E	13	123
Smith	31°36'S	150°12'W	34	150
Smithson	2°24'N	53°36'E	5	101
Smoluchowski	60°18'N	96°48'W	83	26
Snellius	29°18'S	55°42'E	82	160
Sniadecki	22°30'S	168°54'W	43	129
Soddy	0°24'N	121°48'E	42	105
Somerville	8°18'S	64°54'E	15	122
Sommerfeld	65°12'N	162°24'W	169	24
Sömmering	0°06'N	7°30'W	28	98

Name	Lat.	Long.	Size	Page
Soraya	12°54'S	1°36'W	2	118
Sosigenes	8°42'N	17°36'E	17	99
South	58°00'N	50°48'W	104	28
Spallanzani	46°18'S	24°42'E	32	175
Spencer Jones	13°18'N	165°36'E	85	108
Spörer	4°18'S	1°48'W	27	118
Spurr	27°54'N	1°12'W	13	78
St. John	10°12'N	150°12'E	68	107
Stadius	10°30'N	13°42'W	69	98
Stark	25°30'S	134°36'E	49	146
Stearns	34°48'N	162°36'E	36	68
Stebbins	64°48'N	141°48'W	131	25
Stefan	46°00'N	108°18'W	125	38
Stein	7°12'N	179°00'E	33	108
Steinheil	48°36'S	46°30'E	67	176
Steklov	36°42'S	104°54'W	36	152
Stella	19°54'N	29°48'E	1	80
Steno	32°48'N	161°48'E	31	68
Sternfeld	19°36'S	141°12'W	100	131
Stetson	39°36'S	118°18'W	64	152
Stevinus	32°30'S	54°12'E	74	160
Stewart	2°12'N	67°00'E	13	102
Stiborius	34°24'S	32°00'E	43	159
Stöfler	41°06'S	6°00'E	126	158
Stokes	52°30'N	88°06'W	51	39
Stoletov	45°06'N	155°12'W	42	37
Stoney	55°18'S	156°06'W	45	167
Störmer	57°18'N	146°18'E	69	34
Strabo	61°54'N	54°18'E	55	31
Stratton	5°48'S	164°36'E	70	128
Street	46°30'S	10°30'W	57	174
Strömgren	21°42'S	132°24'W	61	131
Struve	22°24'N	77°06'W	164	74
Subbotin	29°12'S	135°18'E	67	164
Suess	4°24'N	47°36'W	8	96

Name	Lat.	Long.	Size	Page
Sulpicius Gallus	19°36'N	11°36'E	12	79
Sumner	37°30'N	108°42'E	50	65
Sundman	10°48'N	91°36'W	40	93
Susan	11°00'S	6°18'W	1	118
Svedberg	81°32'S	64°38'E	14	199
Sverdrup	88°30'S	152°00'W	35	15
Swann	52°00'N	112°42'E	42	48
Swasey	5°30'S	89°42'E	23	123
Swift	19°18'N	53°24'E	10	81
Sylvester	82°42'N	79°36'W	58	18
Szilard	34°00'N	105°42'E	122	65
T. Mayer	15°36'N	29°06'W	33	77
Tacchini	4°54'N	85°48'E	40	103
Tacitus	16°12'S	19°00'E	39	140
Tacquet	16°36'N	19°12'E	7	80
Taizo	24°42'N	2°12'E	6	79
Talbot	2°30'S	85°18'E	11	123
Tamm	4°24'S	146°24'E	38	127
Tannerus	56°24'S	22°00'E	28	188
Taruntius	5°36'N	46°30'E	56	101
Taylor	5°18'S	16°42'E	42	119
Tebbutt	9°36'N	53°36'E	31	101
Teisserenc	32°12'N	135°54'W	62	53
Tempel	3°54'N	11°54'E	45	99
Ten Bruggencate	9°30'S	134°24'E	59	126
Tereshkova	28°24'N	144°18'E	31	67
Tesla	38°30'N	124°42'E	43	66
Thales	61°48'N	50°18'E	31	31
Theaetetus	37°00'N	6°00'E	24	60
Thebit	22°00'S	4°00'W	56	138
Theiler	13°24'N	83°18'E	7	103
Theon Junior	2°18'S	15°48'E	17	119
Theon Senior	0°48'S	15°24'E	18	119
Theophilus	11°24'S	26°24'E	110	120
Theophrastus	17°30'N	39°00'E	9	81

Name	Lat.	Long.	Size	Page
Thiel	40°42'N	134°30'W	32	53
Thiessen	75°24'N	169°00'W	66	16
Thomson	32°42'S	166°12'E	117	166
Tikhomirov	25°12'N	162°00'E	65	87
Tikhov	62°18'N	171°42'E	83	35
Tiling	53°06'S	132°36'W	38	168
Timaeus	62°48'N	0°30'W	32	29
Timiryazev	5°30'S	147°00'W	53	110
Timocharis	26°42'N	13°06'W	33	78
Tiselius	7°00'N	176°30'E	53	108
Tisserand	21°24'N	48°12'E	36	81
Titius	26°48'S	100°42'E	73	144
Titov	28°36'N	150°30'E	31	67
Tolansky	9°30'S	16°00'W	13	118
Torricelli	4°36'S	28°30'E	22	120
Toscanelli	27°56'N	47°37'W	7	76
Townley	3°24'N	63°18'E	18	102
Tralles	28°24'N	52°48'E	43	62
Triesnecker	4°12'N	3°36'E	26	99
Trouvelot	49°18'N	5°48'E	9	43
Trumpler	29°18'N	167°06'E	77	68
Tsander	6°12'N	149°18'W	181	90
Tseraskiy	49°00'S	141°36'E	56	180
Tsinger	56°42'N	175°36'E	44	35
Tsiolkovskiy	21°12'S	128°54'E	185	146
Tsu Chung-Chi	17°18'N	145°06'E	28	87
Tucker	5°36'S	88°12'E	7	123
Turner	1°24'S	13°12'W	11	118
Tycho	43°24'S	11°06'W	102	174
Tyndall	34°54'S	117°00'E	18	163
Ukert	7°48'N	1°24'E	23	99
Ulugh Beigh	32°42'N	81°54'W	54	55
Urey	27°54'N	87°24'E	38	64
Väisälä	25°54'N	47°48'W	8	76
Valier	6°48'N	174°30'E	67	108

Name	Lat.	Long.	Size	Page
van Albada	9°24'N	64°18'E	21	102
Van Biesbroeck	28°42'N	45°36'W	9	57
Van de Graaff	27°24'S	172°12'E	233	148
Van den Bergh	31°18'N	159°06'W	42	52
van den Bos	5°18'S	146°00'E	22	127
Van der Waals	43°54'S	119°54'E	104	179
Van Gent	15°24'N	160°24'E	43	87
Van Maanen	35°42'N	128°00'E	60	66
van Rhijn	52°36'N	146°24'E	46	49
Van Vleck	1°54'S	78°18'E	31	123
Van Wijk	62°48'S	118°48'E	32	191
van't Hoff	62°06'N	131°48'W	92	25
Vasco da Gama	13°36'N	83°54'W	83	94
Vashakidze	43°36'N	93°18'E	44	47
Vavilov	0°48'S	137°54'W	98	111
Vega	45°24'S	63°24'E	75	177
Vendelinus	16°24'S	61°36'E	131	142
Vening Meinesz	0°18'S	162°36'E	87	128
Ventris	4°54'S	158°00'E	95	127
Vera	26°18'N	43°42'W	2	76
Vernadskiy	23°12'N	130°30'E	91	86
Verne	24°54'N	25°18'W	2	77
Vertregt	19°48'S	171°06'E	187	148
Very	25°36'N	25°18'E	5	80
Vesalius	3°06'S	114°30'E	61	125
Vestine	33°54'N	93°54'E	96	64
Vetchinkin	10°12'N	131°18'E	98	106
Vieta	29°12'S	56°18'W	87	155
Vil'ev	6°06'S	144°24'E	45	127
Virchow	9°48'N	83°42'E	16	103
Virtanen	15°30'N	176°42'E	44	88
Vitello	30°24'S	37°30'W	42	156
Vitruvius	17°36'N	31°18'E	29	80
Viviani	5°12'N	117°06'E	26	105
Vlacq	53°18'S	38°48'E	89	176

Name	Lat.	Long.	Size	Page
Vogel	15°06'S	5°54'E	26	139
Volkov	13°36'S	131°42'E	40	126
Volta	53°54'N	84°24'W	123	39
Volterra	56°48'N	132°12'E	52	34
von Baeyer	81°38'S	61°20'E	13	199
von Behring	7°48'S	71°48'E	38	122
von Békésy	51°54'N	126°48'E	96	48
von Braun	41°06'N	78°00'W	60	56
Von der Pahlen	24°48'S	132°42'W	56	131
Von Kármán	44°48'S	175°54'E	180	181
Von Neumann	40°24'N	153°12'E	78	67
Von Zeipel	42°36'N	141°36'W	83	37
Voskresenskiy	28°00'N	88°06'W	49	55
W. Bond	65°24'N	4°30'E	156	30
Walker	26°00'S	162°12'W	32	129
Wallace	20°18'N	8°42'W	26	78
Wallach	4°54'N	32°18'E	6	100
Walter	28°00'N	33°48'W	1	58
Walther	33°06'S	1°00'E	128	158
Wan-Hoo	9°48'S	138°48'W	52	111
Wapowski	82°55'S	53°31'E	11	199
Wargentin	49°36'S	60°12'W	84	171
Warner	4°00'S	87°18'E	35	123
Waterman	25°54'S	128°00'E	76	146
Watson	62°36'S	124°30'W	62	183
Watt	49°30'S	48°36'E	66	176
Watts	8°54'N	46°18'E	15	101
Webb	0°54'S	60°00'E	21	122
Weber	50°24'N	123°24'W	42	38
Wegener	45°12'N	113°18'W	88	38
Weierstrass	1°18'S	77°12'E	33	123
Weigel	58°12'S	38°48'W	35	186
Weinek	27°30'S	37°00'E	32	141
Weiss	31°48'S	19°30'W	66	157
Werner	28°00'S	3°18'E	70	139

Name	Lat.	Long.	Size	Page
Wexler	69°06'S	90°12'E	51	191
Weyl	17°30'N	120°12'W	108	72
Whewell	4°12'N	13°42'E	13	99
Whipple	89°07'N	118°14'E	15	14
White	44°36'S	158°18'W	39	167
Wichmann	7°30'S	38°06'W	10	116
Widmannstätten	6°06'S	85°30'E	46	123
Wiechert	84°30'S	165°00'E	41	15
Wiener	40°48'N	146°36'E	120	67
Wildt	9°00'N	75°48'E	11	103
Wilhelm	43°24'S	20°24'W	106	173
Wilkins	29°24'S	19°36'E	57	158
Williams	42°00'N	37°12'E	36	45
Wilsing	21°30'S	155°12'W	73	130
Wilson	69°12'S	42°24'W	69	186
Winkler	42°12'N	179°00'W	22	36
Winlock	35°36'N	105°36'W	64	54
Winthrop	10°42'S	44°24'W	17	116
Wöhler	38°12'S	31°24'E	27	159
Wolf	22°42'S	16°36'W	25	138
Wollaston	30°36'N	46°54'W	10	57
Woltjer	45°12'N	159°36'W	46	36
Wood	43°00'N	120°48'W	78	38
Wright	31°36'S	86°36'W	39	153
Wróblewski	24°00'S	152°48'E	21	147
Wrottesley	23°54'S	56°48'E	57	142
Wurzelbauer	33°54'S	15°54'W	88	157
Wyld	1°24'S	98°06'E	93	124
Xenophanes	57°30'N	82°00'W	125	27
Xenophon	22°48'S	122°06'E	25	145
Yablochkov	60°54'N	128°18'E	99	34
Yakovkin	54°30'S	78°48'W	37	171
Yamamoto	58°06'N	160°54'E	76	35
Yangel'	17°00'N	4°42'E	8	79
Yerkes	14°36'N	51°42'E	36	81

Name	Lat.	Long.	Size	Page
Yoshi	24°36'N	11°00'E	1	79
Young	41°30'S	50°54'E	71	160
Zach	60°54'S	5°18'E	70	188
Zagut	32°00'S	22°06'E	84	159
Zähringer	5°36'N	40°12'E	11	101
Zanstra	2°54'N	124°42'E	42	105
Zasyadko	3°54'N	94°12'E	11	104
Zeeman	75°12'S	133°36'W	190	195
Zelinskiy	28°54'S	166°48'E	53	166
Zeno	45°12'N	72°54'E	65	46
Zernike	18°24'N	168°12'E	48	88
Zhang Yuzhe	69°12'S	137°43'W	36	183
Zhiritskiy	24°48'S	120°18'E	35	145
Zhukovskiy	7°48'N	167°00'W	81	89
Zinner	26°36'N	58°48'W	4	75
Zöllner	8°00'S	18°54'E	47	120
Zsigmondy	59°42'N	104°42'W	65	26
Zucchius	61°24'S	50°18'W	64	186
Zupus	17°12'S	52°18'W	38	136
Zwicky	15°24'S	168°06'E	150	148
Rupes, Rupēs				
Rupes Altai	24°18'S	22°36'E	427	140
Rupes Boris	30°30'N	33°30'W	4	58
Rupes Cauchy	9°00'N	37°00'E	120	101
Rupes Kelvin	27°18'S	33°06'W	78	137
Rupes Liebig	25°00'S	46°00'W	180	136
Rupes Mercator	31°00'S	22°18'W	93	156
Rupes Recta	22°06'S	7°48'W	134	138
Rupes Toscanelli	27°24'N	47°30'W	70	76
Catena, Catenae				
Catena Abulfeda	16°54'S	17°12'E	219	139
Catena Artamonov	26°00'N	105°54'E	134	84
Catena Brigitte	18°30'N	27°30'E	5	80
Catena Davy	11°00'S	7°00'W	50	118
Catena Dziewulski	19°00'N	100°00'E	80	84

Name	Lat.	Long.	Size	Page
Catena Gregory	0°36'S	129°54'E	152	126
Catena Humboldt	21°30'S	84°36'E	165	143
Catena Krafft	15°00'N	72°00'W	60	74
Catena Kurchatov	37°12'N	136°18'E	226	66
Catena Leuschner	4°42'N	110°06'W	364	92
Catena Littrow	22°12'N	29°30'E	10	80
Catena Lucretius	3°24'S	126°06'W	271	111
Catena Mendeleev	6°18'N	139°24'E	188	106
Catena Michelson	1°24'N	113°24'W	456	92
Catena Pierre	19°48'N	31°48'W	9	77
Catena Sumner	37°18'N	112°18'E	247	65
Catena Sylvester	81°24'N	86°12'W	173	18
Catena Taruntius	3°00'N	48°00'E	100	101
Catena Timocharis	29°00'N	13°00'W	50	59
Catena Yuri	24°24'N	30°24'W	5	77
Vallis, Valles				
Vallis Alpes	48°30'N	3°12'E	166	43
Vallis Baade	45°54'S	76°12'W	203	171
Vallis Bohr	12°24'N	86°36'W	80	94
Vallis Bouvard	38°18'S	83°06'W	284	153
Vallis Capella	7°36'S	34°54'E	49	120
Vallis Christel	24°30'N	11°00'E	2	79
Vallis Inghirami	43°48'S	72°12'W	148	171
Vallis Krishna	24°30'N	11°18'E	3	79
Vallis Palitzsch	26°24'S	64°18'E	132	142
Vallis Planck	58°24'S	126°06'E	451	192
Vallis Rheita	42°30'S	51°30'E	445	176
Vallis Schrödinger	67°00'S	105°00'E	310	191
Vallis Schröteri	26°12'N	50°48'W	168	76
Vallis Snellius	31°06'S	56°00'E	592	160
Promontorium, Promontoria				
Promontorium Agarum	14°00'N	66°00'E	70	82
Promontorium Agassiz	42°00'N	1°48'E	20	43
Promontorium Archerusia	16°42'N	22°00'E	10	80

Name	Lat.	Long.	Size	Page
Promontorium Deville	43°12'N	1°00'E	20	43
Promontorium Fresnel	29°00'N	4°42'E	20	60
Promontorium Heraclides	40°18'N	33°12'W	50	58
Promontorium Kelvin	27°00'S	33°00'W	50	137
Promontorium Laplace	46°00'N	25°48'W	50	42
Promontorium Taenarium	19°00'S	8°00'W	70	138
Rima, Rimae				
Rima Agatharchides	20°00'S	28°00'W	50	137
Rima Agricola	29°00'N	53°00'W	110	57
Rima Archytas	53°00'N	3°00'E	90	43
Rima Ariadaeus	6°24'N	14°00'E	250	99
Rima Artsimovich	27°00'N	39°00'W	70	76
Rima Billy	15°00'S	48°00'W	70	136
Rima Birt	21°00'S	9°00'W	50	138
Rima Bradley	23°48'N	1°12'W	161	78
Rima Brayley	21°24'N	37°30'W	311	76
Rima Calippus	37°00'N	13°00'E	40	60
Rima Cardanus	11°24'N	71°30'W	175	95
Rima Carmen	19°48'N	29°18'E	10	80
Rima Cauchy	10°30'N	38°00'E	140	101
Rima Cleomedes	27°00'N	57°00'E	80	82
Rima Cleopatra	30°00'N	53°48'W	14	57
Rima Conon	18°36'N	2°00'E	30	79
Rima Dawes	17°30'N	26°36'E	15	80
Rima Delisle	31°00'N	32°00'W	60	58
Rima Diophantus	29°00'N	33°00'W	150	58
Rima Draper	18°00'N	25°00'W	160	77
Rima Euler	21°00'N	31°00'W	90	77
Rima Flammarion	2°48'S	5°36'W	80	118
Rima Furnerius	35°00'S	61°00'E	50	161
Rima G. Bond	33°18'N	35°30'E	168	61
Rima Galilaei	11°54'N	58°30'W	89	95
Rima Gärtner	59°00'N	36°00'E	30	31
Rima Gay-Lussac	13°00'N	22°00'W	40	97

Name	Lat.	Long.	Size	Page
Rima Hadley	25°00'N	3°00'E	80	79
Rima Hansteen	12°00'S	53°00'W	25	116
Rima Hesiodus	30°00'S	20°00'W	256	156
Rima Hyginus	7°24'N	7°48'E	219	99
Rima Jansen	14°30'N	29°00'E	35	80
Rima Krieger	29°00'N	45°36'W	22	57
Rima Mairan	38°00'N	47°00'W	90	57
Rima Marcello	18°36'N	27°42'E	2	80
Rima Marius	16°30'N	48°54'W	121	76
Rima Messier	1°00'S	45°00'E	100	121
Rima Milichius	8°00'N	33°00'W	100	97
Rima Oppolzer	1°42'S	1°00'E	94	119
Rima Réaumur	3°00'S	3°00'E	30	119
Rima Reiko	18°36'N	27°42'E	2	80
Rima Rudolf	19°36'N	29°36'E	8	80
Rima Schröter	1°00'N	6°00'W	40	98
Rima Sharp	46°42'N	50°30'W	107	41
Rima Sheepshanks	58°00'N	24°00'E	200	30
Rima Siegfried	25°54'S	103°00'E	14	144
Rima Suess	6°42'N	48°12'W	165	96
Rima Sung-Mei	24°36'N	11°18'E	4	79
Rima T. Mayer	13°00'N	31°00'W	50	97
Rima Vladimir	25°12'N	0°42'W	14	78
Rima Wan-Yu	20°00'N	31°30'W	12	77
Rima Yangel'	16°42'N	4°36'E	30	79
Rima Zahia	25°00'N	29°30'W	16	77
Rimae Alphonsus	14°00'S	2°00'W	80	118
Rimae Apollonius	5°00'N	53°00'E	230	101
Rimae Archimedes	26°36'N	4°06'W	169	78
Rimae Aristarchus	26°54'N	47°30'W	121	76
Rimae Arzachel	18°00'S	2°00'W	50	138
Rimae Atlas	47°30'N	43°36'E	60	45
Rimae Bode	10°00'N	4°00'W	70	98
Rimae Boscovich	9°48'N	11°06'E	40	99
Rimae Bürg	44°30'N	23°48'E	147	44

Name	Lat.	Long.	Size	Page
Rimae Chacornac	29°00'N	32°00'E	120	61
Rimae Daniell	37°00'N	26°00'E	200	61
Rimae Darwin	19°18'S	69°30'W	143	135
Rimae de Gasparis	24°36'S	51°06'W	93	136
Rimae Doppelmayer	25°54'S	45°06'W	162	136
Rimae Focas	28°00'S	98°00'W	100	153
Rimae Fresnel	28°00'N	4°00'E	90	60
Rimae Gassendi	18°00'S	40°00'W	70	136
Rimae Gerard	46°00'N	84°00'W	100	39
Rimae Goclenius	8°00'S	43°00'E	240	121
Rimae Grimaldi	9°00'S	64°00'W	230	115
Rimae Gutenberg	5°00'S	38°00'E	330	121
Rimae Hase	29°24'S	62°30'E	83	161
Rimae Herigonius	13°00'S	37°00'W	100	116
Rimae Hevelius	1°00'N	68°00'W	182	95
Rimae Hippalus	25°30'S	29°12'W	191	137
Rimae Hypatia	0°24'S	22°24'E	206	120
Rimae Janssen	45°36'S	40°00'E	114	176
Rimae Kopff	17°24'S	89°36'W	41	134
Rimae Littrow	22°06'N	29°54'E	115	80
Rimae Maclear	13°00'N	20°00'E	110	100
Rimae Maestlin	2°00'N	40°00'W	80	96
Rimae Maupertuis	52°00'N	23°00'W	60	42
Rimae Menelaus	17°12'N	17°54'E	131	79
Rimae Mersenius	20°00'S	46°30'W	300	136
Rimae Opelt	13°00'S	18°00'W	70	117
Rimae Palmieri	28°00'S	47°00'W	150	136
Rimae Parry	6°06'S	16°48'W	82	118
Rimae Petavius	25°54'S	58°54'E	80	142
Rimae Pettit	23°00'S	92°00'W	450	133
Rimae Pitatus	28°30'S	13°48'W	94	157
Rimae Plato	52°54'N	3°12'W	87	43
Rimae Plinius	17°54'N	23°36'E	124	80
Rimae Posidonius	32°00'N	28°42'E	70	61
Rimae Prinz	27°00'N	43°00'W	80	76

Name	Lat.	Long.	Size	Page
Rimae Ramsden	33°54'S	31°24'W	108	156
Rimae Repsold	50°36'N	81°42'W	166	40
Rimae Riccioli	2°00'S	74°00'W	400	114
Rimae Ritter	3°00'N	18°00'E	100	99
Rimae Römer	27°00'N	35°00'E	110	80
Rimae Secchi	1°00'N	44°00'E	35	101
Rimae Sirsalis	15°42'S	61°42'W	426	135
Rimae Sosigenes	8°36'N	18°42'E	190	100
Rimae Sulpicius Gallus	21°00'N	10°00'E	90	79
Rimae Taruntius	5°30'N	46°30'E	25	101
Rimae Theaetetus	33°00'N	6°00'E	50	60
Rimae Triesnecker	4°18'N	4°36'E	215	99
Rimae Vasco da Gama	10°00'N	82°00'W	60	94
Rimae Zupus	15°00'S	53°00'W	120	136
Sinus, Sinūs				
Sinus Aestuum	10°54'N	8°48'W	290	98
Sinus Amoris	18°06'N	39°06'E	130	81
Sinus Asperitatis	3°48'S	27°24'E	206	120
Sinus Concordiae	10°48'N	43°12'E	142	101
Sinus Fidei	18°00'N	2°00'E	70	79
Sinus Honoris	11°42'N	18°06'E	109	100
Sinus Iridum	44°06'N	31°30'W	236	42
Sinus Lunicus	31°48'N	1°24'W	126	59
Sinus Medii	2°24'N	1°42'E	335	99
Sinus Roris	54°00'N	56°36'W	202	41
Sinus Successus	0°54'N	59°00'E	132	102
Palus, Paludes				
Palus Epidemiarum	32°00'S	28°12'W	286	156
Palus Putredinis	26°30'N	0°24'E	161	79
Palus Somni	14°06'N	45°00'E	143	81
Albedo Feature				
Reiner Gamma	7°30'N	59°00'W	70	95
Landing Site Name				
Apennine Front	25°54'N	3°42'E	6	79
Baby Ray	9°06'S	15°24'E	0	119

Name	Lat.	Long.	Size	Page
Bear Mountain	20°00'N	30°42'E	0	80
Bench	3°12'S	23°24'W	0	117
Block	3°12'S	23°24'W	0	117
Bowen-Apollo	20°18'N	30°54'E	0	80
Bridge	26°00'N	3°36'E	1	79
Brontë	20°12'N	30°42'E	0	80
Camelot	20°12'N	30°42'E	1	80
Cinco	9°06'S	15°30'E	0	119
Cochise	20°12'N	30°48'E	1	80
Cone	3°42'S	17°24'W	0	118
Crescent	2°54'S	23°24'W	1	117
Doublet	3°42'S	17°30'W	0	118
Dune	26°00'N	3°42'E	0	79
Earthlight	26°06'N	3°42'E	0	79
Elbow	26°00'N	3°36'E	0	79
Emory	20°06'N	30°48'E	1	80
End	8°54'S	15°36'E	0	119
Falcon	20°24'N	30°18'E	0	80
Family Mountain	20°24'N	30°18'E	7	80
Flag	9°00'S	15°30'E	0	119
Flank	3°42'S	17°24'W	0	118
Gator	9°00'S	15°36'E	1	119
Halfway	9°00'S	15°30'E	0	119
Halo	3°12'S	23°24'W	0	117
Head	3°00'S	23°24'W	0	117
Hess-Apollo	20°06'N	30°42'E	1	80
Horatio	20°12'N	30°42'E	0	80
Index	26°06'N	3°42'E	0	79
Kiva	8°36'S	15°30'E	1	119
Lara	20°24'N	30°30'E	0	80
Last	26°06'N	3°42'E	0	79
Light Mantle	20°12'N	30°48'E	4	80
Mackin	20°06'N	30°42'E	0	80
Middle Crescent	3°12'S	23°24'W	0	117
Nansen-Apollo	20°06'N	30°30'E	1	80

Name	Lat.	Long.	Size	Page
North Complex	26°12'N	3°36'E	2	79
North Massif	20°24'N	30°48'E	14	80
North Ray	8°48'S	15°30'E	1	119
Old Nameless	3°42'S	17°30'W	0	118
Palmetto	8°54'S	15°30'E	0	119
Plain	26°12'N	3°36'E	2	79
Plum	9°00'S	15°30'E	0	119
Powell	20°12'N	30°48'E	1	80
Ravine	8°54'S	15°36'E	1	119
Rhysling	26°06'N	3°42'E	0	79
Scarp	20°18'N	30°36'E	8	80
Sculptured Hills	20°18'N	31°00'E	8	80
Shakespeare	20°12'N	30°48'E	1	80
Sharp-Apollo	3°12'S	23°24'W	0	117
Sherlock	20°12'N	30°48'E	0	80
Shorty	20°12'N	30°36'E	0	80
Smoky Mountains	8°48'S	15°36'E	3	119
Snowman	3°12'S	23°24'W	1	117
South Cluster	26°00'N	3°42'E	2	79
South Massif	20°00'N	30°24'E	16	80
South Ray	9°12'S	15°24'E	1	119
Spook	9°00'S	15°30'E	0	119
Spot	9°00'S	15°30'E	0	119
Spur	25°54'N	3°42'E	0	79
St. George	26°00'N	3°30'E	2	79
Statio Tranquillitatis	0°48'N	23°30'E	0	100
Steno-Apollo	20°06'N	30°48'E	1	80
Stone Mountain	9°06'S	15°36'E	5	119
Stubby	9°06'S	15°30'E	1	119
Surveyor	3°12'S	23°24'W	0	117
Taurus-Littrow Valley	20°00'N	31°00'E	30	80
Terrace	26°06'N	3°42'E	0	79
Tortilla Flat	20°12'N	30°42'E	1	80
Trap	9°06'S	15°24'E	1	119
Trident	20°12'N	30°48'E	0	80

Name	Lat.	Long.	Size	Page
Triplet	3°42'S	17°30'W	0	118
Van Serg	20°12'N	30°48'E	0	80
Victory	20°12'N	30°42'E	1	80
Weird	3°42'S	17°30'W	0	118
Wessex Cleft	20°18'N	30°54'E	4	80
West	0°48'N	23°30'E	0	100
Wreck	9°06'S	15°30'E	1	119
Satellite Feature				
Abbe H	58°12'S	177°54'E	25	193
Abbe K	59°36'S	177°18'E	28	193
Abbe M	61°36'S	175°30'E	29	193
Abel B	36°42'S	82°48'E	41	162
Abel C	36°0'S	81°0'E	31	162
Abel L	34°24'S	82°36'E	67	162
Abel M	32°12'S	83°36'E	81	162
Abulfeda D	13°12'S	9°30'E	20	119
Adams B	31°30'S	65°36'E	28	161
Adams D	32°30'S	71°36'E	42	161
Adams P	35°12'S	71°0'E	24	161
Agatharchides A	23°12'S	28°24'W	16	137
Agrippa S	5°18'N	8°54'E	32	99
Airy B	17°36'S	8°30'E	29	139
Aitken C	14°0'S	175°48'E	74	128
Aitken Y	12°0'S	173°12'E	35	128
Aitken Z	15°6'S	173°18'E	33	148
Albategnius B	10°0'S	4°0'E	20	119
Albategnius E	12°54'S	6°24'E	14	119
Alden B	20°30'S	112°36'E	17	145
Alden E	23°12'S	112°24'E	28	145
Alekhin E	67°12'S	124°6'W	38	183
Alhazen A	16°12'N	74°18'E	14	83
Alhazen D	19°42'N	75°12'E	33	83
Al-Khwarizmi B	9°0'N	107°24'E	62	104
Al-Khwarizmi H	6°0'N	109°12'E	50	105
Al-Khwarizmi J	6°12'N	107°36'E	47	105

Name	Lat.	Long.	Size	Page
Al-Khwarizmi K	4°36'N	107°36'E	26	105
Alpes A	51°24'N	0°18'W	11	43
Alter K	16°18'N	106°0'W	22	73
Alter W	20°24'N	109°12'W	52	72
Amici N	11°48'S	172°30'W	39	109
Amici P	12°18'S	174°6'W	31	109
Amici Q	12°0'S	175°42'W	47	109
Amici R	11°24'S	175°12'W	34	109
Amici T	9°42'S	174°0'W	43	109
Amici U	8°42'S	175°30'W	96	109
Anaximander B	67°48'N	60°42'W	78	27
Anaximander D	65°24'N	50°6'W	92	28
Anderson E	16°54'N	173°24'E	28	88
Anderson F	16°18'N	173°36'E	49	88
Anderson L	14°36'N	170°54'E	14	88
Ansgarius B	11°54'S	83°48'E	29	123
Anuchin B	46°42'S	103°18'E	24	178
Anuchin L	50°12'S	101°42'E	15	178
Anuchin N	51°36'S	99°36'E	33	178
Anuchin Q	51°6'S	98°18'E	50	178
Apollonius H	3°24'N	59°36'E	20	102
Appleton D	38°0'N	160°36'E	37	68
Appleton Q	34°18'N	155°18'E	26	67
Appleton R	36°12'N	156°12'E	39	67
Archimedes C	31°36'N	1°30'W	8	59
Archytas U	62°48'N	9°12'E	8	30
Armiński K	17°6'S	154°36'E	22	147
Arnold F	67°30'N	35°12'E	10	31
Arrhenius J	57°36'S	88°18'W	18	185
Artem'ev G	10°18'N	142°48'W	60	91
Atlas A	45°18'N	49°36'E	22	45
Atlas D	50°24'N	49°36'E	25	45
Atlas P	49°36'N	52°42'E	27	45
Avicenna E	40°0'N	91°6'W	25	55
Avicenna G	39°0'N	92°0'W	26	55

Name	Lat.	Long.	Size	Page
Babbage A	59°0'N	55°6'W	32	28
Babbage D	58°36'N	61°0'W	68	27
Babcock H	3°0'N	96°30'E	63	104
Backlund L	18°12'S	103°30'E	56	144
Backlund P	18°54'S	102°0'E	27	144
Baco A	52°48'S	20°12'E	39	175
Baco K	53°54'S	17°36'E	29	175
Baco L	49°30'S	16°42'E	7	175
Baillaud A	75°42'N	48°48'E	56	21
Baillaud D	73°36'N	49°42'E	16	185
Baillaud F	75°42'N	53°42'E	20	21
Bailly B	68°48'S	63°6'W	65	185
Bailly F	67°30'S	69°12'W	16	185
Bailly G	65°36'S	59°6'W	18	186
Bailly O	69°36'S	56°42'W	16	186
Bailly U	71°18'S	75°48'W	20	196
Balboa A	17°24'N	81°54'W	47	74
Balboa B	20°18'N	82°18'W	62	74
Ball A	34°42'S	9°18'W	29	157
Ball C	37°42'S	8°42'W	31	157
Banachiewicz B	5°18'N	78°54'E	24	103
Banachiewicz C	7°0'N	75°24'E	19	103
Barbier D	23°0'S	160°12'E	24	147
Barbier U	22°48'S	155°6'E	38	147
Barnard D	31°24'S	89°18'E	47	162
Barocius B	44°0'S	18°18'E	39	175
Barocius G	42°24'S	21°0'E	27	175
Barocius J	44°54'S	21°30'E	27	175
Barocius M	42°24'S	19°30'E	17	175
Barringer C	26°30'S	148°48'W	19	130
Bartels A	25°42'N	89°36'W	17	74
Bayer D	47°54'S	29°48'W	20	173
Bayer H	53°30'S	32°30'W	27	173
Becquerel F	40°54'N	132°54'E	21	66
Becquerel W	42°12'N	126°54'E	27	48

Name	Lat.	Long.	Size	Page
Bečvář J	3°36'S	126°36'E	45	126
Bečvář Q	2°54'S	124°0'E	28	125
Bečvář X	0°36'S	124°12'E	26	125
Behaim S	16°36'S	81°24'E	25	143
Beijerinck D	12°48'S	153°6'E	14	127
Beijerinck J	14°48'S	153°42'E	40	147
Beijerinck R	14°42'S	149°12'E	28	147
Beijerinck U	12°24'S	149°0'E	18	127
Bel'kovich A	58°42'N	86°0'E	58	32
Bel'kovich K	63°48'N	93°36'E	47	33
Bell J	19°54'N	94°0'W	18	73
Bell K	18°18'N	95°6'W	18	73
Bell L	19°42'N	95°48'W	23	73
Bell Q	20°42'N	97°12'W	23	73
Bell T	21°54'N	98°54'W	52	73
Bell Y	25°24'N	96°42'W	23	73
Belyaev Q	20°36'N	139°24'E	50	86
Bergstrand G	20°0'S	179°24'E	34	148
Bergstrand Q	20°6'S	175°6'E	59	148
Berkner A	27°36'N	104°48'W	22	73
Berkner B	29°18'N	104°6'W	33	54
Berkner Y	27°48'N	106°12'W	31	73
Berlage R	64°0'S	167°36'W	25	182
Bernoulli A	36°24'N	60°54'E	22	63
Bernoulli B	36°54'N	65°36'E	22	63
Bernoulli C	35°18'N	67°12'E	19	63
Berosus F	34°0'N	66°36'E	22	63
Berzelius B	32°36'N	43°6'E	23	62
Bessarion A	17°6'N	39°48'W	13	76
Bessarion B	16°48'N	41°42'W	12	76
Bessarion D	19°48'N	41°42'W	9	76
Bettinus A	64°54'S	48°48'W	26	186
Bettinus B	63°36'S	51°0'W	24	186
Biela A	52°54'S	53°18'E	26	176
Biela B	56°30'S	49°36'E	43	189

Name	Lat.	Long.	Size	Page
Biela C	54°18'S	53°30'E	26	176
Bingham H	7°30'N	116°12'E	26	105
Biot B	20°24'S	49°36'E	28	141
Birkeland M	32°0'S	174°6'E	23	166
Birkhoff K	57°48'N	144°18'W	58	25
Birkhoff L	56°36'N	144°48'W	37	25
Birkhoff Q	56°36'N	150°48'W	43	24
Birkhoff R	57°30'N	153°0'W	27	24
Birkhoff X	62°6'N	149°42'W	77	24
Birkhoff Y	59°54'N	146°36'W	25	25
Birkhoff Z	61°18'N	145°18'W	30	25
Blackett N	39°54'S	116°12'W	23	152
Blancanus C	66°30'S	28°0'W	46	187
Blazhko D	33°0'N	145°12'W	34	52
Blazhko L	29°18'N	147°36'W	44	52
Blazhko R	30°0'N	149°48'W	53	52
Bode A	9°0'N	1°12'W	12	98
Bode B	8°42'N	3°6'W	10	98
Boguslawsky B	73°54'S	61°0'E	63	199
Boguslawsky C	70°54'S	27°42'E	36	198
Boguslawsky F	75°18'S	52°30'E	30	199
Boguslawsky G	71°30'S	34°30'E	21	198
Boguslawsky J	72°12'S	28°54'E	36	198
Boguslawsky N	74°0'S	33°18'E	28	198
Bolyai K	36°18'S	126°48'E	29	164
Bolyai Q	36°6'S	122°30'E	28	164
Boole A	63°36'N	80°36'W	56	27
Boole C	65°24'N	82°30'W	18	27
Boole F	64°12'N	79°24'W	34	27
Boole G	64°48'N	90°54'W	41	26
Boole H	61°36'N	88°54'W	75	27
Borman V	37°24'S	150°36'W	28	150
Bose D	52°42'S	166°6'W	20	167
Bose U	52°48'S	174°36'W	38	167
Boss A	52°18'N	80°18'E	27	46

Name	Lat.	Long.	Size	Page
Boss C	52°12'N	76°24'E	21	46
Boss D	44°54'N	87°24'E	15	47
Boss K	49°36'N	80°42'E	19	46
Boss L	50°54'N	82°18'E	40	46
Boussingault B	65°36'S	46°54'E	54	189
Boussingault E	67°12'S	46°48'E	98	189
Boussingault K	68°54'S	50°54'E	29	189
Boussingault N	71°30'S	62°6'E	15	199
Boussingault S	64°6'S	46°54'E	16	189
Bouvard B	41°42'S	79°42'W	25	154
Bouvard C	37°6'S	77°18'W	16	154
Bouvard D	42°48'S	80°30'W	26	171
Bouvard N	38°36'S	76°30'W	66	154
Boyle Z	51°18'S	177°42'E	52	181
Bragg H	41°42'N	101°0'W	40	54
Bragg M	39°6'N	102°30'W	45	54
Bragg P	40°0'N	104°24'W	30	54
Brashear P	76°48'S	175°42'W	71	194
Brayley B	20°48'N	34°18'W	10	77
Bredikhin B	19°0'N	157°24'W	18	70
Brianchon A	76°42'N	86°18'W	50	18
Brianchon B	72°12'N	89°6'W	31	18
Brianchon T	75°48'N	99°48'W	30	17
Bridgman C	46°42'N	140°12'E	35	49
Bridgman E	44°6'N	141°42'E	29	49
Briggs A	27°6'N	73°42'W	23	74
Briggs B	28°6'N	70°54'W	25	56
Brisbane E	50°0'S	71°12'E	56	177
Brisbane H	50°18'S	64°54'E	43	177
Brisbane Z	52°48'S	72°24'E	64	177
Brouwer H	35°54'S	124°24'W	19	151
Buffon D	40°12'S	131°42'W	20	151
Buffon K	46°18'S	128°0'W	18	169
Buffon V	39°12'S	137°6'W	38	151
Buisson Y	1°24'N	112°36'E	36	105

Name	Lat.	Long.	Size	Page
Bullialdus A	22°6'S	21°30'W	26	137
Bullialdus B	23°24'S	21°54'W	21	137
Bunsen A	43°12'N	88°54'W	39	39
Burckhardt A	30°30'N	58°48'E	28	62
Bürg A	46°48'N	33°6'E	12	44
Buys-Ballot H	19°24'N	179°30'E	22	88
Buys-Ballot Q	19°30'N	172°42'E	58	88
Buys-Ballot Z	22°30'N	174°30'E	58	88
Byrd C	84°42'N	26°48'E	52	14
Byrgius A	24°30'S	63°42'W	19	135
Byrgius D	24°6'S	67°6'W	27	135
Byrgius S	26°12'S	61°24'W	43	135
C. Mayer B	60°12'N	15°36'E	36	30
Cabannes J	62°12'S	167°12'W	34	182
Cabannes M	64°12'S	170°12'W	48	182
Cabannes Q	63°18'S	174°30'W	49	182
Cabeus A	82°12'S	39°6'W	48	197
Cabeus B	82°24'S	53°0'W	61	196
Cajori K	49°6'S	169°48'E	32	181
Campbell N	43°12'N	152°18'E	23	49
Campbell X	47°42'N	149°24'E	24	49
Cannon B	17°30'N	80°0'E	31	83
Cannon E	19°12'N	79°6'E	22	83
Cantor T	37°54'N	113°24'E	23	65
Capella B	9°24'S	36°48'E	10	121
Capella E	7°30'S	37°42'E	16	121
Capella F	9°12'S	35°24'E	14	120
Cardanus C	11°18'N	76°12'W	14	94
Carlini D	33°0'N	16°0'W	9	59
Carnot F	52°30'N	138°54'W	35	37
Carpenter U	70°36'N	57°0'W	26	18
Carver K	46°12'S	128°30'E	60	179
Casatus C	72°12'S	30°12'W	17	197
Casatus D	77°12'S	44°18'W	36	197
Casatus E	79°6'S	53°12'W	41	196

Name	Lat.	Long.	Size	Page
Casatus J	74°18'S	32°48'W	22	197
Casatus K	75°0'S	41°24'W	36	197
Cassegrain H	53°6'S	115°36'E	28	179
Cassegrain K	55°0'S	113°30'E	17	179
Cassini C	41°42'N	7°48'E	14	60
Cassini Z	45°42'S	89°12'W	21	170
Catalán U	45°6'S	90°36'W	20	170
Cavalerius A	4°30'N	69°30'W	14	95
Cavalerius B	6°0'N	71°0'W	39	95
Cavendish E	25°24'S	54°12'W	24	135
Chaffee F	38°48'S	152°30'W	35	150
Chaffee S	39°30'S	156°36'W	19	150
Chamberlin H	59°48'S	99°54'E	27	191
Champollion A	41°6'N	177°6'E	27	68
Champollion Y	40°48'N	174°42'E	22	68
Chandler G	43°18'N	175°48'E	33	50
Chandler P	41°42'N	170°18'E	67	68
Chang Heng C	20°24'N	114°0'E	25	85
Chaplygin K	7°42'S	151°12'E	19	127
Chaplygin Y	2°48'S	149°42'E	29	127
Chapman D	51°24'N	96°48'W	39	39
Chapman M	49°0'N	100°42'W	38	39
Chapman V	51°0'N	103°48'W	21	39
Chappell E	55°48'N	171°30'W	59	36
Chappell T	54°48'N	178°54'E	28	50
Charlier Z	39°42'N	131°36'W	46	53
Chaucer B	6°30'N	137°24'W	27	91
Chauvenet C	10°24'S	138°0'E	48	126
Chauvenet D	10°36'S	139°42'E	14	126
Chauvenet E	11°24'S	140°42'E	27	126
Chauvenet G	12°42'S	141°0'E	26	126
Chauvenet J	13°54'S	139°18'E	77	126
Chauvenet Q	13°18'S	135°24'E	42	126
Chauvenet S	12°18'S	134°24'E	38	126
Chebyshev C	30°18'S	127°12'W	27	151

Name	Lat.	Long.	Size	Page
Chebyshev N	37°42'S	134°24'W	24	151
Chebyshev U	33°18'S	137°0'W	36	151
Chebyshev V	33°30'S	133°36'W	23	151
Chernyshev B	48°30'N	175°42'E	20	50
Chrétien C	44°30'S	165°18'E	63	181
Chrétien W	44°18'S	160°48'E	34	181
Cichus G	35°30'S	23°30'W	23	156
Clairaut A	48°54'S	14°48'E	36	175
Clark F	38°24'S	122°30'E	27	164
Clavius C	57°42'S	14°12'W	21	187
Clavius D	58°48'S	12°24'W	28	187
Clavius G	52°0'S	13°54'W	17	173
Clavius K	60°24'S	19°48'W	20	187
Clavius L	58°42'S	21°12'W	24	187
Cleomedes B	27°12'N	55°54'E	11	82
Cleomedes D	29°18'N	61°54'E	25	63
Cleomedes E	28°36'N	54°24'E	21	62
Cleomedes F	22°36'N	56°54'E	12	82
Cleostratus A	62°42'N	77°18'W	35	27
Cleostratus F	61°30'N	80°24'W	50	27
Cockcroft N	29°6'N	163°42'W	56	51
Colombo A	14°6'S	44°24'E	42	141
Compton E	55°24'N	113°24'E	19	48
Compton R	52°36'N	91°30'E	37	47
Compton W	58°36'N	97°12'E	16	33
Comrie K	22°6'N	112°18'W	73	72
Comrie T	23°6'N	115°18'W	43	72
Comrie V	24°36'N	115°54'W	29	72
Comstock P	20°6'N	122°42'W	26	72
Condorcet F	8°12'N	73°6'E	37	103
Condorcet P	8°42'N	70°24'E	46	102
Condorcet Q	11°24'N	73°18'E	31	103
Condorcet T	11°48'N	65°48'E	15	102
Condorcet W	13°54'N	66°54'E	33	102
Congreve G	0°54'S	163°42'W	17	109

Name	Lat.	Long.	Size	Page
Congreve H	1°12'S	165°12'W	37	109
Congreve L	3°36'S	166°18'W	30	109
Congreve N	3°24'S	168°12'W	31	109
Congreve Q	1°24'S	169°36'W	59	109
Cooper G	52°36'N	178°30'E	20	50
Cooper K	51°6'N	178°6'E	30	50
Coriolis C	1°54'N	173°18'E	19	108
Coriolis H	0°30'S	174°12'E	12	128
Coriolis L	1°54'S	172°42'E	32	128
Coriolis M	1°24'S	171°42'E	31	128
Coriolis S	0°6'N	169°42'E	17	108
Coriolis W	3°6'N	168°0'E	37	108
Coriolis Y	3°36'N	171°12'E	31	108
Coriolis Z	4°12'N	171°30'E	53	108
Coulomb C	57°24'N	110°48'W	34	26
Coulomb J	53°6'N	111°36'W	35	38
Coulomb N	50°36'N	115°48'W	32	38
Coulomb P	50°30'N	117°24'W	38	38
Coulomb V	55°36'N	118°6'W	36	38
Coulomb W	56°30'N	120°24'W	34	25
Cremona B	67°54'N	92°24'W	20	26
Crocco G	47°48'S	152°18'E	42	180
Crocco R	48°18'S	147°30'E	57	180
Crookes D	9°36'S	162°48'W	41	109
Crookes P	11°42'S	165°48'W	21	109
Crookes X	6°36'S	166°12'W	24	109
Crüger A	16°0'S	62°42'W	27	135
Curie C	21°6'S	94°6'E	47	144
Curie G	23°36'S	94°48'E	53	144
Curie M	28°24'S	92°30'E	34	162
Curie P	28°24'S	90°6'E	26	162
Curtius B	63°42'S	4°42'E	41	188
Cusanus C	70°24'N	60°48'E	25	21
Cuvier N	53°24'S	12°6'E	4	175
Cyrano A	18°6'S	158°36'E	26	147

Name	Lat.	Long.	Size	Page
Cyrano E	20°6'S	161°12'E	21	147
Cyrillus A	13°48'S	23°6'E	17	120
Cyrillus F	15°18'S	25°30'E	44	140
Cysatus E	66°42'S	1°18'W	48	187
Daedalus C	4°6'S	178°54'W	68	109
Daedalus G	6°36'S	177°24'W	33	109
Daedalus R	7°42'S	175°12'E	41	128
Daedalus S	6°48'S	172°54'E	20	128
Daedalus U	4°12'S	174°54'E	30	128
Daedalus W	3°30'S	177°30'E	70	128
d'Alembert E	52°48'N	168°12'E	22	50
d'Alembert G	50°54'N	167°30'E	18	50
d'Alembert Z	55°24'N	165°36'E	44	50
Damoiseau A	6°18'S	62°24'W	47	115
Damoiseau E	5°12'S	58°18'W	14	115
Danjon J	12°48'S	125°36'E	23	125
Danjon X	10°0'S	122°48'E	65	125
Dante C	28°18'N	177°6'W	54	51
Dante E	26°42'N	177°0'W	43	69
Dante G	24°54'N	178°36'W	24	69
Dante P	23°36'N	179°24'E	27	88
Dante S	24°54'N	177°18'E	17	88
Dante T	25°48'N	176°36'E	20	88
Dante Y	27°6'N	179°30'E	27	88
D'Arsonval A	8°24'S	125°0'E	17	125
Das G	26°54'S	135°12'W	32	131
Davy A	12°12'S	7°42'W	15	118
Davy G	10°24'S	5°6'W	16	118
Dawson D	66°36'S	131°42'W	39	183
Dawson V	66°36'S	137°0'W	58	183
De Forest N	79°30'S	164°42'W	41	194
De Forest P	80°0'S	176°0'W	18	194
De La Rue S	62°54'N	61°36'E	12	32
De Moraes S	48°54'N	140°42'E	45	49
De Roy N	59°42'S	103°6'W	26	184

Name	Lat.	Long.	Size	Page
De Roy P	58°24'S	102°24'W	35	184
De Sitter F	80°12'N	51°0'E	22	21
De Sitter W	79°24'N	54°6'E	40	21
De Vico A	18°48'S	63°30'W	32	135
De Vico T	18°42'S	61°48'W	41	135
De Vries N	21°30'S	177°18'W	30	129
Debye E	50°24'N	171°0'W	41	36
Debye J	48°24'N	172°36'W	30	36
Debye Q	47°30'N	179°12'W	26	36
Dellinger B	5°30'S	141°6'E	53	126
Deluc A	54°6'S	0°24'W	56	174
Deluc B	52°0'S	0°30'E	38	174
Deluc D	56°24'S	2°24'W	27	187
Deluc G	61°36'S	0°42'E	27	188
Deluc H	54°12'S	2°6'W	26	174
Deluc R	55°24'S	0°36'E	22	174
Democritus B	60°6'N	28°36'E	12	30
Denning B	15°12'S	143°30'E	32	146
Denning R	17°12'S	141°12'E	72	146
Denning U	16°0'S	138°36'E	30	146
Denning V	15°30'S	139°42'E	26	146
Denning Y	14°0'S	142°18'E	52	126
Desargues B	70°42'N	65°0'W	50	18
Desargues M	68°24'N	73°54'W	30	27
Descartes A	12°6'S	15°12'E	16	119
Deutsch L	22°24'N	110°48'E	36	85
Dewar S	3°6'S	163°54'E	23	128
Dirichlet E	12°12'N	147°48'W	26	90
Dobrovol'skiy D	12°12'S	130°36'E	49	126
Dobrovol'skiy M	14°36'S	129°36'E	31	146
Doerfel R	71°18'S	119°24'W	32	195
Doerfel U	68°48'S	117°12'W	34	184
Dollond B	7°42'S	13°48'E	37	119
Donner N	33°12'S	97°6'E	19	162
Donner R	34°24'S	92°18'E	15	162

Name	Lat.	Long.	Size	Page
Donner T	31°6'S	94°48'E	46	162
Donner V	30°30'S	95°42'E	19	162
Donner Z	29°42'S	97°48'E	13	162
Doppler B	11°48'S	159°24'W	37	110
Doppler X	10°18'S	161°18'W	18	110
Douglass C	36°42'N	121°0'W	28	53
Douglass X	38°24'N	123°48'W	23	53
Dreyer C	11°12'N	98°12'E	37	104
Dreyer D	10°48'N	99°48'E	27	104
Dreyer J	8°48'N	98°12'E	29	104
Dreyer K	9°0'N	97°24'E	23	104
Dreyer R	8°30'N	94°0'E	18	104
Dreyer W	11°48'N	95°42'E	30	104
Dryden T	32°48'S	158°36'W	35	150
Drygalski P	81°0'S	99°54'W	30	195
Dufay A	9°30'N	170°30'E	15	108
Dufay B	8°30'N	171°0'E	20	108
Dufay D	6°18'N	170°30'E	32	108
Dufay X	7°12'N	168°30'E	42	108
Dugan X	67°48'N	98°30'E	14	33
Dyson B	63°36'N	117°36'W	45	26
Dyson H	59°30'N	113°42'W	21	26
Dyson M	58°24'N	120°54'W	34	25
Dyson Q	59°48'N	125°42'W	89	25
Dyson X	62°36'N	122°30'W	28	25
Dziewulski Q	20°30'N	98°12'E	32	84
Edison T	24°42'N	97°6'E	48	84
Egede A	51°36'N	10°30'E	13	43
Ehrlich N	39°0'N	173°6'W	19	51
Eichstadt E	23°54'S	78°18'W	18	134
Eijkman D	62°18'S	136°54'W	25	183
Eimmart C	22°24'N	61°12'E	24	82
Eimmart G	25°30'N	64°48'E	14	82
Eimmart H	22°6'N	64°24'E	16	82
Eimmart K	20°12'N	67°36'E	13	82

Name	Lat.	Long.	Size	Page
Einstein A	16°42'N	88°12'W	51	74
Einstein S	15°6'N	91°30'W	20	73
Einthoven G	5°18'S	111°48'E	34	125
Einthoven M	7°30'S	109°36'E	52	125
Einthoven X	3°36'S	108°42'E	45	125
Ellison P	52°48'N	109°36'W	32	38
Elvey G	7°48'N	97°54'W	14	93
Elvey K	6°0'N	98°48'W	22	93
Emden D	64°24'N	171°6'W	47	24
Emden F	63°30'N	171°6'W	20	24
Emden M	61°18'N	177°0'W	25	24
Emden U	64°42'N	177°54'E	38	35
Emden V	65°48'N	177°30'E	35	35
Emden W	66°24'N	178°18'E	25	35
Encke B	2°24'N	36°48'W	12	96
Encke C	0°42'N	36°24'W	9	96
Endymion A	54°42'N	62°48'E	30	46
Endymion B	59°48'N	67°12'E	59	32
Endymion C	58°24'N	60°48'E	32	32
Endymion D	52°24'N	62°24'E	20	46
Endymion E	53°36'N	66°12'E	18	46
Endymion G	56°24'N	55°36'E	15	31
Endymion J	53°30'N	50°42'E	67	45
Engel'gardt C	10°6'N	156°54'W	49	90
Engel'gardt J	2°42'N	155°24'W	19	90
Engel'gardt K	2°24'N	157°48'W	18	90
Engel'gardt N	4°24'N	159°18'W	28	90
Epigenes A	66°54'N	0°18'W	18	29
Epigenes D	68°18'N	0°18'E	10	30
Epimenides A	43°12'S	30°6'W	15	173
Epimenides S	41°36'S	29°18'W	26	156
Erro D	6°48'N	100°30'E	30	104
Erro K	3°48'N	99°36'E	17	104
Erro T	5°36'N	96°54'E	16	104
Erro V	6°18'N	97°48'E	18	104

Name	Lat.	Long.	Size	Page
Espin E	28°18'N	111°18'E	35	65
Euclides C	13°12'S	30°0'W	10	117
Eudoxus A	45°48'N	20°0'E	14	44
Evans Q	11°12'S	136°24'W	137	111
Evdokimov G	33°54'N	150°30'W	48	52
Evershed C	38°6'N	156°42'W	48	52
Evershed D	38°48'N	156°0'W	49	52
Evershed S	34°54'N	162°36'W	45	51
Fabricius A	44°36'S	44°0'E	45	176
Fabry X	49°0'N	96°42'E	28	47
Faraday C	43°18'S	8°6'E	30	174
Faraday G	45°48'S	10°6'E	31	174
Fechner T	59°6'S	122°54'E	14	192
Feoktistov X	33°6'N	139°30'E	23	66
Fermat P	23°36'S	19°18'E	37	140
Finsen C	40°36'S	175°48'W	26	149
Finsen G	43°0'S	175°18'W	33	167
Firmicus B	7°18'N	65°48'E	14	102
Firsov S	3°36'N	109°48'E	96	105
Firsov V	5°6'N	110°42'E	44	105
Fitzgerald B	29°6'N	170°54'W	26	51
Fitzgerald W	28°42'N	173°48'W	51	51
Fitzgerald Y	31°0'N	172°42'W	34	51
Fizeau F	58°12'S	124°30'W	19	183
Fizeau G	59°12'S	124°30'W	54	183
Fizeau Q	59°48'S	136°18'W	28	183
Fizeau S	58°42'S	139°54'W	62	183
Flamsteed A	7°54'S	42°54'W	11	116
Flamsteed B	5°54'S	43°42'W	10	116
Flamsteed C	5°30'S	46°18'W	9	116
Fleming D	17°0'N	114°0'E	25	85
Fleming N	12°42'N	108°48'E	24	105
Fleming W	18°0'N	106°12'E	50	84
Fontenelle A	67°30'N	16°6'W	21	29
Fontenelle B	61°54'N	23°0'W	14	29

Name	Lat.	Long.	Size	Page
Fontenelle D	62°30'N	23°24'W	17	29
Foster H	23°12'N	139°36'W	25	71
Foster L	21°0'N	140°30'W	32	71
Foster P	20°12'N	143°30'W	36	71
Foster S	22°54'N	143°42'W	36	71
Fourier C	28°30'S	51°54'W	14	155
Fowler A	46°18'N	145°0'W	52	37
Fowler C	45°0'N	141°54'W	32	37
Fowler N	40°6'N	146°6'W	39	52
Fowler W	46°0'N	150°12'W	31	37
Fox A	1°30'N	98°18'E	13	104
Fra Mauro A	5°24'S	20°54'W	9	117
Fracastorius B	22°30'S	37°12'E	27	141
Fraunhofer B	41°48'S	67°18'E	36	161
Fraunhofer C	42°54'S	64°42'E	38	177
Fraunhofer E	43°24'S	61°42'E	42	177
Fraunhofer H	40°48'S	61°42'E	43	161
Fraunhofer J	42°24'S	63°36'E	63	177
Fraunhofer U	40°12'S	65°6'E	24	161
Freundlich G	24°30'N	173°30'E	25	88
Freundlich Q	22°42'N	167°24'E	17	88
Freundlich R	23°48'N	167°48'E	20	88
Fridman C	10°30'S	124°30'W	36	112
Froelich M	77°36'N	109°18'W	29	17
Frost N	35°0'N	119°12'W	43	54
Furnerius C	33°42'S	57°48'E	22	160
Furnerius D	37°0'S	55°54'E	16	160
Furnerius H	37°36'S	69°30'E	44	161
Furnerius J	34°48'S	64°12'E	24	161
Furnerius K	38°6'S	68°6'E	36	161
Furnerius Q	39°30'S	67°18'E	30	161
G. Bond B	29°54'N	34°42'E	33	61
Gadomski A	38°36'N	145°54'W	32	52
Gadomski X	37°48'N	148°18'W	35	52
Gagarin T	19°24'S	144°30'E	24	147

Name	Lat.	Long.	Size	Page
Gagarin Z	15°24'S	149°24'E	29	147
Galilaei A	11°42'N	62°54'W	11	95
Galilaei B	11°24'N	67°36'W	15	95
Galois C	12°24'S	150°30'W	22	110
Galois U	13°12'S	154°42'W	35	110
Gambart A	1°0'N	18°42'W	12	97
Gambart B	2°12'N	11°30'W	11	98
Gambart C	3°18'N	11°48'W	12	98
Gamow A	67°18'N	148°48'E	31	34
Gamow B	66°24'N	149°30'E	26	34
Gamow U	66°42'N	137°0'E	39	34
Gamow V	66°18'N	139°42'E	49	34
Gamow Y	67°48'N	143°54'E	27	34
Ganskiy S	10°12'S	94°42'E	22	124
Garavito C	45°0'S	159°0'E	25	181
Garavito D	46°36'S	158°48'E	33	181
Garavito Q	49°36'S	153°36'E	42	180
Gärtner F	57°30'N	30°6'E	14	31
Gassendi A	15°30'S	39°42'W	33	136
Gassendi B	14°42'S	40°36'W	26	136
Gaudibert J	11°6'S	39°6'E	10	121
Gauss B	35°54'N	81°12'E	37	64
Gauss C	39°42'N	72°6'E	29	63
Gauss D	39°18'N	73°48'E	24	63
Gauss W	34°30'N	80°12'E	18	64
Gavrilov A	19°36'N	131°54'E	26	86
Gavrilov K	15°0'N	132°30'E	38	86
Gay-Lussac A	13°12'N	20°24'W	15	97
Geiger R	15°42'S	156°30'E	40	147
Geiger Y	12°42'S	158°6'E	29	127
Gemma Frisius D	34°18'S	10°54'E	28	158
Gerard A	45°6'N	82°18'W	17	40
Gerard G	45°42'N	88°18'W	22	39
Gerasimovich R	24°6'S	125°54'W	55	132
Gernsback H	38°12'S	103°30'E	43	163

Name	Lat.	Long.	Size	Page
Gibbs D	13°6'S	85°54'E	13	123
Gilbert J	4°18'S	72°42'E	38	123
Gilbert K	5°30'S	73°12'E	38	123
Gilbert P	0°54'S	75°36'E	18	123
Gilbert W	1°6'S	78°54'E	19	123
Gill B	61°42'S	69°54'E	31	190
Gill C	62°12'S	67°24'E	30	190
Gill G	63°30'S	68°12'E	32	190
Ginzel G	13°42'N	100°12'E	42	104
Ginzel H	12°42'N	100°6'E	50	104
Ginzel L	13°6'N	97°48'E	28	104
Glaisher B	12°36'N	50°6'E	18	101
Glaisher E	12°42'N	49°12'E	21	101
Glazenap P	5°0'S	136°0'E	57	126
Goddard A	17°0'N	89°36'E	12	83
Goddard B	16°0'N	86°48'E	12	83
Godin B	0°42'N	9°48'E	12	99
Golitsyn J	27°36'S	103°0'W	20	133
Graff A	41°12'S	86°6'W	21	153
Green M	0°54'N	132°54'E	37	106
Green P	1°0'N	131°48'E	21	106
Green R	3°24'N	131°0'E	33	106
Gregory K	0°24'S	128°30'E	26	126
Grigg P	11°24'N	131°6'W	32	91
Grimaldi B	2°54'S	69°12'W	22	115
Grimaldi E	3°42'S	64°24'W	13	115
Grimaldi M	8°0'S	67°0'W	18	115
Grissom K	49°30'S	145°42'W	26	168
Grissom M	49°6'S	147°42'W	38	168
Grotrian X	64°30'S	125°30'E	20	192
Gruithuisen B	35°36'N	38°48'W	9	58
Guericke B	14°30'S	15°18'W	16	138
Guillaume D	46°36'N	170°30'W	26	36
Guillaume F	45°24'N	169°24'W	33	36
Guillaume J	43°42'N	170°36'W	17	36

Name	Lat.	Long.	Size	Page
Gullstrand C	46°48'N	126°36'W	15	38
Gum S	39°48'S	85°0'E	33	162
Gutenberg A	9°0'S	39°54'E	15	121
Gutenberg B	9°6'S	38°18'E	15	121
Gutenberg C	10°0'S	41°6'E	45	121
Gutenberg D	10°54'S	42°48'E	20	121
Gutenberg G	6°0'S	40°0'E	32	121
Hagecius A	58°12'S	47°12'E	61	189
Hagecius E	63°18'S	49°6'E	44	189
Hagecius K	61°12'S	52°0'E	31	189
Hahn A	29°42'N	69°42'E	17	63
Hahn B	31°24'N	77°0'E	15	63
Hahn D	27°30'N	68°36'E	15	82
Hahn E	27°42'N	70°0'E	15	82
Haidinger C	39°0'S	22°6'W	19	156
Hainzel A	40°18'S	33°54'W	53	156
Hainzel L	38°6'S	34°54'W	16	156
Hainzel R	38°42'S	36°24'W	19	156
Hale Q	76°30'S	83°6'E	24	199
Hall K	35°30'N	34°12'E	8	61
Hamilton B	42°36'S	82°6'E	32	177
Hanno A	53°24'S	63°12'E	38	177
Hanno B	52°36'S	68°36'E	36	177
Hanno D	59°6'S	78°18'E	18	190
Hanno E	59°18'S	73°0'E	18	190
Hanno H	57°36'S	74°24'E	57	190
Hanno K	53°30'S	76°54'E	25	177
Hanno W	54°36'S	60°6'E	10	177
Hansen A	13°18'N	74°18'E	13	103
Harding A	40°24'N	75°30'W	14	56
Haret C	57°12'S	172°48'W	30	182
Harkhebi J	37°24'N	103°24'E	40	65
Harkhebi K	35°42'N	100°48'E	27	65
Harkhebi T	40°6'N	95°42'E	16	64
Harkhebi U	40°48'N	97°0'E	18	64

Name	Lat.	Long.	Size	Page
Harpalus C	55°30'N	45°6'W	10	41
Harriot A	35°36'N	114°54'E	63	65
Harriot W	35°0'N	111°42'E	39	65
Hartwig A	5°42'S	79°48'W	10	114
Hartwig B	8°18'S	77°24'W	11	114
Hase D	31°0'S	63°18'E	56	161
Hatanaka Q	26°6'N	124°12'W	20	72
Hayford K	9°36'N	174°12'W	26	89
Hayford P	11°6'N	177°36'W	21	89
Hayford T	13°18'N	179°30'E	31	108
Hayn A	62°54'N	70°30'E	54	32
Hayn B	65°12'N	64°6'E	25	32
Hayn D	65°30'N	62°0'E	20	32
Hayn E	67°6'N	66°24'E	42	32
Hayn F	68°0'N	84°0'E	59	32
Hayn G	67°12'N	85°36'E	21	32
Hayn J	66°42'N	64°12'E	39	32
Hayn L	64°24'N	68°0'E	27	32
Healy J	30°12'N	108°48'W	42	54
Healy N	30°54'N	110°48'W	42	54
Heaviside B	5°30'S	169°18'E	23	128
Heaviside C	5°42'S	171°6'E	28	128
Heaviside D	6°42'S	171°48'E	18	128
Heaviside K	13°18'S	168°30'E	110	128
Heaviside Z	8°48'S	166°48'E	12	128
Hecataeus B	19°30'S	75°36'E	69	143
Hecataeus C	19°0'S	73°12'E	22	143
Hecataeus L	19°6'S	79°0'E	21	143
Hedin A	5°30'N	78°6'W	60	94
Hedin F	4°0'N	74°24'W	19	94
Hedin G	3°48'N	73°24'W	14	94
Heinsius C	40°36'S	17°54'W	23	157
Heinsius O	38°48'S	14°48'W	5	157
Heinsius Q	39°54'S	14°30'W	35	157
Heis D	31°42'N	31°6'W	8	58

Name	Lat.	Long.	Size	Page
Helberg C	23°24'N	100°36'W	70	73
Helberg H	21°48'N	101°12'W	29	73
Helmholtz D	66°18'S	54°18'E	46	189
Helmholtz J	64°48'S	67°48'E	22	190
Helmholtz S	64°18'S	56°36'E	31	189
Henderson G	3°36'N	155°48'E	46	107
Henderson Q	3°24'N	151°0'E	17	107
Hendrix M	48°24'S	158°54'W	21	167
Henry Frères C	24°36'S	59°42'W	37	135
Heraclitus D	50°24'S	5°12'E	52	174
Hermite A	87°48'N	47°6'W	20	14
Herodotus A	21°30'N	52°0'W	10	76
Herodotus K	24°30'N	51°54'W	5	76
Herschel C	5°0'S	3°12'W	10	118
Hertzsprung D	3°18'N	125°24'W	45	91
Hertzsprung H	1°18'S	124°24'W	21	112
Hertzsprung K	0°30'S	127°36'W	27	111
Hertzsprung L	0°24'N	127°48'W	33	91
Hertzsprung M	7°30'S	128°54'W	36	111
Hertzsprung P	0°0'N	129°18'W	22	111
Hertzsprung R	0°6'S	131°48'W	33	111
Hertzsprung S	0°36'N	132°30'W	47	91
Hertzsprung V	5°12'N	133°18'W	39	91
Hertzsprung X	3°48'N	129°6'W	24	91
Hertzsprung Y	8°48'N	131°12'W	23	91
Hess W	52°36'S	171°24'E	28	181
Hess Z	52°0'S	174°0'E	73	181
Hevelius A	2°54'N	68°6'W	14	95
Heymans D	76°48'N	132°18'W	25	17
Heymans F	75°0'N	133°36'W	50	17
Hilbert E	16°30'S	111°48'E	49	145
Hilbert G	19°0'S	114°0'E	50	145
Hilbert L	21°12'S	108°54'E	32	145
Hipparchus C	7°18'S	8°12'E	17	119
Hipparchus G	5°0'S	7°24'E	15	119

Name	Lat.	Long.	Size	Page
Hipparchus J	7°36'S	3°12'E	14	119
Hipparchus L	6°48'S	9°0'E	13	119
Hippocrates Q	69°0'N	148°0'W	35	25
Hirayama C	4°12'S	95°24'E	23	124
Hirayama F	5°48'S	97°12'E	35	124
Hirayama K	8°18'S	94°54'E	39	124
Hirayama L	9°24'S	94°24'E	24	124
Hirayama M	9°12'S	93°30'E	29	124
Hirayama N	7°12'S	93°36'E	17	124
Hirayama Q	8°0'S	91°18'E	40	124
Hirayama S	6°30'S	92°18'E	29	124
Hirayama Y	4°30'S	93°12'E	50	124
Hoffmeister D	16°54'N	140°18'E	21	86
Hoffmeister F	14°42'N	141°0'E	19	86
Hoffmeister N	13°42'N	136°24'E	42	106
Hoffmeister Z	17°48'N	136°42'E	29	86
Hogg E	34°6'N	124°54'E	21	66
Hogg P	32°30'N	121°24'E	26	66
Hogg T	33°54'N	119°0'E	27	65
Hohmann Q	21°48'S	98°6'W	15	133
Holetschek R	29°0'S	147°30'E	69	165
Holetschek Z	26°18'S	150°54'E	30	147
Hommel B	55°18'S	37°0'E	33	176
Hommel C	54°48'S	29°36'E	53	175
Hommel G	58°6'S	27°24'E	30	188
Hommel L	56°6'S	27°54'E	18	188
Hommel Q	56°6'S	38°24'E	29	189
Hooke D	40°42'N	55°48'E	19	62
Horrebow A	59°12'N	40°24'W	25	28
Hortensius A	4°24'N	30°42'W	10	97
Houzeau Q	19°0'S	125°18'W	18	132
Hubble C	19°36'N	85°18'E	50	83
Humboldt B	30°54'S	83°42'E	21	162
Hume Z	3°36'S	90°24'E	14	124
Hutton P	35°42'N	167°24'E	42	68

Name	Lat.	Long.	Size	Page
Hyginus A	6°18'N	5°42'E	8	99
Hyginus D	11°24'N	4°18'E	5	99
Hypatia A	4°54'S	22°12'E	16	120
Ibn Firnas E	7°30'N	125°30'E	42	105
Ibn Firnas L	5°54'N	123°0'E	21	105
Icarus D	4°18'S	171°12'W	68	109
Icarus E	5°12'S	168°48'W	21	109
Icarus H	7°48'S	169°24'W	32	109
Icarus J	7°18'S	170°54'W	32	109
Icarus Q	7°48'S	176°12'W	41	109
Icarus V	3°54'S	176°0'W	36	109
Icarus X	2°12'S	175°30'W	43	109
Ingalls G	25°48'N	150°24'W	55	70
Ingalls M	24°0'N	153°0'W	27	70
Ingalls U	27°18'N	156°12'W	28	70
Ingalls V	27°24'N	155°18'W	27	70
Ingalls Y	29°42'N	154°6'W	23	52
Inghirami A	44°54'S	65°18'W	34	171
Inghirami C	44°6'S	74°30'W	18	171
Inghirami G	51°6'S	74°6'W	29	171
Inghirami H	50°12'S	72°42'W	18	171
Inghirami K	49°36'S	73°54'W	23	171
Innes G	26°42'N	122°18'E	22	85
Innes S	27°36'N	117°18'E	33	85
Innes Z	29°48'N	119°12'E	33	65
Isidorus D	4°12'S	34°6'E	15	120
Isidorus E	5°18'S	32°36'E	15	120
Isidorus F	8°42'S	34°12'E	20	120
J. Herschel F	58°48'N	35°24'W	19	28
Jackson X	25°12'N	164°18'W	17	69
Jacobi A	58°30'S	16°0'E	28	188
Jacobi C	59°48'S	10°36'E	35	188
Jacobi F	58°30'S	9°36'E	42	188
Jansky D	9°30'N	91°12'E	20	104
Jansky F	8°48'N	92°12'E	50	104

Name	Lat.	Long.	Size	Page
Janssen B	43°12'S	34°24'E	22	175
Janssen E	48°48'S	39°54'E	25	176
Janssen J	43°24'S	36°36'E	30	176
Janssen K	46°6'S	42°18'E	16	176
Jeans B	52°24'S	94°48'E	11	178
Jeans S	56°48'S	86°48'E	56	190
Jeans U	54°42'S	86°30'E	57	178
Jeans X	53°30'S	89°24'E	44	178
Jeans Y	51°12'S	90°30'E	17	178
Jenner M	46°0'S	95°30'E	11	178
Joule T	27°42'N	148°12'W	37	70
Jules Verne C	33°12'S	149°42'E	30	165
Jules Verne G	35°6'S	150°0'E	42	165
Jules Verne P	38°0'S	145°6'E	62	165
Jules Verne R	36°54'S	140°54'E	49	165
Jules Verne Y	31°18'S	146°0'E	30	165
Jules Verne Z	32°30'S	146°48'E	20	165
Kane G	59°12'N	25°18'E	10	30
Kant D	11°30'S	18°42'E	52	120
Kapteyn A	14°12'S	71°18'E	31	142
Kapteyn B	15°36'S	71°0'E	39	142
Kapteyn E	8°48'S	69°18'E	31	122
Karpinskiy J	71°30'N	175°6'E	25	23
Kästner A	4°30'S	77°18'E	25	123
Kästner B	6°18'S	80°42'E	20	123
Kästner C	8°0'S	76°54'E	19	123
Kästner E	8°6'S	77°36'E	10	123
Kästner R	6°54'S	82°18'E	17	123
Kästner S	8°0'S	83°12'E	30	123
Kearons U	10°30'S	115°54'W	13	112
Keeler L	13°18'S	163°12'E	71	128
Keeler S	11°24'S	158°0'E	30	127
Keeler U	9°6'S	156°54'E	29	127
Keeler V	8°54'S	158°18'E	53	127
Kekulé K	13°54'N	135°48'W	16	91

Name	Lat.	Long.	Size	Page
Kekulé M	12°12'N	137°24'W	19	91
Kekulé S	15°24'N	143°0'W	21	71
Kekulé V	18°24'N	142°0'W	67	71
Kepínski N	26°36'N	126°12'E	40	86
Kepínski W	30°6'N	124°54'E	25	66
Kepler A	7°12'N	36°6'W	11	96
Kepler C	10°0'N	41°48'W	11	96
Kibal'chich H	2°0'N	144°12'W	40	90
Kibal'chich Q	0°42'S	149°0'W	25	110
Kibal'chich R	0°36'N	150°6'W	29	90
Kidinnu E	36°18'N	124°30'E	60	66
Kies A	28°18'S	22°42'W	16	156
Kinau C	60°36'S	20°30'E	30	188
King J	3°12'N	121°48'E	14	105
King Y	6°30'N	119°48'E	48	105
Kircher E	69°6'S	50°6'W	20	186
Kirkwood Y	72°12'N	157°30'W	19	16
Klaproth A	68°12'S	21°36'W	30	187
Klaproth G	68°36'S	31°12'W	30	186
Klute W	38°12'N	143°0'W	13	52
Koch R	44°30'S	146°18'E	20	180
Kohlschütter N	11°36'N	153°42'E	27	107
Kohlschütter Q	13°12'N	153°0'E	20	107
Kohlschütter W	16°18'N	151°12'E	32	87
Kondratyuk Q	15°42'S	114°42'E	28	145
Korolev B	3°54'S	156°6'W	22	110
Korolev C	1°18'S	153°12'W	68	110
Korolev D	0°48'S	151°30'W	26	110
Korolev E	3°54'S	153°12'W	37	110
Korolev F	4°36'S	152°36'W	31	110
Korolev L	6°0'S	156°42'W	30	110
Korolev M	8°48'S	157°18'W	58	110
Korolev P	8°6'S	159°54'W	17	110
Korolev W	0°24'S	160°18'W	34	110
Korolev X	0°36'N	159°0'W	28	90

Name	Lat.	Long.	Size	Page
Korolev Y	0°42'S	158°12'W	19	110
Kostinskiy B	16°18'N	119°54'E	20	85
Kostinskiy D	16°0'N	122°48'E	32	85
Kostinskiy E	15°6'N	122°24'E	26	85
Kostinskiy W	17°12'N	115°36'E	24	85
Kovalevskaya Q	29°24'N	131°0'W	101	53
Koval'skiy H	22°30'S	102°36'E	37	144
Koval'skiy P	22°24'S	100°18'E	25	144
Koval'skiy Q	23°30'S	98°42'E	35	144
Krafft D	15°6'N	73°18'W	12	74
Krafft E	15°54'N	71°42'W	10	75
Kramers C	55°0'N	125°36'W	60	38
Kramers M	50°24'N	126°54'W	30	38
Kramers S	52°48'N	132°12'W	26	37
Kramers U	54°12'N	132°24'W	38	37
Krasovskiy H	2°42'N	171°24'W	46	89
Krasovskiy J	3°12'N	174°6'W	32	89
Krasovskiy L	0°24'S	174°48'W	58	109
Krasovskiy P	0°48'N	177°18'W	41	89
Krylov A	38°54'N	165°6'W	63	51
Krylov B	37°18'N	163°36'W	40	51
Kulik J	40°36'N	151°24'W	46	52
Kulik K	39°6'N	151°36'W	42	52
Kulik L	40°48'N	153°30'W	33	52
Kurchatov T	38°0'N	138°0'E	27	66
Kurchatov Z	41°0'N	141°48'E	27	67
La Condamine A	54°24'N	30°6'W	18	42
La Condamine B	58°48'N	31°30'W	17	28
La Condamine D	53°30'N	30°48'W	10	42
La Pérouse E	10°12'S	78°30'E	34	123
Lacroix F	40°42'S	61°36'W	15	154
Lacroix K	35°12'S	57°42'W	45	155
Lade T	1°0'S	9°0'E	18	119
Lagalla F	44°36'S	25°18'W	29	173
Lagrange C	29°48'S	64°54'W	23	154

Name	Lat.	Long.	Size	Page
Lalande A	6°36'S	9°48'W	13	118
Lalande B	3°6'S	9°0'W	8	118
Lalande C	5°36'S	6°54'W	11	118
Lalande R	4°42'S	7°0'W	24	118
Lamé T	12°30'S	66°30'E	11	122
Lampland K	33°0'S	132°30'E	47	164
Lampland R	31°42'S	129°12'E	45	164
Landau Q	41°0'N	121°42'W	32	53
Langemak N	12°54'S	119°0'E	126	125
Langemak X	6°36'S	117°30'E	47	125
Langemak Z	5°36'S	119°18'E	27	125
Langevin C	46°24'N	165°30'E	19	50
Langevin K	41°36'N	163°48'E	17	68
Langrenus E	12°42'S	60°36'E	30	122
Langrenus M	9°48'S	66°24'E	17	122
Langrenus P	12°6'S	63°6'E	42	122
Langrenus T	4°36'S	62°30'E	42	122
Langrenus X	12°24'S	64°42'E	25	122
Langrenus Y	7°48'S	66°54'E	27	122
Lansberg B	2°30'S	28°6'W	9	117
Lansberg C	1°30'S	29°12'W	17	117
Laplace A	43°42'N	26°48'W	9	42
Laplace D	47°18'N	25°30'W	11	42
Larmor K	30°18'N	179°0'W	24	51
Larmor Q	28°36'N	176°12'E	22	68
Larmor W	33°54'N	177°36'E	27	68
Larmor Z	33°42'N	179°48'W	49	51
Laue U	28°48'N	101°24'W	56	54
Lauritsen A	24°48'S	96°36'E	35	144
Lauritsen Z	26°0'S	96°12'E	52	144
Lavoisier A	36°54'N	73°12'W	28	56
Lavoisier C	35°48'N	76°42'W	35	56
Lavoisier E	40°54'N	80°24'W	49	55
Lavoisier J	37°30'N	86°30'W	22	55
Lavoisier S	39°6'N	83°6'W	24	55

Name	Lat.	Long.	Size	Page
Le Gentil A	74°36'S	52°24'W	33	196
Le Monnier A	26°54'N	32°30'E	21	80
Le Verrier D	39°42'N	12°18'W	9	59
Leavitt Z	42°42'S	139°12'W	65	168
Lebedev C	45°0'S	111°0'E	34	179
Lebedev D	44°36'S	112°30'E	34	179
Lebedinskiy A	10°54'N	163°42'W	38	89
Lebedinskiy B	10°30'N	163°12'W	37	89
Lebedinskiy P	6°0'N	165°0'W	51	89
Leeuwenhoek E	28°12'S	176°42'W	117	129
Legendre D	31°30'S	75°12'E	58	161
Legendre E	33°48'S	78°30'E	28	161
Legendre F	33°48'S	76°24'E	40	161
Lehmann E	37°30'S	54°54'W	48	155
Leibnitz R	39°18'S	176°18'E	19	166
Lemaître C	59°24'S	145°36'W	27	183
Lemaître S	61°36'S	156°18'W	34	182
Lents C	3°18'N	101°36'W	23	93
Lents J	3°42'S	97°18'W	16	113
Letronne F	9°12'S	46°6'W	8	116
Leucippus Q	25°54'N	118°48'W	84	72
Leucippus X	33°24'N	118°48'W	36	54
Leuschner L	1°6'S	108°48'W	18	112
Leuschner Z	5°18'N	109°18'W	18	92
Levi-Civita S	24°6'S	138°48'E	43	146
Lewis R	20°18'S	116°6'W	26	132
Lexell A	36°54'S	1°24'W	34	157
Lexell B	37°18'S	3°24'W	23	157
Licetus F	46°0'S	1°0'E	32	174
Lick N	9°42'N	47°54'E	23	101
Liebig G	26°6'S	45°48'W	20	136
Lilius A	55°24'S	8°48'E	41	174
Lilius B	53°0'S	3°48'E	29	174
Lilius E	50°6'S	2°54'E	38	174
Lilius F	49°24'S	1°42'E	43	174

Name	Lat.	Long.	Size	Page
Lilius K	53°36'S	2°12'E	23	174
Lindblad F	70°36'N	94°18'W	42	17
Lindblad Y	73°0'N	101°12'W	28	17
Lippmann B	52°36'S	110°54'W	29	169
Lippmann E	55°24'S	107°36'W	23	170
Lippmann L	57°36'S	112°30'W	54	184
Lippmann Q	57°0'S	118°42'W	27	184
Lippmann R	57°12'S	121°18'W	37	183
Lipskiy V	1°12'S	178°42'E	36	128
Lipskiy X	0°24'N	178°54'E	24	108
Lobachevskiy M	8°0'N	112°48'E	41	105
Lockyer G	45°42'S	33°18'E	24	175
Lockyer H	44°30'S	32°30'E	31	175
Lodygin C	15°54'S	144°30'W	30	130
Lodygin F	17°36'S	142°48'W	47	131
Lodygin R	18°18'S	149°12'W	30	130
Lohrmann A	0°42'S	62°42'W	12	115
Longomontanus A	52°48'S	24°0'W	29	173
Longomontanus C	53°24'S	19°0'W	31	173
Longomontanus D	54°18'S	22°54'W	29	173
Lorentz P	31°48'N	98°30'W	38	55
Lorentz R	33°24'N	99°12'W	33	55
Lorentz T	34°36'N	100°18'W	20	54
Love G	6°30'S	131°18'E	54	126
Love H	6°54'S	130°24'E	29	126
Lovell F	36°42'S	138°12'W	24	151
Lowell W	10°0'S	107°0'W	18	113
Lubbock N	1°30'S	39°42'E	26	121
Lucretius C	3°42'S	114°24'W	20	112
Lucretius U	7°42'S	123°36'W	24	112
Lundmark B	37°42'S	153°12'E	30	165
Lundmark C	35°48'S	155°36'E	25	165
Lundmark F	39°24'S	157°12'E	26	165
Lundmark G	40°30'S	155°30'E	35	165
Lyman Q	68°36'S	156°42'E	56	193

Name	Lat.	Long.	Size	Page
Lyman T	64°6'S	157°42'E	59	193
Lyman V	62°36'S	154°12'E	37	193
Lyot A	49°0'S	79°36'E	38	177
Lyot C	50°24'S	80°24'E	17	177
Lyot L	54°24'S	83°6'E	70	177
Lyot M	53°18'S	86°12'E	24	178
Lyot S	46°0'S	85°36'E	26	178
M?sting A	3°12'S	5°12'W	13	118
Mach H	14°54'N	144°6'W	40	71
Maclaurin A	3°0'S	67°36'E	29	122
Maclaurin C	1°6'S	69°36'E	26	122
Maclaurin E	3°30'S	65°42'E	20	122
Maclaurin K	0°54'S	66°54'E	34	122
Maclaurin L	1°24'S	71°42'E	30	122
Maclaurin M	4°48'S	69°24'E	42	122
Maclaurin N	3°48'S	68°24'E	29	122
Maclaurin P	6°0'S	69°24'E	29	122
Macrobius M	25°0'N	41°0'E	42	81
Macrobius S	23°18'N	49°36'E	26	81
Macrobius W	24°48'N	44°36'E	26	81
Magelhaens A	12°36'S	45°0'E	32	121
Maginus C	51°42'S	9°24'W	42	174
Maginus N	48°30'S	9°0'W	24	174
Mairan A	38°36'N	38°48'W	16	58
Maksutov U	40°6'S	170°54'W	21	149
Malapert B	79°6'S	2°24'W	37	197
Malapert C	81°30'S	10°30'E	40	198
Malapert K	78°48'S	6°48'E	36	198
Mallet A	45°54'S	53°48'E	28	176
Mallet B	46°36'S	52°0'E	32	176
Mallet D	46°0'S	57°0'E	42	176
Mallet J	48°42'S	55°54'E	52	176
Mallet K	47°36'S	57°0'E	43	176
Malyy G	21°42'N	106°54'E	28	84
Mandel'shtam F	5°12'N	166°12'E	17	108

Name	Lat.	Long.	Size	Page
Mandel'shtam G	4°30'N	166°24'E	29	108
Mandel'shtam N	3°18'N	161°36'E	25	107
Mandel'shtam R	4°30'N	159°48'E	57	107
Mandel'shtam T	5°42'N	160°24'E	37	107
Mandel'shtam Y	9°6'N	161°48'E	32	107
Manzinus E	68°54'S	25°24'E	18	188
Manzinus U	68°36'S	34°30'E	21	189
Marci B	25°12'N	166°18'W	28	69
Marci C	24°18'N	165°24'W	26	69
Marconi H	11°0'S	147°30'E	41	127
Marconi L	11°42'S	145°18'E	38	127
Marinus A	39°54'S	73°12'E	27	161
Marinus B	39°36'S	74°48'E	59	161
Marinus C	38°0'S	73°30'E	37	161
Marinus R	38°0'S	75°18'E	44	161
Mariotte P	29°54'S	139°42'W	30	151
Mariotte R	30°6'S	141°36'W	33	150
Mariotte Z	22°54'S	139°0'W	47	131
Marius A	12°36'N	46°0'W	15	96
Marius B	16°18'N	47°18'W	12	76
Marius C	14°0'N	47°36'W	11	76
Marius D	11°24'N	45°0'W	9	96
Maskelyne A	0°6'N	34°0'E	29	100
Mason C	42°54'N	33°48'E	12	44
Maunder A	3°12'S	90°30'W	15	113
Maupertuis A	50°36'N	24°42'W	14	42
Maurolycus F	40°36'S	12°12'E	25	158
Maury A	36°0'N	41°48'E	21	62
McKellar T	15°6'S	173°0'W	45	129
McKellar U	13°54'S	174°30'W	37	109
McLaughlin A	51°36'N	92°24'W	35	39
McLaughlin B	50°12'N	91°12'W	43	39
McLaughlin C	48°30'N	91°54'W	60	39
McLaughlin P	45°0'N	94°36'W	34	39
McLaughlin U	47°12'N	97°0'W	30	39

Name	Lat.	Long.	Size	Page
McLaughlin Z	52°36'N	92°48'W	21	39
McNally T	22°18'N	129°0'W	19	71
Mechnikov C	9°54'S	148°0'W	35	110
Mechnikov D	10°12'S	147°12'W	53	110
Mechnikov F	11°18'S	145°0'W	30	110
Mechnikov U	10°36'S	150°54'W	30	110
Mees A	15°42'N	95°12'W	36	73
Mees J	12°18'N	94°42'W	26	93
Mees Y	15°42'N	96°36'W	85	73
Meggers S	24°0'N	119°48'E	42	85
Mendel J	51°36'S	107°24'W	58	170
Mendel V	46°42'S	116°42'W	66	169
Mendeleev P	2°42'N	139°24'E	29	106
Mercurius C	47°30'N	59°24'E	26	45
Mercurius E	49°42'N	73°18'E	29	46
Mercurius M	50°54'N	73°54'E	40	46
Mersenius D	23°6'S	46°48'W	34	136
Meshcherskiy K	9°36'N	126°48'E	17	106
Meshcherskiy X	16°0'N	124°12'E	39	85
Messala A	36°36'N	53°48'E	26	62
Messala B	37°24'N	59°54'E	18	62
Messala D	40°30'N	67°48'E	28	63
Messala E	40°0'N	64°54'E	40	63
Messala F	38°54'N	64°24'E	32	63
Messala G	39°6'N	68°36'E	29	63
Messier A	2°0'S	47°0'E	13	121
Mezentsev M	68°42'N	126°48'W	74	25
Michelson G	5°42'N	118°48'W	27	92
Michelson H	4°36'N	116°48'W	35	92
Michelson V	8°0'N	124°24'W	20	92
Milanković E	78°0'N	177°12'W	46	16
Milichius A	9°18'N	32°0'W	9	97
Miller A	37°42'S	1°48'E	39	158
Miller C	38°12'S	0°18'W	36	157
Millikan B	49°48'N	123°30'E	21	48

Name	Lat.	Long.	Size	Page
Millikan Q	43°54'N	118°36'E	33	48
Millikan R	46°0'N	117°42'E	49	48
Mills B	10°42'N	156°54'E	24	107
Mills K	6°48'N	157°0'E	26	107
Milne K	32°30'S	113°6'E	65	163
Milne L	33°42'S	112°42'E	26	163
Milne M	35°42'S	112°6'E	54	163
Milne N	35°30'S	110°48'E	37	163
Mineur V	26°12'N	163°6'W	26	69
Mineur X	27°6'N	162°42'W	31	69
Minnaert C	64°12'S	176°0'W	15	182
Minnaert W	63°24'S	174°6'E	24	193
Mitra A	20°48'N	154°6'W	46	70
Mitra J	15°54'N	153°12'W	46	70
Mitra Y	21°30'N	155°12'W	26	70
Mohorovičić R	19°54'S	167°42'W	42	129
Moigno A	64°48'N	29°42'E	16	30
Moiseev S	8°42'N	100°42'E	29	104
Moiseev Z	11°12'N	103°24'E	80	104
Montgolfier P	46°6'N	160°54'W	36	36
Montgolfier W	49°18'N	164°24'W	37	36
Montgolfier Y	50°30'N	161°18'W	40	36
Moore F	37°24'N	175°0'W	24	51
Morozov F	5°24'N	130°0'E	60	106
Morozov Y	7°18'N	127°0'E	45	106
Morse N	20°12'N	176°6'W	25	69
Morse T	22°0'N	179°30'W	34	69
Moseley C	22°18'N	88°30'W	18	74
Moseley D	22°54'N	87°36'W	17	74
Moulton H	61°30'S	100°36'E	44	191
Mutus C	61°12'S	27°12'E	32	188
Mutus D	58°24'S	23°18'E	22	188
Mutus F	66°12'S	34°6'E	42	189
Mutus L	61°48'S	24°54'E	20	188
Mutus V	62°54'S	31°18'E	24	189

Name	Lat.	Long.	Size	Page
Mutus W	66°36'S	40°0'E	21	189
N?ther E	67°36'N	105°6'W	47	26
N?ther T	66°18'N	121°30'W	44	25
N?ther U	67°36'N	123°24'W	36	25
N?ther V	68°54'N	122°24'W	26	25
Nagaoka U	19°54'N	151°24'E	30	87
Nagaoka W	20°0'N	153°0'E	29	87
Nansen A	82°48'N	63°0'E	46	21
Nansen C	83°12'N	55°30'E	34	21
Nansen F	84°42'N	60°0'E	62	14
Nassau F	24°42'S	179°12'W	112	129
Nassau Y	22°30'S	176°48'E	38	148
Naumann B	37°24'N	60°36'W	10	56
Nearch A	60°6'S	40°6'E	43	189
Nearch B	60°54'S	35°48'E	43	189
Necho P	6°48'S	122°0'E	75	125
Neper D	9°12'N	80°48'E	40	103
Nernst T	35°48'N	96°54'W	25	55
Neujmin P	28°30'S	124°12'E	38	164
Neujmin Q	30°0'S	121°48'E	17	164
Neujmin T	27°6'S	122°0'E	24	145
Neumayer A	75°0'S	73°36'E	31	199
Neumayer M	71°36'S	78°30'E	31	199
Neumayer N	70°24'S	78°42'E	36	199
Newton A	79°42'S	19°42'W	64	197
Newton G	78°12'S	18°18'W	67	197
Niepce F	72°30'N	113°30'W	44	17
Nijland A	36°12'N	134°24'E	26	66
Nijland V	34°30'N	131°36'E	35	66
Nikolaev J	31°42'N	155°30'E	18	67
Nobel B	17°18'N	99°30'W	24	73
Nobel K	13°6'N	100°12'W	20	93
Nobel L	12°30'N	100°54'W	38	93
Nöggerath G	50°18'S	45°48'W	21	172
Nonius L	33°30'S	3°30'E	31	158

Name	Lat.	Long.	Size	Page
Numerov G	71°42'S	151°54'W	26	194
Numerov Z	68°6'S	160°0'W	44	182
Nušl S	31°12'N	164°6'E	42	68
Nušl Y	34°18'N	166°54'E	51	68
Obruchev M	40°30'S	162°12'E	46	166
Obruchev V	36°36'S	158°18'E	39	165
O'Day T	30°24'S	154°24'E	24	165
Oken A	43°12'S	71°18'E	36	177
Oken F	44°24'S	71°30'E	21	177
Olbers B	6°48'N	74°6'W	16	94
Olbers M	8°0'N	81°12'W	33	94
Olbers N	9°0'N	79°42'W	22	94
Olbers Y	6°30'N	83°36'W	21	94
Olcott E	20°54'N	119°48'E	59	85
Olcott L	18°18'N	118°36'E	36	85
Olcott M	17°54'N	117°36'E	46	85
Olivier N	56°42'N	137°6'E	63	34
Olivier Y	61°54'N	136°30'E	47	34
Oppenheimer F	34°42'S	161°30'W	35	149
Oppenheimer R	37°18'S	170°24'W	26	149
Oppenheimer U	34°18'S	167°54'W	38	149
Oppenheimer V	32°0'S	172°42'W	32	149
Oresme K	43°54'S	170°0'E	24	181
Oresme V	40°30'S	165°36'E	51	166
Orlov Y	22°48'S	175°6'W	126	129
Ostwald Y	13°36'N	121°0'E	24	105
Palitzsch A	26°54'S	65°48'E	31	142
Palitzsch B	26°24'S	68°24'E	39	142
Palmieri A	32°12'S	48°24'W	21	155
Paneth A	65°18'N	94°6'W	47	26
Paneth K	61°42'N	92°54'W	31	26
Paneth W	65°0'N	101°12'W	28	26
Pannekoek A	0°54'S	141°0'E	28	126
Pannekoek D	2°36'S	143°30'E	28	126
Pannekoek R	5°24'S	138°18'E	71	126

Name	Lat.	Long.	Size	Page
Pannekoek T	4°6'S	138°12'E	25	126
Paraskevopoulos H	49°42'N	147°12'W	48	37
Paraskevopoulos N	47°12'N	150°48'W	26	37
Paraskevopoulos Q	48°36'N	152°18'W	35	37
Paraskevopoulos S	49°6'N	154°54'W	67	37
Paraskevopoulos U	50°24'N	154°42'W	30	37
Paraskevopoulos X	53°36'N	152°12'W	26	37
Paraskevopoulos Y	53°6'N	150°24'W	46	37
Parenago W	27°48'N	109°42'W	49	72
Parkhurst D	32°48'S	105°24'E	27	163
Parkhurst Q	35°0'S	101°36'E	37	163
Parkhurst Y	29°54'S	102°48'E	49	163
Parrot C	18°30'S	1°12'E	31	139
Parsons D	38°30'N	168°36'W	54	51
Parsons L	33°36'N	170°0'W	31	51
Parsons N	34°12'N	173°12'W	43	51
Pascal A	72°54'N	74°36'W	28	18
Pascal F	75°36'N	75°36'W	27	18
Pascal G	73°0'N	65°42'W	14	18
Paschen G	14°18'S	135°24'W	29	131
Paschen H	16°0'S	135°36'W	27	131
Paschen K	17°54'S	138°54'W	57	131
Paschen L	16°24'S	139°30'W	38	131
Paschen S	14°30'S	142°0'W	48	131
Paschen U	13°12'S	143°0'W	29	111
Pasteur A	7°0'S	105°42'E	25	124
Pasteur B	8°12'S	105°48'E	20	124
Pasteur D	8°48'S	108°48'E	36	125
Pasteur Q	13°36'S	101°30'E	24	124
Pasteur S	12°12'S	102°0'E	29	124
Pasteur T	11°36'S	100°6'E	41	124
Pasteur U	9°48'S	101°30'E	37	124
Patsaev G	17°6'S	136°48'E	28	146
Patsaev K	18°48'S	134°30'E	53	146
Pavlov G	29°6'S	145°24'E	43	165

Name	Lat.	Long.	Size	Page
Pavlov M	32°18'S	141°48'E	74	165
Pavlov V	26°42'S	138°0'E	38	146
Peirescius C	46°12'S	71°30'E	41	177
Peirescius D	48°6'S	71°54'E	43	177
Peirescius G	48°6'S	67°42'E	25	177
Pentland A	67°24'S	13°30'E	44	188
Pentland C	65°0'S	16°18'E	37	188
Pentland D	63°12'S	14°6'E	35	188
Perel'man S	24°18'S	104°24'E	26	144
Perepelkin P	12°24'S	127°18'E	25	126
Perrine E	42°48'N	124°54'W	40	38
Perrine G	42°6'N	124°36'W	58	53
Perrine L	39°12'N	127°12'W	37	53
Perrine T	42°24'N	130°12'W	34	38
Petavius B	19°54'S	57°6'E	33	142
Petermann D	77°6'N	65°48'E	31	21
Petropavlovskiy M	34°30'N	114°42'W	22	54
Petrov B	62°18'S	90°30'E	31	191
Petzval C	60°18'S	107°48'W	52	184
Phillips A	27°6'S	73°36'E	13	143
Phillips B	23°18'S	70°30'E	40	142
Phillips D	25°0'S	70°48'E	61	142
Phocylides A	54°36'S	51°36'W	19	172
Phocylides F	54°48'S	57°24'W	23	172
Phocylides J	54°6'S	62°42'W	22	171
Phocylides K	52°12'S	48°54'W	14	172
Piazzi C	37°6'S	62°36'W	28	154
Piazzi P	38°48'S	67°18'W	20	154
Pico B	46°30'N	15°18'W	12	42
Pico E	43°0'N	10°18'W	9	43
Pictet A	45°0'S	7°54'W	34	174
Pikel'ner G	49°6'S	128°18'E	24	179
Pikel'ner S	48°42'S	120°12'E	62	179
Pingré C	58°24'S	68°18'W	23	185
Pingré K	55°12'S	77°42'W	13	171

Name	Lat.	Long.	Size	Page
Pingré L	53°48'S	85°48'W	17	170
Pingré N	58°6'S	83°42'W	19	185
Pingré S	60°18'S	82°0'W	70	185
Plana C	42°42'N	27°6'E	14	44
Planck C	53°24'S	141°18'E	43	180
Planck J	62°54'S	145°18'E	26	192
Planck K	65°0'S	146°12'E	23	192
Planck L	66°54'S	141°48'E	23	192
Planck W	56°0'S	131°12'E	17	179
Planck X	54°18'S	129°30'E	25	179
Planck Z	56°24'S	135°12'E	72	192
Plaskett V	82°30'N	118°30'E	49	22
Plato H	55°6'N	2°0'W	11	43
Plummer N	27°42'S	156°18'W	42	130
Plutarch C	23°6'N	71°0'E	11	82
Plutarch D	24°18'N	75°42'E	15	83
Plutarch F	23°30'N	73°30'E	12	83
Plutarch H	24°24'N	72°42'E	11	83
Poczobutt J	56°36'N	96°48'W	24	26
Pogson F	42°0'S	114°36'E	35	179
Poincaré C	54°24'S	169°0'E	20	181
Poincaré Q	59°18'S	160°54'E	26	193
Poincaré X	53°48'S	161°54'E	19	181
Poincaré Z	53°42'S	164°54'E	35	181
Poinsot E	80°12'N	129°48'W	25	17
Poinsot P	77°12'N	149°42'W	27	16
Polzunov J	23°36'N	117°24'E	30	85
Polzunov N	23°42'N	113°48'E	35	85
Poncelet A	79°30'N	74°42'W	31	18
Poncelet B	78°36'N	62°18'W	32	18
Poncelet C	77°24'N	73°42'W	67	18
Poncelet D	77°42'N	70°0'W	23	18
Pontanus C	30°0'S	15°30'E	23	158
Pontécoulant A	57°42'S	62°54'E	19	190
Pontécoulant B	57°54'S	58°30'E	39	189
Pontécoulant G	57°12'S	60°6'E	36	190
Pontécoulant J	61°36'S	64°18'E	39	190
Popov D	17°48'N	102°36'E	15	84
Popov W	19°6'N	97°48'E	25	84
Posidonius J	33°48'N	30°42'E	22	61
Posidonius P	33°36'N	27°30'E	15	61
Poynting X	23°18'N	136°12'W	22	71
Prager C	1°24'S	132°24'E	40	126
Prager E	3°0'S	133°6'E	14	126
Prager G	4°36'S	134°0'E	76	126
Priestley K	59°0'S	110°30'E	35	191
Proclus A	13°24'N	42°18'E	15	101
Proclus D	17°30'N	41°0'E	13	81
Proclus U	15°12'N	48°0'E	13	81
Purkyně S	1°48'S	90°36'E	34	124
Purkyně V	0°48'S	92°42'E	24	124
Pythagoras D	64°30'N	72°0'W	30	27
Quetelet T	42°48'N	137°36'W	46	37
R?ntgen B	35°42'N	88°6'W	16	55
Racah B	10°30'S	178°24'W	27	109
Racah J	16°30'S	177°24'W	37	129
Racah K	16°48'S	178°36'W	52	129
Racah N	17°0'S	179°0'E	35	148
Racah T	13°48'S	177°30'E	21	128
Racah U	13°12'S	177°12'E	25	128
Racah W	12°30'S	178°54'E	39	128
Raimond K	13°18'N	158°12'W	34	90
Raimond Q	11°36'N	161°42'W	32	89
Rayet P	43°18'N	114°0'E	17	48
Rayet Y	47°12'N	113°0'E	14	48
Rayleigh C	31°24'N	85°42'E	22	64
Razumov C	40°48'N	112°24'W	48	54
Reichenbach A	28°18'S	49°0'E	34	160
Reichenbach T	29°18'S	45°42'E	64	160
Reiner A	5°12'N	51°24'W	10	96
Reinhold B	4°18'N	21°42'W	26	97
Repsold B	53°12'N	75°48'W	38	40
Repsold G	50°30'N	80°36'W	44	40
Repsold N	49°0'N	78°12'W	14	40
Repsold R	49°48'N	72°12'W	13	40
Repsold T	47°42'N	79°54'W	13	40
Rhaeticus A	1°48'N	5°12'E	11	99
Rheita B	39°6'S	52°48'E	21	160
Rheita E	34°12'S	49°6'E	66	160
Riccioli C	0°36'N	73°0'W	31	94
Riccioli F	8°36'S	73°54'W	28	114
Riccioli G	1°18'S	71°0'W	15	115
Riccioli H	1°6'N	74°54'W	18	94
Riccioli K	2°12'S	77°30'W	43	114
Riccius C	36°12'S	28°48'E	24	159
Riccius E	39°54'S	26°24'E	22	159
Richardson E	31°54'N	103°36'E	22	65
Richardson W	33°30'N	98°18'E	23	64
Riedel Q	49°54'S	141°42'W	25	168
Riedel Z	47°24'S	139°42'W	30	168
Riemann B	41°36'N	85°12'E	24	64
Riemann J	37°24'N	90°12'E	39	64
Ritter B	3°18'N	18°54'E	14	100
Ritter C	2°48'N	18°54'E	14	100
Ritz B	13°42'S	92°48'E	26	124
Roberts R	69°36'N	178°24'E	59	35
Rocca A	13°48'S	70°0'W	63	115
Rocca B	12°36'S	67°24'W	25	115
Rocca C	10°42'S	70°12'W	19	115
Rocca E	11°48'S	69°24'W	43	115
Rocca T	9°42'S	71°0'W	16	115
Römer A	28°6'N	37°6'E	35	61
Römer C	27°42'N	37°0'E	8	81
Römer P	26°30'N	39°36'E	61	81
Römer R	24°12'N	34°36'E	42	80

Name	Lat.	Long.	Size	Page
Rosenberger A	53°30'S	47°0'E	49	176
Rosenberger B	51°54'S	46°6'E	33	176
Rosenberger C	52°6'S	42°6'E	47	176
Rosenberger D	57°30'S	42°54'E	50	189
Rost A	56°30'S	36°42'W	39	186
Rost D	56°36'S	30°54'W	29	186
Rost M	55°30'S	31°24'W	26	173
Rothmann C	28°36'S	25°6'E	19	159
Rowland J	53°6'N	155°30'W	49	37
Rowland K	51°24'N	157°6'W	25	36
Rowland M	51°54'N	162°24'W	58	36
Rowland N	55°24'N	163°42'W	30	36
Rowland R	53°42'N	169°30'W	24	36
Rowland Y	59°6'N	163°0'W	54	24
Rozhdestvenskiy U	85°18'N	151°54'E	44	14
Rumford A	25°12'S	169°12'W	30	129
Rumford F	28°54'S	165°18'W	13	149
Rumford Q	30°42'S	171°36'W	29	149
Rumford T	28°36'S	172°6'W	108	149
Russell B	26°24'N	78°12'W	19	74
Sacrobosco Q	21°36'S	17°30'E	42	139
Saenger P	2°42'N	101°42'E	41	104
Saenger X	6°18'N	101°48'E	18	104
Šafařík A	12°36'N	177°12'E	19	108
Šafařík S	10°0'N	174°24'E	14	108
Saha B	1°30'N	104°30'E	34	104
Saha C	1°24'N	107°48'E	64	105
Saha J	4°0'S	105°18'E	52	124
Saha M	2°12'S	102°36'E	18	124
Saha N	4°6'S	101°30'E	49	124
Saha W	0°36'S	101°24'E	34	124
Sanford T	32°42'N	143°18'W	43	52
Sanford W	33°42'N	140°12'W	38	52
Santbech A	24°12'S	42°18'E	25	141
Sarton L	47°0'N	120°0'W	48	38

Name	Lat.	Long.	Size	Page
Sarton Z	51°36'N	120°36'W	29	38
Saussure A	43°48'S	0°30'W	19	174
Schaeberle U	25°30'S	113°54'E	24	145
Scheiner B	59°30'S	33°18'W	29	186
Schickard H	43°30'S	62°12'W	16	171
Schjellerup H	68°30'N	167°24'E	21	35
Schjellerup N	66°36'N	154°18'E	38	35
Schjellerup R	68°42'N	152°12'E	54	35
Schlesinger A	50°6'N	137°12'W	32	37
Schlesinger B	51°24'N	134°54'W	66	37
Schlesinger M	45°12'N	138°30'W	45	37
Schliemann A	1°12'N	155°24'E	64	107
Schliemann B	2°6'N	156°12'E	32	107
Schliemann T	2°0'S	152°48'E	21	127
Schlüter A	9°12'S	82°24'W	37	114
Schlüter P	0°6'N	85°6'W	20	94
Schlüter V	4°24'S	86°48'W	12	114
Schlüter Z	2°48'S	83°42'W	11	114
Schneller H	39°54'N	160°12'W	35	51
Schneller S	40°48'N	166°18'W	37	51
Schomberger A	78°48'S	24°24'E	31	198
Schomberger C	77°12'S	15°42'E	43	198
Schomberger D	73°30'S	24°36'E	24	198
Schomberger L	80°36'S	17°30'E	17	198
Schorr A	20°30'S	88°24'E	64	143
Schorr B	16°30'S	88°30'E	26	143
Schorr C	13°30'S	88°12'E	13	123
Schrödinger B	68°24'S	141°18'E	25	192
Schubert E	4°0'N	78°36'E	27	103
Schubert F	3°12'N	77°54'E	35	103
Schubert G	4°6'N	75°12'E	56	103
Schubert H	1°24'N	76°6'E	31	103
Schubert K	2°18'N	75°54'E	29	103
Schubert N	1°48'N	72°42'E	75	103
Schuster N	3°24'N	145°48'E	27	107

Name	Lat.	Long.	Size	Page
Schuster P	1°54'N	144°24'E	16	107
Schuster Q	1°0'N	143°24'E	45	106
Schuster R	3°30'N	144°48'E	40	107
Schwabe E	64°0'N	43°24'E	19	31
Schwabe F	66°24'N	50°0'E	20	31
Schwabe G	65°30'N	42°12'E	15	31
Schwarzschild A	78°42'N	124°0'E	50	22
Schwarzschild D	71°54'N	132°24'E	24	22
Schwarzschild K	67°30'N	125°0'E	45	34
Schwarzschild L	69°18'N	122°6'E	45	34
Schwarzschild Q	66°18'N	108°54'E	19	33
Schwarzschild S	67°48'N	104°42'E	17	33
Scott E	81°6'S	35°30'E	28	198
Seares B	75°42'N	149°42'E	26	23
Seares Y	77°54'N	139°30'E	37	23
Sechenov P	9°48'S	143°48'W	23	110
Segers M	44°30'N	127°36'E	54	48
Segers N	44°0'N	127°30'E	27	48
Seidel M	35°18'S	152°0'E	28	165
Seneca B	27°12'N	77°24'E	28	83
Seneca C	26°18'N	75°6'E	22	83
Seneca D	26°36'N	81°18'E	18	83
Seneca E	29°12'N	79°36'E	16	63
Seneca F	29°30'N	81°54'E	15	64
Seyfert A	30°30'N	114°54'E	53	65
Sharonov D	13°30'N	175°24'E	17	108
Sharonov X	14°6'N	172°42'E	36	88
Sharp A	47°36'N	42°36'W	17	41
Sharp B	47°0'N	45°18'W	21	41
Shayn B	34°30'N	173°30'E	35	68
Shayn F	33°0'N	175°30'E	38	68
Shayn H	31°24'N	175°30'E	38	68
Sheepshanks C	57°0'N	18°6'E	11	30
Shi Shen P	71°42'N	97°0'E	22	22
Shi Shen Q	74°12'N	96°18'E	45	22

Name	Lat.	Long.	Size	Page
Short A	76°54'S	0°30'W	34	197
Shternberg C	20°54'N	114°18'W	29	72
Shuckburgh A	43°6'N	55°30'E	19	45
Siedentopf F	22°6'N	138°30'E	42	86
Siedentopf G	20°30'N	138°24'E	61	86
Siedentopf H	20°54'N	137°12'E	42	86
Siedentopf M	19°0'N	135°30'E	31	86
Siedentopf Q	20°42'N	133°42'E	42	86
Simpelius B	75°12'S	10°12'E	50	198
Simpelius D	71°36'S	8°36'E	54	198
Simpelius G	71°48'S	23°0'E	24	198
Simpelius K	74°48'S	15°42'E	23	198
Sinas E	9°42'N	31°0'E	9	100
Sirsalis A	12°42'S	61°18'W	49	115
Sirsalis B	11°6'S	63°42'W	16	115
Sirsalis C	10°18'S	63°48'W	22	115
Sirsalis J	13°24'S	59°48'W	12	115
Sisakyan E	41°24'N	110°42'E	19	65
Sklodowska A	14°42'S	96°30'E	44	144
Sklodowska Y	13°12'S	95°24'E	17	124
Slipher S	49°12'N	158°42'E	26	50
Snellius A	27°24'S	53°48'E	37	141
Snellius B	30°6'S	53°6'E	29	160
Sniadecki J	24°42'S	166°54'W	27	129
Sniadecki Q	23°0'S	170°6'W	77	129
Soddy Q	0°30'S	120°12'E	24	125
Sommerfeld N	62°18'N	162°12'W	39	24
Sommerfeld V	66°54'N	170°18'W	32	24
Sosigenes A	7°48'N	18°30'E	12	100
South B	57°30'N	44°54'W	14	28
Spencer Jones H	12°6'N	167°54'E	17	108
Spencer Jones J	9°42'N	168°0'E	12	108
Spencer Jones K	10°24'N	167°0'E	29	108
Spencer Jones W	15°12'N	163°18'E	50	88
St. John X	13°54'N	147°24'E	30	107

Name	Lat.	Long.	Size	Page
St. John Y	13°48'N	149°0'E	21	107
Stark Y	24°24'S	134°0'E	31	146
Stebbins C	67°42'N	133°36'W	39	25
Stebbins U	65°24'N	147°36'W	44	25
Stein C	8°54'N	178°48'W	27	89
Stein M	3°48'N	178°48'E	28	108
Stein N	2°12'N	178°30'E	16	108
Steinheil G	45°36'S	49°54'E	19	176
Steno N	31°18'N	161°24'E	20	68
Steno Q	29°18'N	157°48'E	29	67
Steno T	32°42'N	159°42'E	37	67
Stetson E	39°24'S	117°0'W	38	152
Stetson G	39°54'S	117°12'W	23	152
Stevinus D	34°48'S	50°54'E	22	160
Stiborius A	36°54'S	35°30'E	32	159
Stöfler D	43°48'S	4°18'E	54	174
Stöfler H	40°18'S	1°42'E	27	158
Stöfler J	42°12'S	2°24'E	76	174
Stoletov C	46°18'N	153°36'W	36	37
Störmer C	58°18'N	150°36'E	61	35
Störmer H	54°48'N	150°12'E	32	49
Störmer T	56°48'N	141°42'E	27	34
Störmer Y	60°18'N	144°48'E	26	34
Strabo B	64°36'N	55°30'E	23	31
Strabo C	67°6'N	59°18'E	17	31
Strabo L	64°12'N	53°24'E	26	31
Strabo N	64°48'N	57°48'E	25	31
Stratton K	7°24'S	165°48'E	41	128
Street M	47°42'S	14°36'W	49	173
Strömgren A	17°48'S	131°42'W	51	131
Strömgren X	17°24'S	134°36'W	42	131
Struve B	19°0'N	77°0'W	14	74
Struve C	22°54'N	75°18'W	11	74
Struve G	23°54'N	73°54'W	14	74
Struve H	25°12'N	83°18'W	21	74

Name	Lat.	Long.	Size	Page
Struve L	20°42'N	76°0'W	15	74
Suess B	5°42'N	47°18'W	8	96
Sundman J	8°54'N	90°12'W	10	93
Sundman V	11°54'N	93°30'W	19	93
Sylvester N	82°24'N	67°18'W	20	18
Szilard H	32°30'N	108°24'E	50	65
Szilard M	31°6'N	106°36'E	23	65
T. Mayer A	15°18'N	28°18'W	16	77
T. Mayer C	12°12'N	26°0'W	15	97
Tannerus E	56°6'S	19°36'E	26	188
Tannerus F	55°0'S	22°6'E	36	175
Taruntius F	4°0'N	40°30'E	11	101
Taruntius H	0°18'N	49°54'E	8	101
Taruntius X	7°42'N	53°0'E	23	101
Taylor A	4°12'S	15°24'E	38	119
Teisserenc C	33°6'N	134°42'W	47	53
Teisserenc P	30°6'N	137°6'W	25	53
Teisserenc Q	31°6'N	137°18'W	30	53
Ten Bruggencate D	8°6'S	136°54'E	43	126
Ten Bruggencate H	10°0'S	135°36'E	33	126
Ten Bruggencate Y	6°42'S	134°0'E	57	126
Tesla J	37°12'N	126°42'E	18	66
Thebit A	21°30'S	4°54'W	20	138
Theophilus B	10°30'S	25°12'E	8	120
Thiel T	40°24'N	136°36'W	31	53
Thiessen Q	73°54'N	174°36'W	39	16
Thomson J	35°54'S	169°36'E	44	166
Thomson V	30°42'S	162°12'E	13	166
Thomson W	30°12'S	163°18'E	17	166
Tikhomirov J	20°54'N	165°42'E	29	88
Tikhomirov K	21°18'N	163°54'E	23	88
Tikhomirov R	24°6'N	160°18'E	21	87
Tikhomirov T	25°24'N	158°48'E	26	87
Tikhomirov X	27°18'N	160°36'E	24	87
Tikhomirov Y	28°18'N	160°18'E	20	68

Name	Lat.	Long.	Size	Page
Tiling D	52°0'S	131°12'W	34	169
Timiryazev B	2°18'S	145°42'W	23	110
Timiryazev L	8°12'S	146°24'W	18	110
Timiryazev S	6°0'S	149°24'W	53	110
Timiryazev W	3°0'S	150°0'W	32	110
Tiselius L	4°36'N	177°24'E	12	108
Tisserand A	20°24'N	49°24'E	24	81
Titov E	29°6'N	153°54'E	22	67
Torricelli A	4°30'S	29°48'E	11	120
Torricelli C	2°42'S	26°0'E	11	120
Tralles A	27°30'N	47°0'E	18	81
Tralles B	27°18'N	50°36'E	11	81
Trumpler V	29°48'N	164°0'E	36	68
Tsander B	9°36'N	147°0'W	55	90
Tsander R	3°24'N	152°12'W	36	90
Tsander V	7°54'N	153°30'W	37	90
Tseraskiy K	53°0'S	144°36'E	45	180
Tseraskiy P	51°18'S	139°36'E	33	180
Tsinger W	58°6'N	173°48'E	53	35
Tsinger Y	58°6'N	175°6'E	31	35
Tsu Chung-Chi W	18°30'N	143°18'E	24	86
Tycho A	39°54'S	12°0'W	31	157
Tycho D	45°36'S	14°0'W	27	173
Tycho W	43°12'S	15°18'W	19	173
Ulugh Beigh A	34°6'N	79°18'W	41	56
Valier J	6°18'N	174°54'E	26	108
Van de Graaff C	26°36'S	172°48'E	20	148
Van de Graaff F	26°48'S	174°36'E	20	148
Van den Bergh Y	33°6'N	159°42'W	43	52
Van der Waals K	45°48'S	122°0'E	55	179
Van der Waals W	41°18'S	117°6'E	46	163
Van Gent D	16°18'N	161°42'E	35	87
Van Gent N	13°30'N	160°0'E	32	107
Van Gent P	12°36'N	159°24'E	47	107
Van Gent X	16°24'N	159°42'E	38	87

Name	Lat.	Long.	Size	Page
Van Maanen K	33°12'N	129°6'E	23	66
van Rhijn T	52°12'N	140°0'E	35	49
van't Hoff F	61°30'N	126°12'W	41	25
van't Hoff M	56°48'N	132°6'W	36	25
van't Hoff N	57°54'N	132°18'W	46	25
Vasco da Gama A	12°42'N	80°0'W	23	94
Vasco da Gama B	15°42'N	83°0'W	27	74
Vasco da Gama C	11°24'N	84°54'W	44	94
Vasco da Gama R	10°0'N	83°24'W	59	94
Vasco da Gama S	12°36'N	82°48'W	28	94
Vendelinus E	17°54'S	61°0'E	21	142
Vendelinus F	18°30'S	65°0'E	32	142
Vening Meinesz C	1°12'N	163°48'E	46	108
Vening Meinesz W	1°30'N	161°0'E	39	107
Ventris A	4°24'S	158°12'E	26	127
Ventris B	2°24'S	158°12'E	18	127
Ventris C	3°12'S	158°54'E	48	127
Ventris D	3°24'S	160°18'E	21	127
Ventris N	7°6'S	157°36'E	63	127
Vernadskiy U	23°42'N	126°30'E	37	86
Vernadskiy X	25°54'N	129°0'E	64	86
Vertregt J	21°30'S	174°18'E	17	148
Vertregt K	20°6'S	172°0'E	27	148
Vesalius D	2°12'S	116°54'E	50	125
Vesalius H	3°54'S	119°0'E	36	125
Vesalius J	4°48'S	119°6'E	25	125
Vesalius M	5°42'S	114°30'E	31	125
Vestine A	36°12'N	94°48'E	17	64
Vestine T	33°54'N	91°6'E	49	64
Vetchinkin F	10°0'N	134°0'E	30	106
Vetchinkin K	9°36'N	132°18'E	22	106
Vetchinkin P	7°42'N	130°18'E	17	106
Vetchinkin Q	9°36'N	130°42'E	23	106
Vieta B	30°30'S	60°12'W	40	154
Vil'ev V	5°18'S	142°54'E	44	126

Name	Lat.	Long.	Size	Page
Virtanen B	17°48'N	177°48'E	24	88
Virtanen C	17°18'N	178°12'E	20	88
Virtanen J	14°0'N	177°54'E	21	88
Virtanen Z	16°30'N	176°36'E	31	88
Vitello A	34°6'S	41°54'W	21	155
Vitello D	33°12'S	41°0'W	18	155
Vitello N	32°6'S	36°6'W	5	156
Viviani N	3°30'N	116°30'E	16	105
Vlacq D	48°42'S	36°12'E	34	176
Volkov J	14°24'S	132°24'E	32	146
Volta D	52°30'N	83°18'W	20	40
Volterra R	56°12'N	129°36'E	31	34
von Békésy F	52°54'N	137°18'E	18	49
von Békésy T	52°12'N	121°54'E	29	48
Von der Pahlen E	24°30'S	128°48'W	32	131
Von der Pahlen H	27°6'S	127°30'W	35	131
Von Kármán L	47°42'S	177°54'E	29	181
Von Zeipel J	40°48'N	139°18'W	39	53
Voskresenskiy K	28°48'N	84°6'W	34	55
W. Bond B	64°54'N	7°36'E	15	30
Walker W	24°36'S	164°18'W	44	129
Wan-Hoo (Van-Gu) T	10°0'S	140°24'W	21	111
Wargentin A	47°6'S	59°6'W	21	172
Wargentin F	51°30'S	66°6'W	20	171
Watt D	50°18'S	55°12'E	32	176
Watt L	52°36'S	57°36'E	32	176
Watt M	53°6'S	59°54'E	42	176
Webb H	2°6'S	59°30'E	10	122
Webb J	0°36'S	64°0'E	24	122
Wegener K	43°18'N	111°54'W	32	38
Wegener W	47°30'N	116°6'W	53	38
Weigel B	58°48'S	41°6'W	37	186
Wexler H	70°30'S	96°42'E	14	200
Wexler U	68°12'S	82°0'E	51	190
White W	42°6'S	162°42'W	24	167

Name	Lat.	Long.	Size	Page
Wiechert J	85°36'S	177°0'W	34	15
Wiechert P	85°30'S	150°30'E	37	15
Wiener F	41°12'N	150°0'E	47	67
Wiener K	39°18'N	147°48'E	101	67
Wiener Q	39°30'N	145°0'E	30	67
Wilhelm D	41°48'S	17°42'W	32	157
Wilhelm O	43°6'S	17°12'W	17	173
Wilsing C	19°0'S	153°0'W	33	130
Wilsing W	18°30'S	159°48'W	36	130
Wilsing Z	20°54'S	155°12'W	30	130
Wilson A	71°18'S	53°30'W	15	196
Wilson E	72°30'S	55°0'W	24	196
Winkler L	40°0'N	178°24'W	31	51
Winlock M	32°18'N	106°0'W	68	54
Woltjer P	43°24'N	161°30'W	33	36
Wood S	43°48'N	123°36'W	35	38
Wurzelbauer D	36°18'S	17°36'W	38	157
Wyld C	0°42'N	100°30'E	28	104
Wyld J	3°48'S	99°24'E	24	124
Xenophanes A	60°6'N	84°48'W	42	27
Yablochkov U	61°54'N	120°48'E	30	34
Young C	41°30'S	48°12'E	30	160
Young D	43°30'S	51°48'E	46	176
Young F	44°48'S	51°48'E	23	176
Zach B	58°36'S	3°0'E	32	188
Zach D	62°6'S	7°54'E	32	188
Zagut B	32°6'S	18°42'E	32	158
Zanstra A	4°30'N	125°12'E	36	105
Zeeman E	74°12'S	123°54'W	29	195
Zeeman G	74°18'S	107°24'W	45	195
Zeeman U	73°48'S	148°12'W	26	194
Zeeman X	71°30'S	138°6'W	26	194
Zeeman Y	72°48'S	137°36'W	33	194
Zeno A	44°30'N	70°0'E	44	46
Zeno B	44°0'N	71°0'E	37	46
Zeno F	42°24'N	80°0'E	17	46
Zeno H	41°24'N	74°24'E	17	63
Zeno K	42°48'N	66°36'E	18	46
Zeno X	43°36'N	76°54'E	17	46
Zernike T	18°30'N	166°54'E	17	88
Zernike W	19°36'N	166°48'E	27	88
Zernike Z	20°54'N	168°0'E	30	88
Zhiritskiy F	24°54'S	121°36'E	75	145
Zhiritskiy Z	23°12'S	120°24'E	22	145
Zhukovskiy Q	6°12'N	168°48'W	23	89
Zhukovskiy U	8°30'N	173°12'W	29	89
Zhukovskiy W	9°48'N	170°18'W	31	89
Zhukovskiy X	10°30'N	171°6'W	30	89
Zhukovskiy Z	10°0'N	166°48'W	34	89
Zöllner D	8°18'S	17°42'E	24	119
Zsigmondy A	62°48'N	102°36'W	63	26
Zsigmondy Z	62°6'N	104°54'W	23	26
Zucchius B	61°48'S	54°18'W	25	186
Zwicky R	18°18'S	163°24'E	28	148
Zwicky S	16°18'S	162°36'E	44	148